SUSTAINABLE MANAGEMENT AFTER IRRIGATION SYSTEM TRANSFER

Experiences in Colombia – The RUT Irrigation District

Promoter: Prof. E. Schultz PhD, MSc
                        Professor of Land and Water Development
                        UNESCO-IHE Institute for Water Education
                        The Netherlands

Awarding Committee: Prof. Dr. Fernando Gonzalez Villareal
                        National University of Mexico, Mexico City
                        Mexico

                        Dr. Carlos Garces Restrepo
                        Research in Irrigation and Drainage (IPTRID), FAO, Rome
                        Italy

                        Prof. Dr. Ir. S.E.A.T.M. van der Zee
                        Wageningen University
                        The Netherlands

                        Prof. Dr. Ir. N.C. van de Giesen
                        Delft University of Technology, the Netherlands

                        Prof. P. van der Zaag PhD, MSc
                        UNESCO-IHE Institute for Water Education
                        The Netherlands

SUSTAINABLE MANAGEMENT AFTER IRRIGATION SYSTEM TRANSFER

Experiences in Colombia – The RUT Irrigation District

DISSERTATION

Submitted in fulfilment of the requirements of
the Academic Board of Wageningen University and
the Academic Board of UNESCO-IHE Institute for Water Education
for the Degree of DOCTOR
to be defended in public
on Tuesday 5 September 2006 at 15:00 hour in Delft, the Netherlands

by

NORBERTO URRUTIA COBO

Born in Morales - Cauca, Colombia

CRC Press
Taylor & Francis Group
Boca Raton London New York

CRC Press is an imprint of the
Taylor & Francis Group, an **informa** business
A TAYLOR & FRANCIS BOOK

CRC Press
Taylor & Francis Group
6000 Broken Sound Parkway NW, Suite 300
Boca Raton, FL 33487-2742

© 2006 by Taylor & Francis Group, LLC
CRC Press is an imprint of Taylor & Francis Group, an Informa business

No claim to original U.S. Government works

ISBN-13: 978-0-415-41693-1 (pbk)
ISBN 978-90-8504-495-6 (Wageningen University)

**Visit the Taylor & Francis Web site at
http://www.taylorandfrancis.com**

**and the CRC Press Web site at
http://www.crcpress.com**

Contents

Acknowledgement

This research was possible thanks to the contribution of several institutions and people.

At the first place I like to thank the Netherlands Fellowship Programme of the Netherlands Organization for International Cooperation in Higher Education (NUFFIC) for funding my PhD research, the UNESCO-IHE Institute for Water Education for the opportunity to pursue my postgraduate studies, the CINARA Institute for the economic support given during some stages of the research and my employer the Universidad del Valle, which assigned this research to me.

I like to express my sincere gratitude to my promoter Prof. E. Schultz, PhD, MSc for his support and directions given during the various stages of my research. I'm also very thankful to my mentor Herman Depeweg for his permanent guidance, support, collaboration and patience; he and his dear wife created a friendly and familiar environment during my stays in the Netherlands.

My thanks go also to the field, technical and administrative staff of the Rio Recio, Coello, Saldaña, Zulia, and María La Baja Irrigation Districts for their valuable inputs during the progress of the research. Special acknowledgement goes to my friends in the RUT Irrigation District, to the members of the Board of Directors of RUT, to the irrigation inspectors, technical and administrative staff, and especially to the farmers; their contributions were very important to become acquainted with and to comprehend fully the diversity of factors influencing the management of the RUT Irrigation District.

My sincere thanks and appreciation go to my friends Gerardo Galviz and Antonio Arias, they deeply influenced and guided my academic life. I'm also very grateful to my colleagues in the Universidad del Valle, to my friends in Colombia and the Netherlands, and to my family, especially my father, sisters and the memory of my mother. But most of all, I wish to thank my dear wife Angelina and my children Beto and Nata for their understanding, support, confidence and love.

Last but no least I thank to God for his powerful help.

Summary

Colombia is a tropical country located in South America. It has a total area of 114 million ha. In Colombia two irrigation sectors are distinguished: the small-scale irrigation and the large-scale irrigation sector. The small-scale irrigation sector is developed on lands located on sloping areas, where food crops and cash products such as corn, potato and specially vegetables are cultivated. In these areas the technique of cultivation is not well developed and therefore agriculture with advanced technology is not yet found there. Small farmers with a low income, a high level of illiteracy and malnutrition, form the main part of the traditional sector. The large-scale irrigation sector corresponds to large landholdings, located on the flat areas and with soils of good quality, that are destined for the cultivation of cash crops such as sugar cane, cotton, sorghum, fruit trees, etc. In this type of cultivation mechanization prevails and advanced irrigation technologies are applied.

The area of the large-scale irrigation sector that corresponds to the public sector is grouped in 24 irrigation districts covering 264,802 ha; 132,918 ha with an irrigation and drainage infrastructure and 131,884 with only drainage. The area covered by the irrigation districts has a total capacity of around 500 m^3/s; the irrigation network is 1,905 km long, of which 37% is for the main canals, 45% for the secondary canals and 18% for the tertiary canals; the drainage network is 2,132 km long, of which 44% is for the main drains, 39% for the secondary drains and 17% for the tertiary drains; and the road network is 3,382 km long. About 9% of the land is used for permanent crops, 51% for non-permanent crops and 40% for grass. About 62% of the landholdings is smaller than 5 ha and cover about 9% of the total area, 17% of the properties is in the range of 5-10 ha and covers 13% of the area, about 16% of the landholdings is in the range of 10-50 ha, while 3% of the farms is larger than 50 ha and cover 38% of the total area.

Potential for irrigated agriculture

At the global scale, Colombia is considered as a country with abundant natural resources, especially in relation to the land and water resources that present a great potential for the development of agriculture under irrigation.

Soils. The Agustin Codazzi Geography Institute (IGAC) carried out the soil study for Colombia in 1973. Two soil types were distinguished. The type-A soils correspond to the soil classes I, II and III and are suitable for mechanization, intensive agriculture and livestock, and for the development of irrigated agriculture. These soils are mainly located in the Caribbean plains and in the Orinoquia and Amazon regions. Most of the type-B soils correspond to class IV; their ground surface varies from flat to concave-flat requiring among others land improvement like flood protection, drainage, and leaching. Once they are improved, their potential use is similar to the soil type A. The type A and B soils together constitute a potential area of 10.6 million ha for mechanized and irrigated agriculture.

Water resources. According to the World Water Balance and Water Resources for the Earth of UNESCO, Colombia has a specific surface water yield of 0.59 l/s per ha, and the country occupies the fourth place at world level. This specific yield is six and three times larger than the values for the continental world and Latin America, 0.10 and 0.21 l/s per ha, respectively.

Colombia has an average annual rainfall of 1,900 mm. On 70% of the area the average annual rainfall is 2,000 mm or more and on almost one third of the area the rainfall is more than 4,000 mm (Pacific Coast and East plains). These values are above the average of 1,600 mm per year for Latin America, and above the global average of 900 mm per year. In contrast, the Caribbean Coast has little rainfall and is considered as a dry region.

Colombia has more than 1,000 perennial rivers, a figure that is larger than for Africa (60). It has 10 interior rivers, which have an average annual discharge greater than 1,000 m^3/s. Colombia has a natural drainage network of 740,000 micro-basins in view of the high rainfall, topography and geology. This natural drainage is distributed in 4 regions or main river basins.

Irrigation development. Colombia contributes less than 3% of the irrigated land area in Latin America, occupying the sixth place among the South American countries in view of irrigated lands over the total cultivated lands (12.5%). Colombia has followed a similar tendency as the Latin American region for its irrigation development. The area increased rapidly up to the end of 1960s, but the growth considerably dropped during the last two decades. During the period 1990 – 2000 the public sector concentrated its resources on the rehabilitation, completion and enlargement of the existing irrigation districts.

Expectations for the irrigated agriculture. According to the National Water Study of 1984 Colombia has 14.4 million ha suitable for agriculture, of which 6.6 million ha (45.8%) are suitable for mechanized agriculture and the development of irrigated agriculture through irrigation and drainage infrastructure. This figure could be increased to 10.6 million ha after the improvement of some soils, which at present limit the conditions for agricultural use. Of these 6.6 million ha only 11.4% has an infrastructure for irrigation and drainage, of which 38% has been developed by the public sector and the remainder by the private sector.

The Colombian Government established the "Decade Land Improvement Program for 1991 – 2000" to improve 535,500 ha with an irrigation and drainage infrastructure. During the different periods the Government has adapted the land improvement policies with the purpose to follow the preliminary land improvement program that was established for this period by the National Planning Department. However, the land improvement programs had a slow development.

Policy for the land improvement sector. The historic development of the land improvement in Colombia has been largely influenced by the policy as established by the different governments, which have implemented their policies through the formulation of respective legal frameworks. The policy for the land improvement sector has changed by mandate (directive) after mandate, which did not permit a consistent development on the long term.

The political outlines of the Colombian Government for the sub-sector of land improvement were enclosed in the Economic-Social Development Plan for the period 1991 – 2000. In 1991 the Colombian Government considered that the public and private investments are good mechanisms to contribute to the modernization of the agricultural production. The investments in land improvement projects would preferably have to be promoted and developed jointly by the public and private sector. To facilitate the investment recovery the Government tried to promote community participation during the design process of the projects, which after construction would be transferred to the water users associations for administration, operation and maintenance. A new institutional framework was proposed in which the Government would change its role from being a simple executor to more an investment promoter with a major involvement of the community. The law 41 (1993) set up the rules by which the land improvement sector was organized.

After 7 years of land improvement programs within the new institutional framework, strong criticism rose in view of the role of the governmental agency due to a ineffective management of the economic resources and an almost zero-development of the land improvement programs launched from 1990 by different governmental programs. The criticisms were among others related to the management of external resources for the land improvement program, the costs for the functioning of the agency, and the quality of the agency's performance.

As a result, a change in the institutional framework was introduced at the beginning of 2003 and the existing Governmental Agency for the Land and Improvement Sector and other public institutions within the public agricultural sector were suppressed and integrated in a new organization called Colombian Institute for Rural Development (INCODER). The Agreement 03 of February 2004 was considered as a key element within the new institutional framework by which organizational guidelines were established for the users' organizations including the basic regulation for the functioning of the small, middle and large irrigation districts. However, this agreement contained controversial points causing a new regulation for the administration, operation and maintenance of the irrigation districts including the transfer forms (delegation, administration, and concession) and considering that INCODER must not have direct interference in the water users organizations, but only in the management of the infrastructure (Resolution 1399/2005).

Irrigation management transfer

At the middle of the 1970s the Colombian Government started a program to transfer the irrigation management to water users associations. The transfer of irrigation districts in Colombia has been carried out during three periods. In 1976, the irrigation districts Coello and Saldaña were transferred by request of the water users. The Rio Recio, Roldanillo-La Union-Toro (RUT), Samaca, and San Alfonso Irrigation Districts were transferred during the period 1989-1993 and finally, during 1994-1995, most of the remaining irrigation districts were transferred following the policy established by the Colombian Government. At the moment, 16 irrigation districts out of 24 have been transferred and the remaining districts are still under the administration of the Government Agency.

However, serious problems were detected during and after the transfer process concerning disputes on water rights and legal aspects, inadequate operation and maintenance of the hydraulic infrastructure, disagreement about the role of the Government Agency and water user associations, environmental degradation, problems related to economic and financial aspects, insufficient community participation, lack of leadership, and insufficient training and irrigation management. For some irrigation districts, these problems and constraints resulted in an inefficient performance, deterioration of the physical infrastructure and conflicts related to the operation, maintenance and administration of the system causing an unsustainable development of the irrigation districts.

In the framework of this research the background of the transfer program in Colombia, its implementation, current status and impacts were reviewed. In this way the irrigation districts of Rio Recio, Coello, Saldaña, Zulia and Maria La Baja were visited and fieldwork and interviews with technical staff and farmers were carried out. Staff from the Government Agency for the Land Improvement Sector, and from private and public entities at local and regional level was also contacted. A review of some experiences and characteristics of the transfer of the irrigation management in other countries with similar characteristics as Colombia was also carried out.

The experiences with the transfer of irrigation management in Colombia showed that the governmental financial burden was significantly reduced by the shift of the total costs for administration, operation and maintenance to the water users associations. The partial and in many cases the total elimination of the subsidies and the reorganization of the new institutional structure for the Land Improvement Sector facilitated the implementation of the decentralization and delegation policy for the sector. The impact of the delegation of functions to regional and local institutions can not yet be evaluated, because the institutions only recently started to take up their tasks after the delegation; however, scepticism exists among the water users' organizations concerning the effective action of this delegation.

The development of the transfer program showed a clear difference between the first phase of the transfer program and the following stages. During the first phase the management transfer was promoted by the users themselves and showed a clear opposition of the Government Agency; while for the following phases the motivation came mainly from the Government and was interpreted by the users as a governmental strategy to face the fiscal crisis of the 1990s and to obtain the acquired compromises with the international banking.

About 30 years after the start of the transfer program the researched water users organizations offered diverse grades of development. Some of them, presented a very well developed organizational stability (Rio Recio, Coello, Saldaña Irrigation Districts) based on financial self-sufficiency as a result of the provision of adequate and reliable irrigation water services; other districts as Maria La Baja showed a serious organizational crisis that resulted in a governmental intervention. The Zulia Irrigation District presented a great organizational model with a dynamic character that was based on an integral and participatory approach around profitable agricultural activities that were focused on rice production and on a clear vision of the future development of the organization. Subsequently, it can be said that in the post-transfer process and from organizational point of view three crucial phases seem to take place: namely adaptation, maturity or full development, and consolidation.

An increase in efficient water use was another objective of the management transfer to the water users associations. However, observations during the research showed that the water use in the irrigation districts is a critical issue that will hamper the environmental sustainability of the irrigation systems. The financial self-sufficiency as a result of the water selling for the provision of irrigation water, especially for crops demanding large quantities of water like rice seemed to be a factor which did not contribute to the efficient use of irrigation water.

From an environmental point of view it has to be mentioned that the growing deforestation process affects the watersheds of the main surface waters of most of the researched irrigation districts. This situation affects amongst others the quality and costs of the maintenance and operation activities and the availability of the water resource.

Lastly, the experience of the irrigation management transfer in Colombia allowed for the consideration of a wider concept of the role of an irrigation system. An irrigation system must not be considered as a simple hydraulic infrastructure that only provides irrigation water to the farmers, but it should be recognized as an important component of a production system, whose final objective is to contribute to the improvement of the living conditions of the farmers through irrigated agriculture under criteria of profitability, equity, efficiency, and an integral and participative management approach.

Sustainable management of transferred irrigation systems; conceptual framework

A conceptual framework for the sustainable management of irrigation districts transferred to water users associations is formulated and developed based on general concepts and approaches of sustainable water management and especially the sustainable use of land and water resources and the management of land and water systems. This framework is supported by certain general principles, approaches and definitions, some of which were stated in numerous events and conferences concerning improvement and preservation of the natural resources with special emphasis on land and water resources.

The sustainability of an irrigation and drainage system refers to the physical and organizational infrastructure for management, administration, operation and maintenance during the lifetime of the system according to established specifications and giving the users the expected benefits without causing damage to the environment. The concept of sustainability involves three elements that play an important role in the definition of the conceptual framework for sustainable management of an irrigation system. The three components are Community (C), Environment (E), and Science & Technology (S&T) and they are interconnected within the existing institutional and legal contexts. This interrelationship determines the sustainability of an irrigation system.

Community (C) represents the different interest groups in the irrigation and drainage sector; in which the farmers are the main beneficiaries. The community with its specific socio-economic-cultural characteristics and organizational forms interconnects with the Environment and Science & Technology components in order to contribute to the sustainable management of the systems. *Environment (E)* provides the natural resources for irrigated agriculture as production system. Climatic conditions, soils, ecosystems, and water resources are the main inputs provided by the environment component. Factors that affect the relationship between Environment and Community in two directions form risk factors. *Science and Technology (S&T)* represents the scientific and technological knowledge, methods, tools, and experiences that are available and can be used for the community with the purpose to intervene in the natural environment in order to satisfy its needs and desires. The Science & Technology constitutes the mechanism by which the risk factors for the relation between Environment and Community can be controlled or overcome. The technological alternatives for the provision of an irrigation and drainage service must be in line with the cultural, socio-economic, technical, and financial conditions of the Community. The level of service specifications, cost of the service, willingness to pay for the service provision, organizational forms, technical capacity, participation for operation and maintenance, decision-making, conflict resolution, and resource mobilization, play an important role in the relation Science and Technology and Community.

RUT Irrigation District

The RUT Irrigation District was taken as case study for this research. The historic background; description and characteristics of the physical, management and socio-economic aspects of the district; and finally the characteristics and impacts after the transfer process are presented. The RUT Irrigation District is located in the South-west of Colombia, in the northern part of the Valle del Cauca Department. It lies within the three municipalities that give their name to it, namely Roldanillo, La Union, and Toro. The altitude of the district ranges between 915 and 980 m+MSL (Mean Sea Level). Before the construction of the RUT Irrigation District the area was regularly subject to floods from both the Cauca River and the streams descending from the Western Mountain Range. As a consequence an area of 1,500 and 3,500 ha was under permanent and periodic floods respectively. As a result of the high water an area of about 2,500 ha had a high moisture content that only allowed the growth of grass, and an area of 4,000 ha that could only be used for extensive cattle breeding.

In general terms, the irrigation system is primarily a flood protection and irrigation/drainage scheme. The long and narrow bowl-shaped area is surrounded by a protection dike, running along the East border with the Cauca River and a flood interceptor canal on the West side. A main drain divides the area almost in half, running through the lowest elevations. Parallel to the river dike is the main irrigation canal, which is served by three pumping stations with a total capacity of 13.8 m^3/s. The flood interceptor canal also acts as an irrigation canal for the water users who use small centrifugal pumps to serve their individual needs. There is a complementary network of both irrigation and drainage canals throughout the area. The main drain discharges freely into the Cauca River during low stages and the drainage water is pumped during the high stages of the river.

The RUT Irrigation District is one of the most important irrigation systems in Colombia. The district has excellent characteristics for the development of agriculture, namely the good quality of the available natural resources (climate, soils, and water), its strategic geographical location, and its potential for further agro-industrial development in view of national and international markets. Nowadays, the RUT Irrigation District shows a growing crisis, which is a serious threat for the sustainability of the whole system, when the necessary measures are not taken on short notice. External and internal factors; with their social, economic, environmental, technical, and institutional characteristics have contributed in different grade to this situation.

From its early beginning, the management of the RUT Irrigation District has met a strong opposition from the local community that considered the irrigation scheme as one imposed by the National Government. As many projects of the 1960s and 1970s, also this project was developed with minimal participation of the local community and under almost total responsibility of the Government, who considered the implementation and development of a land improvement project completely as mere physical interventions with limited attention to the social, cultural, and economic requirements within an acceptable social context. Therefore, the farmers showed a very low sense of togetherness and they believed that it was the responsibility of the Government to carry out all the administration, operation and maintenance activities. During the administration of the district by the Government (1971-1989), this attitude did not have a serious impact, because the government agency was in charge of the whole management that included the operation and maintenance of the system with financial resources provided by the government. However, the paternalistic attitude of the land improvement sector, the negative impact of the canal alignments on their farms and agricultural practices, insufficient compensation for their land acquired by the Government for construction works, and lack of credibility of the investment project did not improve the motivation and attitude of the farmers.

After the transfer in 1989, the Water Users Association ASORUT became fully responsible for the administration, operation and maintenance activities covering all the costs without any subsidy from the Government. The actual transfer was mainly focused on activities related to the organization of the water users association and on considerations concerning water charges and repair of the intake and main canal. Even today, the farmers believe that the transfer has been too rapid without sufficient assistance from the Government side and that the main objective of the Government was an as-fast-as possible management transfer of the irrigation district in order to satisfy the wishes of international financial entities that were interested in the land improvement sector. The main focus of the present management is the provision of irrigation water, from which they will obtain the necessary revenues for the financing of the organization. As the present organization is acting in a comparable way as the previous government agency, the stakeholders believe that the transfer process brought only a change of actors within the management.

Some time later ASORUT faced the following main negative characteristics: low collection of water charges, temporal and spatial suspension of the maintenance activities for the irrigation network, a low quality of the provided irrigation services, an inefficient use of the irrigation infrastructure and irrigation water, conflicts amongst water users and between them and the administrative and technical staff of the organization, and finally the failure of an agreement concerning the payment of the water tariffs.

The present management attitude clearly shows that the management only provides irrigation water without questioning whether the water contributes to the success of the production system of the farmers. The present situation is a result of the growing discrepancy in interests of the two parties: the organization is focused on the provision of irrigation water that is paid through a tariff and the farmers expect actions from the organization that will support them in the development of projects that will improve their standard of living.

This management approach and some external factors caused several problems on, for instance the interceptor canal, the water quality, salinization processes, maintenance of the infrastructure, and development projects with international cooperation for the modernization of the district. The present crisis can be seen from the poor conditions of ASORUT, especially in the technical, economic, social, environmental, and institutional parts of the organization.

The operation of the district is nowadays much more complex and the water supply has been affected by changes in the cropping pattern, by lack of irrigation schedules and cropping plans. The farmers decide themselves about the crops, the planting date and the moment that they request irrigation water. Sugar cane covers one-third of the area and requires much more water than the traditional crops. Modern agriculture practices use fruit trees that require a water supply with short irrigation intervals.

The major environmental problems are related to the water quality. The domestic and industrial waste-water and the deforestation of the adjacent catchment areas are affecting the water quality of the irrigation canals and the Cauca River. At several places the salinization process is intensified by the reuse of drainage water, inefficient irrigation practices, lack of a drainage system at farm level, and salinization caused by groundwater.

One factor that seriously influences the critical situation is the fact that there are two types of farmers in the area, namely the small and middle farmers and the large ones. Their minimal economic resources together with a low technology and small land area, the small and middle farmers can not, like the large ones, introduce and grow crops with a high profitability, like fruit crops. This has caused a lot of tension and disagreement between the two groups.

Sustainable Management for the RUT Irrigation District; implementation of a conceptual framework

An analytical framework for integrated water resources management has been used to formulate a well developed proposal for an integrated and sustainable management plan for ASORUT. The assessment of the present situation, the formulation of this management plan and the required interventions have been derived from interviews with farmers, administrative, technical, and field personnel from ASORUT, representatives of the municipalities and representatives of public and private institutions closely related to the land improvement sector.

Proposals for integrated and sustainable management scenarios will be based on the potential internal and external factors of the district and their influence on the management process. Two scenarios can be developed for the district, namely a Direct Governmental Intervention Scenario and a Participatory and Communal Approach: a New Role for ASORUT.

Both scenarios start from the current critical condition, but the first scenario is based upon the assumption of a continuous deterioration due to the present low management capability to produce the necessary changes that can lead to a gradual recovery. It is expected that some negative factors will continue to build up, for instance an increase in non-profitable agricultural practices resulting in a low willingness and capability of the farmers to pay the water fees; an increase of outstanding bills as the water users are not able to pay for the services resulting in a decreasing financial capacity of the organization. Other factors include changes in land use, especially the expansion of extensive crops like sugar cane and of other, non-agricultural uses; changes in the land tenure and lack of incentives for investments, especially in cash crops, agro-industry and marketing; increase of the presence of illegal groups that will cause unemployment and poverty. In this scenario ASORUT will have a very low management capability and its credibility will be at a very low level for the farmers, government and supporting public and private entities and that will result in a serious isolation of the organization within the national, regional and local context. In this scenario the gap between the two groups of farmers will not be bridged, but even deepened by the large difference in socio-economic conditions.

This situation will form the start for the next phase of this scenario in which the government takes over the total control of the district and will undertake the necessary actions for a new organizational model. In this case the government may decide for a concession model, which has recently been considered during the changes of the legal framework for the land improvement sector (August, 2005). According to the law a Concession Contract contains the provision, operation, exploitation, organization and partial or total management of a public service; or the construction, exploitation and partial or total conservation of a work or good for the public use or service. It also includes all the necessary activities to be developed by the concessionaire for the appropriate provision or functioning of the work or public service and the remuneration for the control and care of the work or service by the concessionary entity is arranged by rights, tariffs, tax, and valuation, amongst others.

This scenario will result in a management organization that will concentrate its services as irrigation water supplier on the large farmers, which present a strong socio-economic entity due to their most profitable agriculture practices. This management model will result in economic, financial and technical sustainability in the short and middle term, but its social sustainability will be seriously endangered due to the deepening of the differences between the small-middle farmers and large ones; the expected benefits will result in social inequalities as the large farmers will be strengthened and the small-middle ones will be excluded from any benefit.

In the second scenario the community consisting of the water users in the district will play the main role; the water users association as representative of the users will head the development process. The fear of the community that the government could give the system management to another private organization forms the main incentive to change the present management situation without the interference of the government. The development process towards a sustainable management will follow an integral and participatory approach involving all the internal and external actors: namely the farmers as first beneficiaries; the technical and administrative staff; representatives of each municipality located in the influence area of the irrigation district; the land improvement sector; other supporting private and public entities. As a legal organization, ASORUT will act on behalf of all the farmers before public and private institutions in order to develop programs in a mutual way. Within this participatory scenario the organization will have a clear statement about its mission, which is aimed at the improvement of the quality of life of the farmers through a sound management of the district under principles of equity, competitiveness, sustainability and multi-functionality in line with the legal framework established by the government. This means that the activities will have to be broad and be focused on improved socio-economic conditions and living standard of its beneficiaries in an integrated and sustainable way.

In this scenario ASORUT will have a well developed ability to formulate, plan and implement development-coherent programs for the short, medium, and long term. These programs will include socio-economic, technical, and environmental components and will be based on an integral, participative, and communal approach, meaning the participation of all the directly and indirectly involved actors. The programs will be based upon the promotion and practice of irrigated agriculture under criteria of profitability, efficiency, competitiveness, and sustainability, and at the same time the protection and conservation of the natural resources.

Some major expectations from the implementation of this participatory and communal approach might include: active participation and an improved sense of belonging of the farmer's side due to a better standard of living as a result of new, successful commercial projects; the farmers will have a better opinion concerning the organization; they will see that the organization pursues the well-being of all of them and as a consequence conflicts will reduce; commercial agriculture and corresponding business will reduce poverty and unemployment and increase the standard of living and will create a favourable environment for the conservation of the natural resources; the management capability of the organization will get stronger resulting in an efficient water supply to all the farmers, but also in the formulation and implementation of profitable agricultural projects; the community in the district will obtain more authority in a gradual way and will develop a mutual decision-making process in which the major differences between the two groups of farmers can be brought down resulting in a harmonic development; ASORUT will be recognized at local, regional and national level by all the public and private entities in view of its sound management and it will have a high ability for inter-institutional relationships; the physical infrastructure (pumping stations and other installations) will be used in a more efficient way and ASORUT will support and enhance the marketing of agricultural products.

As a final conclusion it can be said that this scenario is the most suitable one for the district, the water users and farmers in view an integrated and sustainable management. Its implementation and successful development will create new conditions that will empower the community and allow them to participate in an active way in the decision-making process, while they take their responsibilities as they are the central actors for the necessary change processes.

Physical and non-physical interventions considering the new management scenario

Before the implementation of the new management scenario it is important that the present physical infrastructure of the RUT Irrigation District is brought to a higher level. The most important interventions are related to the pumping stations together with the implementation of new sand traps; improvement of the marginal irrigation canal, conveyance canal, canal 1.0 and the interceptor canal to reduce seepage losses and costs for maintenance; the modernization of three water level control structures in the interceptor canal to assure a correct operation of the canal. The new management would have to pay special attention to a proper water management that will result in a more efficient water use, a proper use of the hydraulic infrastructure and a correct operation of the pumping stations in order to reduce the high energy costs. In this way the management will regain its authority.

The interventions from an environmental point of view should be mainly focussed on the water quality, the proper disposal of solid waste, the protection and conservation of the micro-watersheds, the control of the salinization process, and the formulation of an elaborated plan for the environmental management. The operational function of the new organization will have to focus on the use, reuse and control of water for irrigation in order to fulfil the specific needs and demands of the farmers.

One of the major problems to be solved by the new management concerns the financing of the administration, operation and maintenance activities. That financing should come from the establishment of a new tariff structure. Discussions with the farmers have shown that they are willing to pay for the service that supplies irrigation water, because until now agriculture is for them the only profitable economic activity. The tariff would have to be supported by incentives for an efficient water use and reuse; and fines by an inefficient use of irrigation water. Activities to improve the economic condition of the farmers will help to increase their willingness to pay for water, especially in the case of the small and middle farmers.

ASORUT needs to consider other possibilities for additional economic resources different to those coming from water fees. On the short term new financial resources can be obtained from the participation in the identification and implementation of agricultural production projects; from obtaining new credit lines; provision of technical services for the development of agricultural activities; and marketing of agricultural products. On the mid- and long-term, ASORUT can obtain more economic income from other, but parallel money-making projects, like agro-tourism, small rural enterprises, agro-industrial processing; and training programs for other irrigation districts.

Productive agriculture in the future would have to be based on to the existing natural conditions within the area. Some of the crops, like grains, fruits, vegetables, cotton and grass offer good perspectives in view of international markets (fruits and vegetables) and can get some added value through agro-industrial processes within the area. Corn for example offers good possibilities to meet the local and national demand and can provide raw material for the national industry. Sugar cane forms the basis for an important production chain in the Valle del Cauca Department, but the crop is not very suitable for the prevailing conditions of ASORUT and it had a very negative impact on the physical infrastructure.

ASORUT would have to promote privately negotiated future contracts, especially for small and middle farmers that will assure favourable selling prices, volume and quality of the agricultural products. These contracts can be for example with large department stores, large farmers who need a certain production volume and with the agro-processing industry or the National Agricultural Stock Exchange.

The new ASORUT needs to try to recover and strengthen the sense of togetherness of the farmers by adapting objectives of the organization that are in line with the interests and expectations of the farmers, who for instance expect the development of profitable agricultural practices and the improvement of their standard of living in an efficient and equitable way. Transparency, accountability, clear actions, administrative efficiency, and equity are requirements demanded by farmers to create confidence in the organization.

Expectation for a New Role for ASORUT

A process that clearly reflected a bottom-up approach was followed in the framework developed as part of this research. The different stakeholders and other actors involved got the opportunity to present their views and ideas in several participatory workshops and direct discussions. In this way, the key factors for the present critical situation of ASORUT, the wishes and expectations of the farmers and other actors, and the required interventions have been identified.

Two scenarios for the management of ASORUT have been developed and after a thorough analysis, one model, namely the "Scenario for a Participatory and Communal Approach: a New Role for ASORUT" is considered as the most suitable one in view of an integrated and sustainable management of the RUT Irrigation District.

The transition stage will start from the present low organization level of ASORUT that is a result of different socio-economic and political events during the lifetime of the organization and for which ASORUT had not yet the adequate answers and adaptation capability to tackle the new circumstances and acquired responsibilities after the transfer of the irrigation management. At this moment, ASORUT suffers under a process of destruction at different levels that will practically end in a state of total collapse of the organization. This meagre state of the organization is reflected for instance in the poor socio-economic condition of the farmers, especially the small and middle ones, in the poor management capability of the organization and in a low credibility in the eyes of the farmers and the society at large.

The questions to be answered are: how is it possible to implement the selected scenario under the current situation of ASORUT; how long will it take to arrive at the desired management condition and what are the requirements for the envisaged development? A vision of the near future under the new management model shows that ASORUT will be an organization based on equity, efficiency, sustainability and competitive criteria with the mission to promote the welfare of all the beneficiaries (farmers), to improve their socio-economic conditions through profitable agricultural practices and to be involved in all the stages of the production chain: production, post-harvest management, added value process (agro-industry), and marketing.

Therefore, the stakeholders of ASORUT and the different actors involved have to make very vital and wide-reaching decisions on the short term to safeguard the organization and the physical infrastructure. In the presented scenario the current situation is considered as a "transition stage", which bridges the present degenerating organization and the new one that will be characterized by a dynamic behaviour and a capability to interpret and satisfy the wishes of the stakeholders in view of the sought improvement of their socio-economic living conditions. Moreover, this dynamic attitude of the new organization will be able to maintain and develop stable and lasting contacts with supporting public and private organizations in the land improvement sector, and will take advantage of all the available opportunities for further development as they are offered within the new management framework and the recent government policy for the land improvement. Passing from the current status (transition stage) to the desired condition will take considerable time, which is based on the fact that the ongoing deterioration process of the organization started a long time ago. The formulation and implementation of some small productive agricultural pilot projects that will involve small groups of small and middle farmers will be used as a strategic approach.

This knowledge is based upon successful experiences in the Zulia Irrigation District, in the north-east of Colombia, which has now reached a degree of sustainable development where the farmers perform the central role, meaning that they lead and determine the development policy of the organization. In order to embark on and to reach the new management conditions for ASORUT the key actors will have to play an important role. The Board of Directors would have to include representatives of the farmer's community along with the Government Agency represented by the Departmental Agriculture Secretary. Other supporting public and private entities will join their resources and efforts in a concerted way to contribute to the final realization of the new management model.

1

Introduction

1.1 General

Colombia is located in the Northwest corner of South America at latitude 5° North. Colombia is a mountainous country with a population of 44 million inhabitants. The country has relatively abundant water resources including more than 1,000 perennial rivers. According to the study "World Water Balance and Water Resources of the Earth" carried out by UNESCO (1978) Colombia has a specific surface water yield of 0.59 l/s per ha, which means the fourth place in the world. It has both tropical and temperate climates and an average rainfall of 1,900 mm/year. A marked bimodal rainfall distribution in April/May and October/November makes the need for irrigation primarily a supplemental one.

Colombia has a total area of 114 million ha, of which not more than 33 million ha (30%) are considered suitable for agricultural production (14 million ha for crops and 19 million for livestock). About 5.8% of the total area (6.6 million ha) is considered to be suitable for mechanized agriculture and modern irrigation. This figure could be increased to 10.6 million ha after the improvement of some soils, which at present limit the conditions for agricultural use. Of these 6.6 million ha only 11.4% (750,513 ha) has an infrastructure for irrigation and drainage, of which 38% (287,454 ha) has been developed by the public sector and the remainder by the private sector.

Agriculture in Colombia represents an important sector in the economy of the country in spite of its changing pace during the last decade. Before the 1990s its contribution to the Gross National Product was 20% varying to 16.3, 13.7 and 14.2% for the years 1991, 1998 and 2002 respectively. The yearly average growth for the period 1990-1998 and 1998-2002 was 1.6% and 1.3% respectively; presenting during the latter period even a negative value (-0.05% in 1999). Agriculture generates employment for about 3.5 million persons, which is about 28.2% of the rural population (about 12.4 million inhabitants) and it contributes to about 18% of the export of all the merchandises (National Administrative Department for Statistics, 2001).

The Economic Liberalization and the violence raised in the rural area are factors that have a serious impact on the development pace of the agricultural sector. The Economic Liberalization decreased the cropped area with 800,000 ha during the period 1990-1997, caused an increase in the un-employment rate from 3.5% (1991) to 5.8% (1997), and a loss of more than 228,000 jobs. For the same period, the growth per year of the export and import has been 10.7% and 50.1% respectively. For the year 2003 the agricultural sector showed a growth of 4.4% as a consequence of an increase of the cropped area with 42,568 ha (1999-2000) and with 141,700 ha in 2003 (season A), generating 147,000 new employments (National Administrative Department for Statistics, 2003). Finally, the report "Memorias 2003 - Manejo Social del Campo" (Ministerio de Agricultura y Desarrollo Rural, 2004), presented for the agricultural sector the following growth indicators for the year 2003: a total growth of 4.4%, a growth of the agricultural productivity of 2.8%, and of the rural employment of 9.2%. On the other hand, during the first semester of 2004, the sector showed a total growth of 4% and an increase of the cropped area with 44,877 ha.

Most of the area under irrigation is located in the low-lying, warm climate areas and in the inter-Andean valleys at altitudes up to 1,000 m+MSL (Mean Sea Level). Of the 10 million ha suitable for rainfed production, about 7 million ha are in the sloping, Andean foothills, which are primarily used for permanent crops to minimize erosion problems and the other 3 million ha are mainly in the flatlands along the Caribbean coast.

In Colombia two irrigation sectors are distinguished: the small irrigation and the large irrigation sector. The small irrigation sector, also called, the traditional sector, is developed on lands located on sloping areas, where food crops and other cash products are cultivated. In these areas the technique of cultivation is not well developed and therefore agriculture with advanced technology is not yet found there. Small farmers with a low income, a high level of illiteracy and malnutrition, form the main part of the traditional sector. Increases of sustainable production and rural income in the small farmer sector, where rural poverty is concentrated, are constrained by the small landholding sizes, which impose a physical limit to production increases. In these areas the climate also forms a major risk for crop growing. This type of agriculture is characterized by the following aspects: the farmer is owner of the land which has a small size (1 – 3 ha); the agriculture is generally rain dependent (rainfed); the number of harvests per year is not more than two, depending upon the rainfall pattern; the yield of the crops is low; and erosion and soil losses are serious.

Due to the expansion of cash crops in the flat areas, the sloping areas will become very important for the production of food crops. The statistics for the agricultural sector clearly show that a large part of the food products come from these sloping areas. For this reason, the Colombian Government promoted and developed a project called "Program of Small Irrigation - SIP". The aim of this project, which was planned for the years 1991-1996, was the strengthening of the agricultural activities on sloping areas through irrigated agriculture. Before the actual start of this project, a successful pilot project had been developed during the period 1986 - 1990.

The large irrigation sector corresponds to large landholdings, located on the flat areas and with soils of good quality, that are destined for the cultivation of cash crops such as sugar cane, cotton, sorghum, fruit trees, etc. In this type of cultivation mechanization prevails and advanced irrigation technologies are applied (surface, sprinkler and trickle irrigation). In this sector, the extension of cultivable land plays an important role. The objective is to extend cash crop cultivation through the development of irrigated agriculture. For this reason, the policy of the Colombian Government for the period 1991 - 2000 was to increase the area with irrigation and drainage infrastructure with about 1,365,000 ha. The goal of the Government for that period was the development of irrigated agriculture on about 535,500 ha (being almost two times the improved land of the public sector) and representing 8% of the potential land of mechanized agriculture and modern irrigation.

The area of the large irrigation sector that corresponds to the public sector is grouped in 24 irrigation districts covering 264,802 ha; 132,918 ha with an irrigation and drainage infrastructure and 131,884 with only drainage. More than two decades ago the Central Government started a program to transfer the irrigation and drainage infrastructure to water users associations. The program started in 1976 with the transfer of the Coello and Saldaña Irrigation Districts. In 1989 the RUT and Rio Recio Irrigation Districts were transferred and the transfer of most of the other irrigation districts took place during the period 1990 – 1995. At the moment, 16 irrigation districts have been transferred and the remainder is still under the administration of the

Government Agency INAT (National Institute for Land Improvement), later INCODER (Colombian Institute for Rural Development).

1.2 Problem Identification

Serious problems have been detected during and after the transfer process. FEDERRIEGO, the Federation of Water Users Associations, mentioned six aspects that are considered to be the major problems faced during and after the transfer (Quintero, 1997). These problems include disputes on water rights and legal aspects, inadequate operation and maintenance of the hydraulic infrastructure, disagreement about the role of the Government Agency and water user associations, environmental degradation, problems related to economic and financial aspects, and lack of leadership. For some irrigation districts, these problems and constraints resulted in an inefficient performance, deterioration of the physical infrastructure and conflicts related to the operation, maintenance and administration of the system causing an unsustainable development of the irrigation districts.

1.2.1 Water Rights and Legal Issues

The management of an irrigation and drainage district depends to a large extent on the present water laws and regulations. The existing water legislation and the related water rights, administrative procedures, jurisdiction and jurisprudence form the legal basis. The National Code for Renewable Resources and Environmental Protection (Law 2811 of 1974) and the Law 99 of 1993 regulate the water resources. The first law states that the environment is a collective possession and the State and people have to participate in the management and protection of the environment. The renewable natural resources belong to the nation without discrimination of the legitimate rights acquired by private individuals.

In the case of the water resources, the right to use the water can be acquired by law, permit, concession or association. By law, all the inhabitants of the country have the right of free use of the natural water resources in order to satisfy their primary needs; while only by concession the water resources can be used for economic activities and its use depends on the availability of the water and on the purpose of the water use; as is the case for the irrigation districts.

The code establishes that the Government is responsible for among others, the authorization and control of the improvement of the water resources, the coordination of the water management activities of the official organizations and water users associations and the control of the use of private water resources in order to avoid any environmental deterioration. Moreover, the code describes the taxes for the utilization of water and the water resources, namely a tax for the renewable use of water by environmental services (returning tax) and a tax for the utilization of the water resources (use tax). The returning tax includes payment for the services provided for the elimination or control of harmful impacts caused by profitable activities, while the use tax is meant as payment for the utilization of the water with the aim to develop economic activities.

Alternatively, the Law 99 of 1993 established the Ministry of Environment that resulted in a reorganization of the public sector for the management and conservation of the renewable natural resources and it is responsible for the organization of the so-called National Environmental System. The Ministry of Environment is in charge of the development of the policies and regulations for the reclamation, conservation, protection, management, use and improvement of the renewable natural resources and the environment in view of a sustainable development.

The Law 99 also included the establishment of the Autonomous Regional Corporations with the objective that they execute the policies, programs, plans and projects related to the environment and the renewable natural resources. They are the highest environmental authority within their region and they are responsible for among others, the provision of concessions, permits, authorizations and environmental licenses for the use of the renewable natural resources or for the introduction of activities that might affect the environment. They also have to establish the acceptable and allowable limits for the emission, spilling, transport, and storage of substances that can affect the environment and they have to evaluate, control and monitor the water uses and other renewable natural resources. They have the right to establish the amount and to collect the contributions, tax, tariffs and fines for the use of the natural resources. However, the water tariffs have been established without a careful analysis and they create some trouble for the administration of the irrigation districts.

Another major legal problem is the ownership of the infrastructure after the transfer of the irrigation districts. The Colombian Government owns the infrastructure, goods and equipment of the existing irrigation districts, which were transferred before 1993; the established water users associations have only the right to administer them. The districts that were transferred after 1993 own the infrastructure, but they have to pay back all the investment costs during an amortization period of 20 – 30 years.

1.2.2 Financial Sufficiency and Future Investments

Insufficient financial resources for the administration, operation and maintenance of the irrigation districts endanger the sustainability of the irrigation and drainage infrastructure and might result in the deterioration of the infrastructure. The available budget for these activities does not cover the total requirements and the irrigation districts do not generate sufficient income by their services provided (INAT, 1996).

According to INAT, most of the irrigation districts have a low fee-collection level, which is mainly due to the lack of up-to-date records of the water users within a district, a low capability of the water users to pay the required fees, a poor and inefficient water delivery to these water users; moreover several water users are not familiar with the functions and services of the irrigation districts.

Moreover, the irrigation districts have insufficient economical resources for the modernization and rehabilitation of their irrigation schemes. The water users associations have increased the fee, but the amount collected is less than that required for operation and maintenance. Therefore some maintenance activities are deferred and are causing the need for modernization. This situation is critical both for the schemes that do not receive subsidies from the government as those that receive subsidies.

The irrigation districts should create reserve funds to be used for the replacement of machinery, equipment and vehicles. At the moment these funds are insufficient for the procurement of these items or to secure these activities in the future.

The debt of the water users to the water users associations is increasing due to the fact that the water users are not able or willing to pay for the services of the association. A solution for this situation is hampered by the existing bureaucratic procedures of the Government Agency. Also the farmers are not able to increase their payment to the districts or to follow up a compromise drawn between them and the district.

Training for financial management is needed for the administrative departments of the water users associations. Transparent management of the economic and financial resources is a prerequisite to create confidence among the users.

1.2.3 Operation and Maintenance

The operation of the irrigation schemes has been affected by a progressive deterioration of the hydraulic infrastructure caused by inappropriate maintenance, a cropping pattern different from the one assumed during the planning stage of the irrigation districts, by improper and inefficient irrigation schedules, and by an underused infrastructure. As a consequence the systems are operated in a deficient way and the use of the available land and water resources is not efficient.

Deficiency in budgets caused postponement of the required maintenance programs. Many water users associations intend to solve this problem by a request for rehabilitation, a program that is financed by the Government. To support their request the present maintenance programs are of low quality.

The reduction in specialized technical staff has affected the quality of the services for the operation and maintenance of the irrigation schemes. Less expensive and less experienced staff has replaced the initial, well-trained staff.

The performance of the irrigation schemes is hard to evaluate due to a lack of interest of the Government Agency to collect and process the required information of the irrigation schemes.

1.2.4 Role of Government Agency and Water Users Associations

The Government Agency has to redefine its role with respect to a future transfer program. Among the issues to be taken into account are:
- The responsibilities of the parties involved in the transfer process, both during the transfer and after it;
- The required modifications of the transfer procedures taking into account the past experiences;
- A training plan to support the water users associations after the transfer;
- Definition concerning budgetary compromises, to pursuit subsidies, to establish training plans, modernization programs and schedules for replacement of equipment;
- Research and technology transfer.

1.2.5 Environmental Aspects

Among the environmental aspects can be mentioned:
- *Catchment deterioration.* Plans for conservation and improvement of the catchment areas are needed, otherwise decreasing discharges and increasing sediment loads will affect the functioning of irrigation schemes;

- *Water quality deterioration.* Plans to improve the water quality are needed, mainly in view of the use of fertilizers and pesticides;
- *Contamination of irrigation water.* The water supply for irrigation, households and industry is affected by the contamination of industrial wastewater. In the specific case of the RUT Irrigation District the contamination of the irrigation water by wastewater coming from the municipalities is an explicit environmental aspect. Technical alternatives for the re-use of treated wastewater from the cities for irrigation have been suggested in the research program.

1.2.6 Leadership

Actions are needed to motivate all the water users to participate in the decision-making process required for a sound and sustainable management of the irrigation districts. At the moment less than 50% of the water users are present in the general assembly. This means that only a minority of the water users is responsible for the decisions and the decision-making process. A higher percentage of attendance is required to have a sound dialogue between the water users and the staff of the irrigation district regarding farmer's perceptions of what physical construction and institutional development need to be made for a sustainable management.

1.3   Objectives

The main objective of this study is to present and evaluate a methodology for an integrated and sustainable development of the irrigation districts after their transfer to a water users association. This methodology has been applied to the RUT Irrigation District under its present conditions. The specific objectives include:
- Study of the past and ongoing transfer programs in other countries with similar conditions to Colombia;
- The analysis and evaluation of the transfer process in Colombia including the study of positive and negative impacts of the transfer process with special emphasis on the RUT Irrigation District;
- The development of a methodical framework for an integrated and sustainable management of the irrigation districts within optimal social, economic, legal, technical and environmental conditions. This framework will be applied for the specific conditions of the RUT Irrigation District;
- Contribute to the selection of the best sustainable development scenario for the area by using the framework;
- Contribute to the management of other systems after transfer, as well as of the future transfer of other projects in Colombia.

1.4   Scope and Relevance

The scope of the research will be the formulation of a methodological framework for an integrated and sustainable management of the irrigation districts after the transfer from the government to the water user associations. The basic components involved in the concept for sustainable development such as financial, technical, socio-economic, legal, and environmental sustainability will be taken into account. The implementation of the proposed methodological framework will be focused on a specific irrigation management transfer, namely in the RUT Irrigation District. The results of the implementation will allow for recommendations for the short, middle and long term for the present and future transfer program in Colombia. The recommendations will be at institutional, organizational and operational level.

The present and future program for the transfer of the irrigation management from the Government to the water users associations is part of the governmental policy to decentralize the provision of public services. The objectives of this policy are to improve the efficiency of the service levels, to decrease the financial responsibility of the central government, to supply autonomy for the service providing organizations and for an efficient use of the available resources. In this way the research will contribute to the attainment of the objectives of the transfer program for the land improvement sector. Among the questions to be answered by the research can be mentioned:

- What is the concept for a sustainable management of the irrigation districts after transfer to the water users associations for the specific Colombian circumstances?
- What are the critical parameters and their nature influencing the transfer of the irrigation districts?
- In which way will the water users associations obtain full autonomy for the management of the transferred irrigation systems?
- What are the essential conditions to be fulfilled so that the water users associations become independent from the central government in view of the management of the irrigation districts after transfer?

1.5　Structure of the Thesis

Chapter 1 presents general information on Colombia, the importance of the agricultural sector for the economy, general characteristics of the small and large irrigation sectors existing in Colombia, and a general description of the problems faced by the irrigation districts after the transfer to water users associations.

Chapter 2 presents a short description about the availability of land and water resources in Colombia and their contribution to the development of the irrigated agriculture. It presents also the historic development of irrigation in Colombia and the general characteristics of the institutional framework for the land improvement sector to promote irrigation development.

Chapter 3 describes the background of the transfer program in Colombia, its implementation, current status and impacts in some irrigation districts; and finally, conclusions are given concerning the transfer program in Colombia. Chapter 3 illustrates also some experiences and characteristics of the transfer carried out in other countries with similar characteristics as Colombia.

Chapter 4 develops the framework for a sustainable management of irrigation districts transferred to water users associations. In this way, this chapter compiles the basic concepts and approaches for an integrated water resources management, sustainability and sustainable management of water resources with which a conceptual framework is formulated and developed.

Chapter 5 describes the present situation of the RUT Irrigation District that is taken as case study for this research. It presents the historic background; description and characteristics of the physical, management and socio-economic aspects of the district; and finally the characteristics and impacts after the transfer process.

Chapter 6 presents an evaluation of the existing situation of the RUT Irrigation District, describes the expectations and possible scenarios for the implementation of the framework for an integrated and sustainable management and also the required physical and non-physical interventions; and finally outlines a plan for monitoring and impact evaluation.

Chapter 7 carries out an evaluation about the impacts caused by the implementation of the framework for the sustainable management of the RUT Irrigation District and the possibilities to extend it to other irrigation districts in Colombia.

2

Irrigation Development in Colombia

The public and private sectors have developed the land improvement in Colombia in a shared way. The private sector has improved a major land area (463,059 ha) where the irrigation infrastructure is predominant over the drainage. However, an important part of this area has only simple works for diversion, conveyance and distribution of water and is mainly used for the cultivation of rice and African palm. The area with a more complex irrigation infrastructure has been developed for cash and export crops such as sugar cane, banana, flowers and more recently fruits. The area with sugar cane is almost 200,000 ha and can be found in the flat region of the Valle del Cauca Department where the soils have a very good quality; flowers are grown on the savannah of Bogotá; banana in the Northwest region of Colombia, African palm in the regions of the Pacific Coast, eastern plains, and Caribbean Coast; and fruits in the Valle del Cauca and Tolima Departments. Figure 2.1 shows a map of Colombia.

The public sector has contributed to the development of 287,454 ha, of which 155,570 ha (54.1%) have an irrigation and drainage infrastructure and the remainder (131,884 ha, 45.9%) has only a drainage infrastructure. About 51% (79,467 ha) of the area with irrigation and drainage is located in the East-Centre region of the country, 26.6% (41,351 ha) in the Caribbean Coast, and the remainder in the West, Orinoquia and Amazon regions and the hillside areas. The Caribbean Coast and West-Centre regions have the largest area with a drainage infrastructure, 93,022 ha (70.5%) and 33,443 ha (25.4%) respectively (National Planning Department, 1991). The total area covered by the irrigation districts has a total capacity of 499 m³/s; the irrigation network is 1,905.2 km long, of which 37% is for the main canals, 45% for the secondary canals and 18% for the tertiary canals; the drainage network is 2,132.6 km long, of which 44% is for the main drains, 39% for the secondary drains and 17% for the tertiary drains; and the road network is 3,382.4 km long. About 9% of the land is used for permanent crops, 51% for non-permanent crops and 40% for grass. About 62% of the landholdings is smaller than 5 ha and cover about 9% of the total area, 17% of the properties is in the range of 5 – 10 ha and covers 13% of the area, while 3% of the farms is larger than 50 ha and cover 38% of the total area.

At the moment about 1,500 million hectares are currently cultivated lands in the world; 1,100 million hectares are without a water management system, 270 million hectares have an irrigation infrastructure, and 130 million hectares have only a drainage system (Schultz, 2002). Of the total area under irrigation 6.6% (15.1 million ha) is found in Latin America. Irrigation in Latin America is very concentrated: Mexico (36%), Brazil (17%), Chile (11%), Argentina (9%), Peru (7%), Cuba (5%), Ecuador (5%), Venezuela (3%) and the remainder is distributed among the other countries. Colombia also contributes 5%. Colombia occupies the sixth place among the South American countries in view of irrigated lands over the total cultivated lands (23.4%). Chile and Suriname have the highest percentages of irrigated lands over the total cultivated lands, 82.4% and 74.6%, respectively. (ICID, 2001)

Colombia has followed a similar tendency as the Latin American region for its irrigation development. The area increased rapidly up to the end of 1960s, but the growth considerably dropped during the last two decades. The average annual growth is for the period 1945 - 1960, 3.1%; 1961 - 1970, 6.6%; 1971 - 1980, 3.2%; and 1981 - 1990, 1.5% respectively (National Planning Department, 1991); the growth for the period 1990 – 2000 has been almost zero. The small growth during the last mentioned period was mainly due to investments of the private sector since the public sector concentrated its resources on the rehabilitation, completion and enlargement of the existing irrigation districts.

Figure 2.1 Map of Colombia (Periódicos Asociados Ltda, 2003)

2.1 Land and Water Resources

At world level, Colombia can be considered as a country with abundant natural resources, especially in relation to the land and water resources that present a great potential for the development of agriculture under irrigation.

2.1.1 Land Resources

The soil study for Colombia (1973) that was carried out by IGAC (Agustin Codazzi Geography Institute) followed the same method as used by the United States Agriculture Department (USDA). This method considers 8 land classes, where class-I is the best one and class-VIII is the worst. The classes I-IV refer to those lands that are appropriate for crops and other uses, and the classes V–VIII have some constraints in view of their agricultural use.

Table 2.1 Inventory of Land Classes in Colombia (Agustin Codazzi Geography Institute, 1973)

Class	Studied Area (1,000 ha)	Studied Area (%)	Potential Use
I	172	0.3	All crops
II	979	1.9	All crops
III	5,418	10.3	Almost all crops
IV	4,398	8.3	Restricted crops
V	1,605	3.0	Grass and woods
VI	13,654	25.9	Grass and woods
VII	18,202	34.5	Grass and woods
VIII	7,743	14.7	Natural vegetation
Water and Urban Area	600	1.1	
Subtotal	52,774	100	
Not studied area	61,401		
Total	114,175		

The soil study initially covered 53 million ha, which is 46% of the total area of the country. The remainder (61.4 million ha) includes the major part of the existent woodlands (40 million ha). Next, the study area was enlarged and included also the Colombian Amazon. Table 2.1 shows the results of the soil inventory according to the land classes.

The soil study distinguishes two soil types, namely soil type A and B. The type-A soils correspond to the soil classes I, II and III and are suitable for mechanization, intensive agriculture and livestock, and for the development of irrigated agriculture. These soils are mainly located in the Caribbean plains and in the Orinoquia and Amazon regions. Most of the type-B soils correspond to class IV, their ground surface varies from flat to concave-flat requiring among others land improvement like flood protection, drainage, and leaching. Once they are improved, their potential use is similar to the soil type A. Table 2.2 shows the distribution of the soil types A and B in the country. Table 2.2 shows that 6.6 million ha are suitable for mechanized agriculture (5.8% of total area) and the area could be increased to 10.6 million ha through the improvement of some of the type B soils.

On the other hand, 2.7 million ha are located in dry areas and their agricultural development might only be possible by supplying irrigation water. Other regions show wet and dry periods characterized by low rainfall and an irregular distribution.

Table 2.2 Distribution of Soil Types per Region in Colombia (Institute of Hydrology, Meteorology and Land Improvement (HIMAT), 1985)

Regions	Soil type					
	A (1,000 ha)	A (%)	B (1,000 ha)	B (%)	A and B (1,000 ha)	A and B (%)
Caribbean Plains	2,474	37.6	1,628	41.0	4,102	38.9
Orinoquia	1,039	15.8	----------	-----	1,039	9.8
Amazon	124	1.9	361	9.1	486	4.6
Remainder	2,952	44.8	1,973	49.8	4,925	46.7
Total	6,589	62.44	3,962	37.6	10,552	100

2.1.2   Water Resources

Colombia has an average annual rainfall of 1,900 mm. However, in about 70% of the area the average annual rainfall is 2,000 mm and the rainfall is more than 4,000 mm on almost a third of the area (Pacific Coast and East plains). These values are above the average of 1,600 mm per year for Latin America, and above the global average of 900 mm per year (Marin, 1992). In contrast, the Caribbean Coast has little rainfall and is considered as a dry region. According to the World Water Balance and Water Resources for the Earth (UNESCO, 1978) Colombia has a specific surface water yield of 0.59 l/s per ha, and the country occupies the fourth place after Russia, Canada and Brazil. This specific yield is six and three times greater than the values for the continental world and Latin America, 0.10 and 0.21 l/s per ha, respectively (Marin, 1992). Colombia has more than 1,000 perennial rivers, a figure that is larger than for Africa (60). Table 2.3 shows the surface water availability related to the altitude above mean sea level, the percentage of the area in relation to the area of the country and the percentage of the population living in the specific area.

Table 2.3 Surface Water Availability with Respect to Altitude, Area and Population (Jimenez and Carvajal, 1998)

Altitude (m+MSL)	Surface Area (%)	Population (%)	Water Availability (%)
> 3,000	9	1	4
1,000 – 3,000	35	66	44
< 1,000	56	33	62

From Table 2.3 follows that the water availability (62%) in the regions located under 1,000 m+MSL is high, that the area covers 56% of the total area and that 33% of the population lives in this region. The region between 1,000 – 3,000 m+MSL plays also an important role, it has a relatively high water availability/area-ratio and 38% of the agricultural activities, livestock and a significant part of the industry take place in this area.

Colombia has 10 interior rivers, which have an average annual discharge greater than 1,000 m³/s. The Caqueta River is the greatest river with 10,000 m³/s, but is surpassed by the Orinoco and Amazon River, which run along the borders with Venezuela and Brazil, respectively. Colombia has a natural drainage network of 740,000 micro-basins in view of the high rainfall, topography and geology. This natural drainage is distributed in 4 regions or main river basins as presented in Table 2.4.

Figure 2.2 shows the location of the basins before mentioned plus the basin of the Catatumbo River.

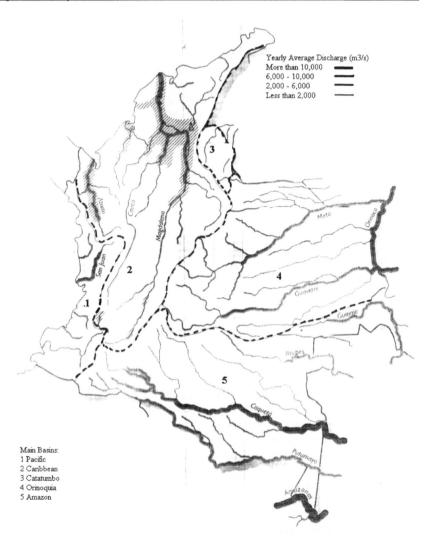

Figure 2.2 Main river basins (Periódicos Asociados Ltda, 1997)

Table 2.4 Regional Distribution of Surface Water (Jimenez and Carvajal, 1998)

Rivers in the Region	Average annual discharge (m^3/s)	Average annual discharge ($10^9\,m^3$)
Amazon	22,185	699
Orinoquia	21,399	675
Caribbean	15,430	487
Pacific	6,903	218
Total	65,917	2,079

In addition, 1.8% of the surface water is stored in lakes, lagoons, swamps and reservoirs, which occupy an area of 2,680,000 ha and have an artificial storage (reservoirs) of 6,773 million m^3.

The groundwater resources of the country have not yet been studied in detail. In the Valle del Cauca Department the agricultural sector is the main user of the groundwater and uses 88% of pumped groundwater, followed by the sector for drinking water supply and the industrial sector. The average installed capacity of tube wells is about 127 m^3/s, which produce a total annual water volume of 1,000 – 1,200 million m^3. The operation regime is not continuous over the year as the main user (agricultural sector) operates the wells mainly during the dry season. Table 2.5 shows the water demand for the different sectors, with the major demands from the hydropower and agricultural sector with 2,000 and 1,000 m^3/s, respectively.

Table 2.5 Water Demand for different Sectors (Jimenez and Carvajal, 1998)

Sector	Water Demand (m^3/s)	%
Hydropower	2,000	61.9
Agriculture	1,000	30.5
Industrial	42	1.3
Human Consumption	100	3.0
Other	142	4.3
Total	3,284	100

2.2   Potential for Irrigated Agriculture

According to the National Water Study (National Planning Department, 1984) Colombia has 14.4 million ha suitable for agriculture, of which 6.6 million ha (45.8%) are suitable for mechanized agriculture and the development of irrigated agriculture through irrigation and drainage infrastructures. Only 750,473 ha (11.4%) have an irrigation and drainage infrastructure contributing to the private sector with 62%. The most recent studies for land improvement identified about one million ha for the development of irrigated agriculture with an irrigation and drainage infrastructure. Table 2.6 and 2.7 show the present and potential area for irrigated agriculture and the plans for irrigation projects for the period 1991 – 2000 (Marin, 1992).

Table 2.6 Present and Potential Area for Irrigated Agriculture (Marin, 1992)

Region	Potential Area (1,000 ha)	Area with Works			
		Total (1,000 ha)	%	Public Sector (1,000 ha)	Private Sector (1,000 ha)
Caribbean Coast	2,474	216	8.8	134	82
West	1,081	254	23.5	11	243
East-Centre	1,718	195	11.3	113	82
Orinoquia	1,1645	58	5	1	56
Amazonia	151	5	3.6	5	0
Sub-Total	6,589	728	11.0	265 [1]	463 [2]
Small Irrigation [3]		22		22	
Total	6,589	750	11.4	287	463

[1] : Includes 132,000 ha with only drainage works
[2] : Estimated: 80% of the area has only irrigation works
[3] : Distributed over the country, it is not included within the potential area

Table 2.7 Land Improvement Projects Identified for the Period 1991 – 2000 (Marin, 1992)

2.1.1.1 Status	Gross Area (1,000 ha)
Design Phase	43
Feasibility Phase	202
Pre-feasibility Phase	725
Identification Phase	395
Total	1,366

The implementation of the land improvement program for the period 1991 – 2000 will be discussed in the next sections.

2.3   Historical Development

Irrigation in Colombia dates from several centuries before the arrival of the Spanish conquerors in the fifteenth century. When they arrived in the country, especially in the Southern part, they found a hydraulic infrastructure with a certain degree of complexity used for the capture, conveyance and storage of water for irrigation purposes.

Up to the early 1900s most irrigation systems in the country were provided by the private sector, after that the Government decided that water services should be provided by the public sector and later they concluded that such services should be furnished by agencies of the central Government. These interventions in the provision of water services were based on the belief that strong government interventions in the economy were required to ensure economic growth.

The first public sector irrigation and drainage project was Prado and Sevilla that covers 40,000 ha and is located in the Northern part of the country; it was built at the end of the nineteenth century. The United Fruit Company built an estate for banana on it, which is the largest and still functioning estate of the country (National Planning Department, 1991). In 1936, the National Ministry of Economy built the irrigation districts La Ramada and Fuquene-Cucunuba in the Cundinamarca Department, Firavitoba in the Boyaca Department and Bugalagrande Channel in the Valle del Cauca Department. In the 1940s, the Institute for Improvement of Water and Electric Promotion built the irrigation and drainage districts Samaca (Boyaca Department), Rio Recio (Tolima Department), the drainage district Alto Chicamocha (Boyaca Department) and the hydraulic works of the Gualanday Canal (Tolima Department). In addition, the Caja Agraria built the irrigation and drainage districts Coello and Saldaña and the enlargement of the Rio Recio District at the end of the 1940s.

In 1958, feasibility studies started for the land reclamation and development of what is now the RUT (Roldanillo - La Union - Toro) Irrigation and Drainage District. The studies were carried out by CVC (Autonomous Regional Corporation of the Cauca Valley). In 1962, with only 40% of the construction works completed, lack of funds stopped the work. In 1964, under INCORA (National Institute for Agrarian Reform), the flood control and land development works continued until 1970 when the construction was concluded. Between 1963 and 1972, INCORA started the construction of 15 new irrigation and drainage districts located in 8 departments.

At the beginning of the 1980s, the land improvement projects presented several problems: some projects were not finished due to lack of funds, progressive deterioration of the hydraulic infrastructure and equipment and conflicts with the users.

For this reason the Government Agency devoted its resources to the improvement of the existing schemes during this decade.

The Colombian Government established through HIMAT the Small Irrigation Program (1986-1990) that included 307 small irrigation projects and covered 25,293 ha, where about 11,118 families lived. In view of the importance of this program, the development of 850 projects was foreseen for the period 1991 – 1996, covering 60,900 ha with about 43,000 families. This area is about 11% of the proposed area (535,500 ha) to be developed and the investment costs are 18% of the expected total investment costs. Table 2.8 and figure 2.3 show the location of the irrigation districts built and operated by the public sector.

Table 2.8 Irrigation Districts Built and Operated by the Public Sector (National Planning Department, 1984)

Region	Area (1,000 ha)		
	Irrigation and Drainage	Drainage	Total
Caribbean Coast	41	93	134
East-Centre	79	33	112
West	11	0	11
Orinoquia	1	0	1
Amazonia	0	5	5
Small Irrigation	23	0	23
Total	155	131	286

2.4 Policy and Legal Framework

The historic development of the land improvement in Colombia has been largely influenced by the policy as established by the different governments, which have implemented their policies through the formulation of respective legal frameworks. The policy for the land improvement sector has changed by mandate (directive) after mandate, which did not allow a consistent development on the long term.

2.4.1 Outline of Policy

The political outlines of the Colombian Government for the sub-sector of land improvement were enclosed in the Economic-Social Development Plan for the period 1991 – 2000. The following summary presents the main outlines. Irrigation investment contributes in several ways to the growth of the agricultural production:
- Increases the arable area by the enlargement of the agricultural land;
- Land is used more intensively;
- Modern crop varieties and inputs are used for crop diversification and the adaptation of new crop varieties where the climate conditions are not favourable for their development.

Irrigation investments generate important benefits among others:
- Generation of direct and indirect employment;
- Increase of the farmers' income. It is estimated that the income from irrigated agriculture is four times larger than from rainfed agriculture;
- Promotion of price stability for consumers and producers;
- Reduction of the freezing effects on the crops.

Figure 2.3 Locations of land improvement projects of the public sector (National Planning Department, 1991)

In 1991 the Colombian Government considered that the public and private investments are good mechanisms to contribute to the modernization of the agricultural production. For this reason, the intensification and increased efficiency of the land use was the main objective for agricultural development. To reach that objective the

Government proposed to renew the investments in new irrigation projects, which had been suspended a decade ago.

From past experiences the following constraints can be identified for an efficient investment in irrigation and drainage projects:

- *Institutional.* The main actions of the central and regional agencies have been implementing activities; they did not efficiently use the investment resources, because of their scarce contact with the community, which would have to benefit from these developments. These agencies have developed projects of different scale without taking into account the advantages of each institution and without applying an adequate methodology for the prioritization of the investment. The regional agencies have developed small and middle size projects, but their executive capacity was constrained by the scarce economic resources and inadequate policy for tariffs and investment recovery. In addition these agencies did not carry out an efficient control and supervision of the private irrigation expansion, which resulted in adverse effects on the environment. The Ministry of Agriculture has not established a Water Division in charge of the technical aspects of the land improvement projects. The executive institutions were under the control of different central organizations: the regional agencies under the National Planning Department and the government agency under the Ministry of Agriculture. As a consequence INAT was created in June 1994 to perform the tasks and activities of a Water Division;

- *Users Participation.* The government agencies, insufficiently taking into account the interest and participation of the community carried out all the activities for the implementation of the project cycle. This situation had a negative impact on the functioning of the projects, the investment recovery and the operation and maintenance of the irrigation districts. The transfer of the irrigation management to the water users associations has been hampered by existing high subsidies for the most inefficient schemes, especially those where the water is captured by pumping or those with only drainage. During the years 1982 – 1988 the subsidies given by the Government were about 77% of the total costs for administration, operation and maintenance. In 1990, the government agency transferred the administration of 4 districts to the water users associations and subsidized the remainder of the 20 districts according to the type of irrigation and drainage, namely gravity, pumping or only drainage received 63%, 76% and 91% respectively;

- *Financial Aspects.* For the public sector the main constraint has been the scarcity of financial resources due to the low recovery of the investments. On the other hand, the financial constraints have also affected the investments of the private sector. The existing credit facilities do no correspond to the amount of money and instalments required for the development of this type of projects.

- *Technical Aspects.* Most of the irrigation districts show inflexible design conditions, which do not allow a variable water delivery and/or the production of other crops than those planned at the beginning of the project. The non-existence of devices to measure the amount of water delivered to the farmers has caused an inefficient use and unrealistic billing of water. In addition the supporting services to increase the productivity like technical assistance, training and applied research have been scarce. Monitoring and evaluation of the performance of the irrigation districts have not been carried out. Therefore adjustments and improvements to the development of the schemes have not been done in order to reach the objectives proposed during the planning phase.

- *Environmental Aspects.* Deforestation and inappropriate land use have caused the deterioration of the catchments areas. For this reason the water sources show decreased discharges during the dry season and increased sediment loads during the wet season. These aspects cause a decrease of the conveyance capacity of the channels and high maintenance cost for the sediment removal.

2.4.2 Role of the State for the Investment in Land Improvement

Private investments in projects for land improvement are considered to be important for the development of an efficient and profitable agriculture. In some irrigation projects, such as those for improved groundwater use, the incentive of private profitability is sufficient to obtain efficient investments of the resources. However, the Government will have to permit and control the groundwater use to guarantee that the various users equally benefit from this resource. Moreover, the Government would have to promote that the water supply is by gravity because most of the private sector has not the resources and means to develop another type of irrigation.

The investments in land improvement projects would preferably have to be promoted and developed jointly by the public and private sector. To facilitate the investment recovery the Government will have to promote community participation during the design process of the projects, which after construction will be transferred to the water users associations for administration, operation and maintenance.

The Government can support and encourage pre-feasibility and feasibility studies with the purpose to promote these investments. Moreover, the Government can promote, execute and finance the investments and can carry out jointly with the communities the identification, formulation, construction, community organization, training and supporting services for the development of projects. In this case, the private sector will receive technical support from the Government to develop some phases of the project cycle or to develop them completely according to the established rules.

2.4.3 Physical Goals

To reach the above-mentioned aims the Colombian Government established the "Decade Land Improvement Program for 1991 – 2000" to improve 535,500 ha with an irrigation and drainage infrastructure. This figure is two times the land area improved by the public sector at the moment and is 8.1% of the land potential to be improved. Table 2.9 shows the preliminary land improvement program for this period.

Table 2.9 Preliminary Land Improvement Program, Period 1991 – 2000 (National Planning Department, 1991)

Project	Area (1,000 ha)	Total Investment (millions US$)
Rehabilitation, improvement and enlargement	144.5	184
Medium and large size	86.2	210
Small irrigation	60.9	197
Other to define	243.9	527
Total	535.5	1,118

The governmental agency (HIMAT, later INAT in 1994) was responsible for the promotion and development of the large irrigation projects and also for the involvement of the community in all the development phases of the projects, while the regional agencies were responsible for the medium and small projects. The small irrigation program expected to develop 850 projects that would improve 60,000 ha for about 43,000 benefiting families and recovering 50% of the investment. The average project size was 70 ha.

During the different periods the Government has adapted the land improvement policies with the purpose to follow the preliminary land improvement program that was established for the period 1991 – 2000 by the National Planning Department. In this way, the policy for the period 1994 –1998 established a program for land improvement with the strategic name "Social Jump", that was looking for a modern, profitable, and productive agriculture and the improvement of the living conditions of the rural population. In this way, the Colombian Government renewed the construction of the hydraulic infrastructure, which had been suspended approximately 25 years ago.

The objective of the program was the improvement of 145,160 ha with a cost of US$ 578.8 million for the design and construction of irrigation projects. It was expected that about 15,182 families would benefit from the activities. Moreover, about 93% of the improved area involved large irrigation projects and the remainder middle ones. Approximately 35% of the budget was allocated for the construction of irrigation schemes, 2% for the design and construction, 37% for the designs, and the remainder for feasibility studies. During this period, only the construction of the large Alto Chicamocha Irrigation District started in 1997 and 2,000 ha of the potential area of 6,022 ha has been in service in 1998. Table 2.10 shows the general characteristics of the land improvement program for the middle and large irrigation projects.

Additionally, a program for small irrigation schemes had also been considered and included the construction of new small irrigation projects and the rehabilitation of more than 150 ones. In this way, 73 small irrigation projects would be built at the beginning of 1997, covering 7,692 ha and benefiting 4,399 families. The finance would be from the World Bank and the National Fund for Land Improvement (FONAT). In summary, the overall goal for the above mentioned period was the improvement of 200,000 ha, to be divided in 110,000 ha for middle and large irrigation projects, 40,000 ha for small ones, and 50,000 ha for the rehabilitation of irrigation schemes.

The governmental policy for the land improvement for the first two years of the period 2002 – 2006 has been derived from the report "Memorias 2003-2004. Manejo Social del Campo", prepared by the Ministry of Agriculture and Rural Development, which presented an up-to-date balance of the land improvement projects. According to this report, the budget for the year 2003 was Col$ 39,516 million, of which 67% was used for the irrigation infrastructure, 8% for the supervision of the works executed, and the remainder for support of the development of the land improvement projects within the cooperation agreement between the Colombian Government and the Executive Secretary of the Andres Bello Agreement (SECAB). The investments in irrigation infrastructure have been used for the construction of 21 small irrigation schemes (44%), for the rehabilitation of 28 small ones (9%), and the rehabilitation of 21 middle irrigation schemes (47%). Table 2.11 shows the figures for this period.

According to the Colombian Institute for Rural Development (INCODER), the schedule for land improvement for the year 2003 also included the following:

-   Reformulation of the Santo Tomas del Uvito irrigation project, which was suspended due to its high costs. The economic resources assigned to this project were addressed for the rehabilitation and complementation of the Manatí, Repelón, and Santa Lucía irrigations schemes, which altogether cover 7,000 ha;

- Finalization of the designs and start of the works for Ranchería Irrigation Scheme;
- Enlargement of the Coello Irrigation Scheme involving a new area of 21,000 ha;
- Construction of the Ariari Irrigation Project;
- Rehabilitation of the Río Zulia Irrigation Scheme and;
- Design for the Golondrinas irrigation scheme located in Tolima Department and covering 14,000 ha.

Table 2.10 Land Improvement Program for the Period 1994 - 1998 (Based on National Institute for Land Improvement, 1995)

Status	Project (Location Department)	Scale	Area (ha)	Families	Cost (million US$)
C	Magara (Santander)	LI	10,000	230	12
	Alto Chicamocha (Boyacá)	LI	6,022	3,100	45
	Zanja Honda - Triangulo (Tolima)	LI	24,700	6,000	170
	Guamo (Tolima)	LI	5,210	160	22
	Cabrera Tres Pasos (Huila)	LI	5,570	350	30
	Sub-total		51,502	9,840	279
D&C	San Juan del Cesar (Guajira)	MI	3,000	162	13.8
D	Ranchería (Guajira)	LI	16,000	754	120
	Santo Tomas del Uvito (Atlántico)	MI	4,500	393	22
	Hatico-Tamarindo (Tolima)	MI	2,400	200	10
	Patia (Cauca)	LI	6,543	476	41
	Ariari (Meta)	LI	23,815	2,000	93
	Sub-total		53,258	3,823	286
F	Montería-Lorica (Córdoba)				
	La Gloria (Sucre)	LI	7,500	182	
	Catatumbo (N. Santander)	LI	11,900	600	
	Aguas Blancas (Cesar)	LI	6,000	250	
	Toronto-La Patria (Córdoba)	LI	12,000	325	
	Sub-total		37,400	1357	
TOTAL			145,160	15,182	578.8

C: Construction   D&C: Design and Construction   D: Design   F: Feasibility
LI: Large Irrigation MI: Middle Irrigation

Table 2.11 Executed Budgets for Land Improvement Projects for the year 2003 (Ministry of Agriculture and Rural Development, 2004)

Components	Budget in %	Characteristics					
I&D infrastructure	67	Status	Type	Area (ha)	#	Families	% Budget
		C	SIP	3,131	21	2,356	44
		R	SIP	2,554	28	1,976	9
		R	MIP	67,271	21	13,300	47
				72,956	70	17,632	
Work supervision	8						
SECAB	25						
Total (million Col$)	39,516						

SECAB: Executive Secretary of Andres Bello Agreement   C: Construction   R: Rehabilitation
SIP: small irrigation projects MIP: middle irrigation projects

The budget that was assigned for the year 2004 was Col$ 72,785 million of which 59% was invested in the irrigation infrastructure, 23% for agrarian reform, 4% for the fishing and aquaculture sector, and the remainder for productive development. The irrigation infrastructure included the design and construction of the dam and main conveyance canals for the Ranchería Irrigation Scheme; and the revision and completion of detailed studies for the Triangulo del Tolima and Ariari Irrigation Projects. Moreover, Col$ 9,796 million that was supplied by the Public Audience Program would be used for investment in the irrigation infrastructure of an area of 26,443 ha with about 3,422 benefiting families. Table 2.12 shows the figures for the year 2004.

From the information in the Tables 2.10 to 2.12 follows that the land improvement programs had a slow development. For instance, consider the cases of the large irrigation projects Ranchería, Triangulo del Tolima and Ariari, which are since 1994 the most important projects within the different governmental programs. Their potential area is about 64,242 ha or 60% of the total target area for the projects under design and construction for the period 1994 – 1998. However, 10 years later, the projects have still the same status, namely design for Triangulo del Tolima and Ariari, and design and partial construction for the Ranchería Project.

Table 2.12 Schedule for Land Improvement Projects for the year 2004 (Ministry of Agriculture and Rural Development, 2004)

Project	Location Department	Cost (Million Col$)	Families	Area (ha)	Characteristics
Ranchería	Guajira	19,981	754	15,820	D&C dam and main conveyance
Triangulo Tolima	Tolima	9,000	6,000	24,607	D: revision and
Ariari	Meta		2,000	23,815	complementation
Río Recio	Tolima	494	360	9,492	R and enlargement
Cravo Sur	Casanare	2,000	164	10,650	D
Guamal San Sebastian	Magdalena	1,402	386	386	C
Beltran	C/marca	1,000	57	185	D&C
La Copa I Etapa	Boyacá	500	2,000	2,500	D&C
Fomeque	C/marca	900	201	300	D&C
Ubaque	C/marca	1,500	200	200	D&C
Tesalia-Paicol	Huila	2,000	54	2,730	D&C
Total		37,277	12,176	90,685	

C: Construction   D&C: Design and Construction   D: Design   R: Rehabilitation

2.4.4 Strategies and Mechanisms

For the execution of the land improvement program for the years 1991 – 2000, some strategies and mechanisms were established such as:

- Creation of special credit facilities to finance the basic studies, designs and construction of individual and associated projects. In addition, the government agencies will give technical support, and legal and juridical support needed to create water users associations and to obtain the water concessions;

- A new institutional framework was proposed based on investment recovery, clear and transparent subsidy policy, criteria for tariff collection, community participation in the development of the project cycle, supporting services and environmental protection;

- In the new institutional framework the Government changed its role from being a simple executor to more an investment promoter with a major involvement of the community;
- The community participation in the development of the projects was carried out through technical committees constituted by specialized professionals contracted by the community;
- The public investment was addressed to the development of profitable projects from a social and economic point of view and with the criterion of social equity. Among the criteria for the selection of the projects the following can be mentioned: social profit, strategic location with respect to export harbours and consumer centres, and the level of interest of the community;
- The water tax would be a proper financial resource for the regional agencies. These agencies fixed the amount and the tax collected would be used for investment in new land improvement works;
- Simple mechanisms were designed which allowed the government agencies to recover the investments including some pressure mechanisms for the case of users, who would be reluctant to pay. Credit facilities were designed to facilitate the investment payment by the users before, during and after the construction works;
- A tariff policy was addressed to reach self-sufficiency in order to finance the total costs for administration, operation, maintenance, and equipment replacement. As a general criterion and looking for efficient water use it has been established that the volume tariff had a more important weight than the fixed one;
- The irrigation district management, existent and future, would be transferred to the water users through delegation contracts. The government agencies would train the water user associations in aspects related to financial administration, accounting and engineering. The creation of the federation of water users associations (FEDERIEGO) would facilitate the communication between the beneficiaries and the government agency.

2.4.5 Legal Framework

The legal framework for the improvement of the water resources is supported by the principles of the National Constitution of 1991, the code of natural resources (Decree 2811, 1974) and the Laws 99 and 373 of 1993 and 1997 respectively). Among the more outstanding principles can be mentioned: protection and conservation of the environment, natural resources, biodiversity, special ecological areas; the recognition of the ethnical and cultural diversity of the country; and joined, participated and decentralized environmental actions (Calle et al., 1998).

The National Constitution states in article 80:

"the Government will plan the management and improvement of the natural resources guaranteeing their development, conservation, recovery or replacement. In addition, it must foresee and control the factors of environmental deterioration; impose the legal penalties and require the amendments of the caused damages".

In turn, the Law 99 establishes:

"for the water resources, the human consumer will have priority over any other user".

The Law 373 establishes clear policies about efficient water use, measurements, maximum and basic consumption, groundwater, users training, and low consumption technologies. The Law 41 of 1993 set up the rules by which the land improvement sub-sector is organized. It defines land improvement as:

"The infrastructure works devoted to endow an area with facilities for irrigation, drainage and flood control to increase the productivity of the farm sector. It is considered as the public sector and defines users as every juridical or natural person".

The Ministry of Agriculture, Superior Council for Land Improvement (CONSUAT), the Government Agency (INAT) and National Fund for Land and Improvement (FONAT) together constitute the land improvement sub-sector. CONSUAT is the advisory entity in charge of the application of the policies for the land improvement sub-sector. INAT is the executive entity to develop the land improvement policy with other public and private entities. FONAT is the administrative entity for the financing of irrigation, drainage and flood control projects.

The transfer process employs a legal rule in the country's constitution and is called the "delegation of administration", by which a public good can be turned over to a private corporate entity (water users association) for administration on behalf of the Government. The delegation of administration created several conflicts between the districts and the Government, which resulted in legal debates and proceedings until the 1990s (Vermillion and Garces-Restrepo, 1996).

Law 41 and its associated decrees 1278 and 2135 determine that after a transfer the control over irrigation districts, finances, operation and maintenance procedures and personnel would be placed in the hands of the water users associations. The new agreements were referred to as "concessional contracts" rather than agreements for "delegation of administration".

After 7 years of land improvement programs within the new institutional framework, strong criticism has risen in view of the role of the governmental agency (HIMAT, later INAT) due to a non-efficient management of the economic resources and an almost zero-development of the land improvement programs launched from 1990 by different governmental programs. These criticisms were among others related to the management of external resources for the land improvement program, the costs for the functioning of the agency, and the quality of the agency's performance; these aspects have been widely mentioned and discussed (Ramirez, 1998).

As an example of the non-efficient management of the economic resources can be mentioned the loans from external sources (credits 2667-CO and BIRF-3113-CO). The first loan with a value of US$ 114 million was approved by the World Bank in 1985 and became effective in 1987. The loan was meant for the rehabilitation of 6 large and middle irrigation districts, and to develop a small irrigation program for about 3,900 small farmers. The aim of the program was to recover the agricultural production in areas affected by natural disasters and to improve the sustainability of the public investments in the irrigation infrastructure through cost recovery for the operation and maintenance of the irrigation districts. The results after 8 years showed that about US$ 16.9 million were not used and were returned to the World Bank; however, the objectives of the investment program and civil works in the irrigation districts were not achieved. On the other hand the government agency (INAT) looked well after its own institutional strengthening by the purchase of vehicles and an overseas training program for the directors. Torres (1999) mentioned a striking detail, namely that at the end of the loan the planned enlargement of the Coello Irrigation District was not fully carried out. However, in 1998 the Water Users Association of the Coello District with the right resources (US$ 6 million) was able to finalize the project within six months.

Similar situations occurred with the second loan with a value of US$ 78.2 million, which was approved in 1990 with the objective to finance a small irrigation program of 60,000 ha for about 4,300 small farmers. After 4 years of zero-execution, the program was reformulated and the target was reduced to 20,000 ha including some middle and large irrigation projects, which were finally not executed. At the closing date of the loan, US$ 28.2 million were returned to the financial agency.

On the other hand, INAT is considered as a demanding institution in view of the high costs required for the execution of its assigned functions. According to the Land Improvement Program, the unit cost for the land improvement was US$ 2,500/ha for the period 1991-2000; however, the national budget assigned for the land improvement projects was US$ 6,500/ha, being the contribution of the Government as its share in the program. These figures are in contradiction with the unit cost of US$ 907/ha for land improvement by the private sector and as reported by the Riopaila sugar cane factory in 1997. Besides, the National Planning Department (1997) presented the ratio of INAT's budget for its functioning and investment during the period 1990 – 1997 and the number of improved hectares as US$ 31,495/ha (price 1996). In the same line, Torres (1999) mentioned that during the period 1994 – 1996, the Colombian Government approved 745 land improvement projects that covered 16,780 ha and that would require an investment of approximately US$ 680/ha to be executed by the Incentive for Rural Capitalization (ICR).

The performance of the government agency has been measured and evaluated in terms of the area improved for irrigation in comparison with the assigned national budget. In a direct way the Government has improved 5,800 ha for the middle and large irrigation projects during the years 1969 – 1997; besides 38,000 ha were improved for the small irrigation projects during the period 1983 – 1997, of which 65% was only working after a rehabilitation program (Torres, 1999). On the other hand, the national program for land improvement for the years 1994 – 1998 had as target 200,000 ha, namely 40,000 ha for small irrigation, 110,000 for middle and large irrigation, and 50,000 for rehabilitation with a total cost of US$ 450 million (price 1994). About 59% of the work was assigned to INAT, but the final results were not proportional to the assignments. Of the Alto Chicamocha Irrigation District (built in 1997) only 2,000 ha of the planned 4,000 ha were in operation in 1998. According to the Law 41 of 1993 for land improvement, the available governmental resources plus the share supplied by the farmers should have been sufficient to reach the governmental goal.

Additionally, the performance of INAT as mediator between the water user associations and the Superior Council for Land Improvement for the development of land improvement projects has also been criticized; INAT submitted to the Superior Council only those projects in which INAT was interested. A study of the World Bank showed that some projects submitted by INAT were not feasible given their high costs per unit area, a very small number of benefited farmers, cropping patterns with low value crops, and a return rate of less than 12%. Besides theses projects presented a poor design and the design criteria were only focused on the engineering aspects; little attention was given to the development at farm level.

Concluding it can be said that the institutional framework, of which the INAT was a component, was a significant factor in the high costs of the land improvement works in the public sector. This fact has caused a very low institutional credibility, no interest for private investments, a serious obstacle for rural economic growth; corruption in the processes of hiring and executing the works, and finally the results of the construction works did not match the users needs (Torres, 1999).

This situation required a change in the institutional framework for the land improvement. As a consequence, changes were introduced at the beginning of 2003 as a first step of the new Government for the period 2002 – 2006. In this way, INAT and other public institutions within the public agricultural sector were suppressed and integrated in a new organization called Colombian Institute for Rural Development, INCODER. This organization was created in July 2003 after allowing the integration of INAT with the Colombian Institute for Agrarian Reform (INCORA), the National Institute for Fishing and Aquaculture (INPA), and the Co-Financial Fund for the Rural Investment (DRI). This integration resulted in the first place in a staff reduction of 56% for the agricultural public sector. INCODER belongs to the Ministry of Agriculture and Rural Development and basically comprises the Direction Council, Managers Office, and Territorial Linking Offices. The Direction Council is integrated by the Minister of Agriculture, Minister of Social Protection, Minister of Territorial Development, Housing and Environment, Director of the National Planning Department, representative of the President, and one representative of the following organizations: agricultural workers, natives, Afro-Colombian people, and Agricultural Sector Union. Here, it is important to mention that the Federation of Irrigation Districts Users Associations (FEDERRIEGO) was represented before in the CONSUAT, but in the new set-up this representation is lost. INCODER, is now charged with the execution of the policy for rural development by facilitating the access to production factors, strengthening the regional entities and their communities and promoting the voicing of the institutional actions in order to contribute to the socio-economic development of the country. Figure 2.4 shows a chart of the organizational structure of INCODER.

With the purpose to reach its institutional mission INCODER has formulated the following general objectives:

- To lead the identification and consolidation of areas for rural development promoted by public, private and mixed initiatives, in order to carry out rural development programs with the common purpose to satisfy specific needs of rural areas and communities;
- To strengthen the participative processes for institutional planning at regional and local level in order to define rural development programs allowing to rural actors the identification of development opportunities according to their context and the arrangement of the required investments;
- To strengthen the inter and intra-sectoral coordination processes which allow for the integration within the rural context;
- To consolidate the delegation and decentralization process to local and regional administration through the support of competent institutions at departmental and municipal level as well as of the rural producers and communities;
- To negotiate and give co-financing resources, subsidies and incentives to support the execution of rural development programs, facilitating the access and use of the production factors to middle and small producers;
- To contribute to the strengthening of fishing activities through the research, arrangement, control, administration, and regulation for the use and sustainable development of these resources.

The new organizational framework does not include the Superior Council for Land Improvement (CONSUAT), which was introduced by the Law 41 of 1993. Now, its function will be executed by the Director Council of INCODER with support of the Infrastructure Office, which has among others to advise the Managers Office and the Territorial Linking Offices in the formulation of plans and programs for the physical infrastructure and basic services with special emphasis on land improvement topics. On the other hand, 9 territorial linking offices were created (Agreement 002 of July 2003) replacing the previous 20 regional offices of INAT. Those new offices were created with the objective to satisfy the service demands in the national and territorial context, being in line with the delegation, diversion and decentralization principles of functions. Their basic function is the direction and coordination of plans, programs, projects, and policy of INCODER in their influence areas.

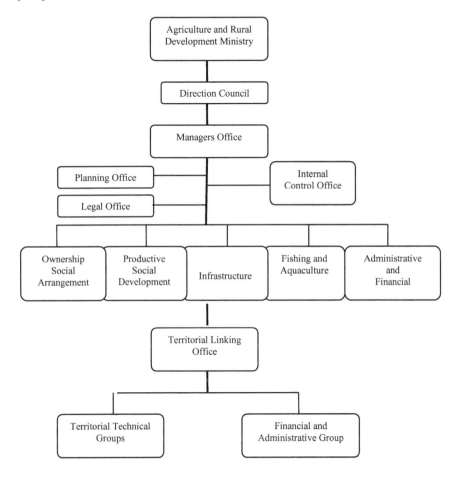

Figure 2.4 Organizational chart of INCODER (Colombian Institute for Rural Development, 2003)

The Agreement 003 of February 2004 is considered as a key element within the new institutional framework by which organizational guidelines are established for the users' organizations including the basic regulation for the functioning of the small, middle and large irrigation districts. The concepts, definitions and regulations given by this agreement in relation to the water users associations are now considered in a more extensive way in comparison with the Law 41 of 1993. A water users association is now defined as a legal person with private rights, with a corporative character and special objectives and it is a non-profitable association; its main objective is the administration of an irrigation district under competitiveness, equity, and sustainability principles. The law not only authorizes the water users associations for the administration, operation and maintenance of the irrigation districts under their responsibility, but also for the promotion of activities in order to reach a major social and economic development within its influence area, and to carry out actions in relation to the defence and conservation of the available water resources. Thus for example, the water users associations can formulate and develop training programs for their users and board members, help users to develop programs to increase the agricultural productivity, to promote the creation of organizations and agro-industrial enterprises, and take care of the integrated development of the organization.

The Agreement also discusses a controversial aspect that is related to the so-called "weighted vote" by which the influence of the user on the decision-making in the general assembly depends on the size of the area that he/she owns. Before, the decision-making was based on one vote per user, now it will be affected by the ratio of the area owned by the user and the total benefited area. For the election process of the Board of Directors carried out at the start of 2004, a large majority of the water users associations did not yet follow the new rule; however, a strong division in opinion in the Coello Users Association resulted in a request for intervention of the Government. The new rule caused great worry amongst the small farmers, because they believe and argue that the new rule will give extra power to the large farmers, especially in the decision-making process at administrative level and in the water resources management. This argument should be carefully considered by the Government taking into account that 79% of the properties covering 22% of the irrigation districts area have an area smaller than 10 ha, while 3.1% of the properties covering 39% of the total area are an area larger than 50 ha (INAT, 1996).

In view of the conflict among the water users of the Coello District, the Delegated Attorney for the Preventive Vigilance of the Public Function declared that the voting rule is unfair and inconvenient for the small land owners (minifundios), and that it helps the large land owners (latifundios), since the rule will result in a major land and water concentration. As a consequence of this situation, the article related to "weighted vote" has been suppressed, because it is not supported by the democratic principle of 1 vote per person. In the same way, the Infrastructure Office of INCODER has evaluated the contents of the Agreement 003, and as result it will propose to the Director Council a total elimination of this agreement. A new regulation has been proposed for the administration, operation and maintenance of the irrigation districts including the transfer forms (delegation, administration, and concession) and considering that INCODER must not have direct interference in the water users organization, but only in the management of the infrastructure (Resolution 1399 of 2005).

3

Management Transfer of Irrigation Systems

3.1   General Aspects of Irrigation Management Transfer

Management transfer of irrigation systems is being undertaken in many countries at present, especially in the developing countries of Asia and Africa, and in Eastern Europe and in some Latin American countries. The justification is in many respects similar to the rationale of privatization in other economic sectors. There are financial, social and technical arguments for the view that state-owned irrigation systems can be managed better, if they are transferred to local organizations, controlled either by their users or by some combination of users and government officials. However, these programmes are not always successful and the new organizations are in many cases not able to perform the management functions as well as expected. The problems that hamper success in these programmes are usually of a financial and social kind, rather than of a technical type. Transfer of irrigation management faces problems that are different from those of other privatizations, because the profitability of irrigated agriculture is under pressure, and so it is difficult to attract adequate and sufficient capital resources for it.

The policy of transfer of irrigation management (sometimes called "turn-over") began in the mid-1970s in a few developing countries, such as the Philippines and Colombia. It gathered pace through the 1980s and 1990s and in 1997 Vermillion identified 24 countries pursuing such policies, 10 of them in Asia, 8 in Africa and 6 in Latin America. To those countries should now be added several in Eastern Europe and Central Asia. By now, there are probably some thousands of organizations of irrigation farmers that have been set up under programmes of this kind and in a large variety of developing countries (Abernethy, 2001).

Irrigation management transfer (IMT) can be defined as the transfer of the responsibility and authority for irrigation system management from government agencies to water users associations, or private sector entities. The words 'turnover', 'handing over', 'devolution' and 'privatization' are often used synonymously with transfer. Irrigation management transfer may include the transfer of water rights from the government to water users associations as in Mexico; or it may only include turning over the partial management responsibilities, such as water delivery, canal maintenance and paying for irrigation services to water users as in Sri Lanka or Philippines, while the final approval of operation and maintenance plans and budgets are subject to government approval as has been the case with the first wave of irrigation management transfer in Colombia (FAO and INPIM, 2002).

The International Commission on Irrigation and Drainage (ICID, Yalta Declaration 2002) describes irrigation management transfer as the process to delegate the management responsibility and authority for irrigation systems, previously held by governmental institutions, to farmers or organizations of water users. It may also include the transfer of ownership of parts of the systems. These transfers are in line with the significant changes from the tradition of centralized planning and production systems into a market economy.

Irrigated agriculture is facing organizational changes world-wide. There is a growing recognition that irrigation water management is a service provided to customers with better results when operated by decentralized organizations: this leads to irrigation management transfer. Irrigation management transfer is a complex process, involving possibly infrastructural changes, institutional changes and legislative requirements. At the beginning of the 1990s started in many Eastern European countries a period of transition, namely the change from a central planning economy to a market economy. During this period several specific economic difficulties and rather critical situations could be observed and a lot of common issues in agriculture and water management could be distinguished. In irrigated agriculture the majority of the Eastern European countries faced similar problems: such as a decrease in agriculture production, deterioration of the irrigation and drainage infrastructure at on-farm level and a breakdown of the previous well-functioning markets. Everywhere in Eastern Europe, the processes of agricultural reform as well as reforms in water management were started and have been developed to solve these problems. Reforms in agriculture and water management in different countries were initiated with respect to the local natural conditions, traditional farm practices, infrastructural particularities as well as institutional and legislative developments (Zhovtonog et al., 2005).

As was mentioned before, irrigation management transfer (IMT) is the full or partial transfer of some or all the responsibilities from the present, governmental irrigation management to local entities, for instance a water users association. For large public irrigation systems the transfer often involves a partial transfer, with the government agency retaining the management and ownership of the water resources and the main distribution canals, while the farmers organization being responsible for the management of the secondary and minor canals and the canals within the distribution blocks.

Other transfer programs, involving smaller irrigation systems, consist of the full transfer of the management of the government to the farmers' organization. Among the responsibilities transferred can be mentioned the allocation and distribution of the irrigation water, operation and maintenance of the hydraulic infrastructure for irrigation and drainage, decision-making on the cropping pattern, irrigation scheduling, establishment of water charges and fee collection, preparation of the budget and administration forms.

The transfer can also include the transfer of the ownership of the whole physical infrastructure (canals, drains, roads, structures and communication system), of the irrigation staff and of the water rights to the local entity. The more comprehensive the transfer, the less the influence of the government on the future management activities of the local water users association will be. Some of the main objectives for a transfer are:
-    ensure sustainability of the irrigation districts;
-    reduce the financial burden on the government;
-    pass responsibility for operation and maintenance to the water users;
-    increase efficiency in the use of water, and improve and sustain system performance;
-    reduce the number of public employees in the irrigation districts.

The common tendency is to create organizations, which are more or less independent from the government. Decentralization of water management helps to involve users in the decision process and to allocate in an easy way funds for operation and maintenance of the infrastructure. In many countries with continuing transformation, the tasks of control and regulation, as well as the governance and executive functions are separated under different organizations. In the case of the process developed in Eastern European countries the main obstacles, which have been faced during the phase of restructuring the irrigation agencies, are the following: failures of the government in finding the correct model for restructuring the irrigation sector; weakness of the legislation leading to an unresolved legal situation of agencies and thereafter a lack of supervision and control; the mounting burden of debts; declining performance of the agencies; the distraction by non-core activities; overstaffing; fragmentation of agencies; and over-priced water charges. Restructuring the governmental agencies involved in irrigation appears an intervention that will have a massive impact on the established bureaucracy and employment security, and that usually meets strong resistance (Zhovtonog et al., 2005).

Merrey (1996) presents five hypotheses about the likely relationship between institutional principles and performance. The hypotheses are based upon autonomy and dependency of the agency and whether the agency manages a single system or multiple systems.

The main hypothesis states that irrigation systems managed by financially and organizationally autonomous system-specific organizations that are accountable to their customers perform better and more sustainable than systems managed by agencies dependent on the government and by agencies responsible for many systems. In other words fully autonomous organizations accountable to their customers and managing a single irrigation system will exhibit the highest performance, will prove to be most adaptive to changing conditions and therefore will prove to be most sustainable. Table 3.1 shows Merrey's hypotheses based upon the relationship between agency and government whether they are autonomous or dependent, managing a single system or multiple systems. Table 3.2 gives some examples of autonomous and dependent agencies managing a single system or multiple systems.

Table 3.1 Merrey's Hypotheses Based upon the Liaison Agency-Government (Merrey, 1996)

		Relationship of Agency to Government		
		Autonomous		Dependent
Agency manages a single irrigation system	1	Achieve highest performance Most adaptive to changing conditions Most sustainable	2	Mixed, but generally low performance Most adaptive to changing conditions Most sustainable
Agency manages multiple irrigation systems	3	Performance will vary among systems, but overall will be lower than (1), but higher than (4) Adaptability and sustainability will vary among systems, but overall will be lower than (1), but higher than (4)	4	Wide range of, but generally low performance Low adaptability and sustainability, with variation among systems based on local factors

Financial autonomy means the condition where an irrigation agency relies on user fees for a significant portion of the resources used for operation and maintenance, and the agency has expenditure control over the use of the funds generated from these charges (Small, 1990). It is believed that in an environment of financial autonomy, irrigation performance will improve both by freeing the operation and maintenance budget from the constraints imposed by the central government's fiscal difficulties, and by increasing the accountability of the irrigation system managers to the water users (Small and Curruthers, 1991).

Perry (1995) mentions that there are three elements, which are a prerequisite for proper irrigation management:

- clearly defined water rights;
- infrastructure capable of providing the service embodied in the water rights;
- assigned responsibilities for all aspects of irrigation system management.

Table 3.2 Some Examples of Autonomous and Dependent Agencies (Merrey, 1996)

		Relationship of Agency to Government		
		Autonomous		Dependent
Agency manages a single irrigation system	1	Mendoza, Argentina Irrigation Districts, USA Irrigation Districts, Colombia Taiwan systems Communals, Philippines Farmers-managed Irrigation System, Nepal Irrigation Districts, Mexico	2	Egypt Haryana, India Punjab and Sind, Pakistan ORMVA's, Morocco
Agency manages multiple irrigation systems	3	National Systems under NIA, Philippines	4	Irrigation Districts, Mexico (pre-transfer) Sri Lanka Nepal Indonesia technical systems West Frontier Province and Baluchistan, Pakistan

Water rights, infrastructure and responsibilities interact and are interdependent; changes in one will lead to changes in the other elements. All other common issues, e.g. proper operation and maintenance, sound institutions are either a subset or combination of the three basic elements.

Many governments are now either formulating new policies or implementing programs to transfer irrigation management to farmers associations or to autonomous, third party organizations, which are accountable to the farmers associations. "Management transfer is being carried out around the world, but little information is available about its results. The main reason for promoting the transfer programs is to reduce the cost of irrigation and drainage management for the government. There is little evidence, however, to suggest that irrigation management transfer policies affect the overall government expenditure in the water or agriculture sectors of developing countries" (Vermilion, 1997).

3.2   Developments in Colombia

Colombia has 287,454 ha with an irrigation and drainage infrastructure developed by the public sector. This area is grouped in 24 irrigation districts of which 16 have been transferred to the water users associations for irrigation management and the remainder is still under the administration of the Government Agency.

3.2.1 Background

In the 1970s the government decided that the private provision of water services is the most efficient means of improving both economic efficiency and social welfare. In the case of irrigation and drainage, this means transferring the responsibility for operation and maintenance from the public agency to water users associations. The transfer of irrigation districts in Colombia has been carried out during three periods. In 1976, the irrigation districts Coello and Saldaña were transferred by request of the water users. During the period 1989 – 1993 the irrigation districts Rio Recio, RUT, Samaca, and San Alfonso were transferred to compromise with the IBRD (International Bank for Reconstruction and Development). Finally, during 1994 – 1995 most of the irrigation districts were transferred following the policy established by the Colombian Government.

3.2.2 Current Status

Table 3.3 shows the development of the transfer program including information about irrigation and drainage areas, name of the water users association, and the transfer date.

Table 3.3 Status of Transfer Program of Irrigation Districts (Quintero, 1997)

District	Department	Area (ha)				
		Total	I&D	D	Total developed	Transfer date
Irrigation Districts Transferred to Water Users Associations						
Alto Chicamocha	Boyaca	16,722	1,373	8,321	15,235	1995
Samaca	Boyaca	2,947	2,934	0	2,934	1992
San Rafael	Boyaca	590	590	0	590	
Maria La Baja	Bolivar	19,600	6,429	2,861	17,300	1995
La Doctrina	Cordoba	3,000	2,000	0	2,570	1994
San Alfonso	Huila	3,028	1,042	8	3,028	1993
Tucurinca	Magdalena	9,866	7,214	0	7,214	1994
Sevilla	Magdalena	8,104	5,978	0	5,978	1994
Aracataca	Magdalena	13,300	10,692	0	10,692	1994
Abrego	N. Santander	2,000	1,200	0	1,200	1995
Zulia	N. Santander	15,843	10,519	0	10,519	1992
Rio Prado	Tolima	8,500	1,286	0	5,520	1995
Rio Recio	Tolima	23,600	10,374	0	18,650	1989
Coello	Tolima	44,100	25,000	1,870	40,000	1976
Saldaña	Tolima	22,500	13,000	0	14,050	1976
RUT	Valle	10,750	9,500	0	9,700	1989
Subtotal		204,450	109,131	13,060	165,180	
% Transferred Area		64	90	13	62	
Irrigation Districts Managed by Government Agency						
Manati Candelaria	Atlantico	29,000	0	21,268	25,206	1997
Repelon	Atlantico	3,800	2,100	250	3,400	1996
Santa Lucia	Atlantico	3,000	1,000	0	2,400	1996
Cerete-Lorica-Monteria	Cordoba	55,000	4,100	47,039	51,139	1997
El Juncal	Huila	5,100	2,997	123	3,100	1996
El Porvenir	Huila	762	241	115	357	1996
Sibundoy	Putumayo	8,500	0	7,855	7,855	1997
Lebrija	Santander	9,131	1,457	7,241	8,698	1997
Subtotal		114,293	11,875	83,891	102,155	
Total		318,743	121,006	96,951	267,335	
% Area to Transfer		36	10	87	38	

I&D: Irrigation and drainage   D: Drainage

From table 3.3 follows that 64% of the area developed by the public sector has been transferred to the water users associations, 90% has an irrigation and drainage infrastructure and only 13% has a drainage one, while 10% of the not transferred area has an irrigation and drainage infrastructure and 87% has only a drainage one. Notice that the transferred area is mainly concentrated in the Tolima Department (49%) and the Atlantic Coast (27%), while the remainder is in the Departments of North Santander (9%), Boyaca (10%) and Valle del Cauca (5%).

Transfer of Coello and Saldaña Irrigation Districts

As was mentioned before, the Coello and Saldaña Irrigation Districts were the first districts that were transferred. They are located in the Tolima Valley, in the central part of Colombia at an elevation of 380 m+MSL. The Valley is located between the central and western mountain ranges of the country, and the Magdalena River runs through it. The Valley has mostly a flat topography with undulating terrain towards both mountain ranges and has primarily alluvial soils, fans, terraces, and narrow valleys with minor rivers. The main soil characteristics are sand and loam in Coello and clay and loam in Saldaña. Yearly precipitation in the Valley is 1,000 to 1,500 mm. A marked bimodal rainfall distribution in April – May and October – November makes irrigation primarily supplemental. The mean temperature is 28 °C, the average relative humidity is 74%, and the yearly average pan evaporation is 1,800 mm.

In the 1930s, land reform in the Tolima Valley replaced the old hacienda system of peasant cultivation with land-ownership for farmers. Irrigation introduced in the 1950s transformed the agriculture in the Valley. Cotton became an important crop during the 1950s and 1960s. It was finally replaced by rice, which has been the main irrigated crop since the 1970s. Maize, sorghum, fruit and vegetables are also irrigated in the Valley now.

The Coello and Saldaña Irrigation Districts became operational in 1953 when the districts after 8 years of construction were completed. Initially they were constructed and managed as a single district. In 1976 they were separated after the management was turned over to a water users association. Table 3.4 shows the main characteristics of the Coello and Saldaña Irrigation Districts.

The irrigation supply in the Coello and Saldaña Irrigation Districts is by gravity. The Coello irrigation system captures water from the Coello River by a weir and lateral intake. Due to sedimentation problems the water is nowadays captured by a diversion channel with a length of 200 m. Next, the irrigation water is conveyed by an unlined conveyance canal with a capacity of 25 m³/s, which is 16.5 km long, The first 8 km has the name of Gualanday Canal and the remainder is called the Coello Canal. The Gualanday Canal suffers from siltation due to the high sediment load that is a result of the ongoing deforestation in the catchment of the Rio Coello Basin. Four tunnels, a sand trap, weirs, and structures for the evacuation of sediments are main hydraulic structures in the Gualanday Canal.

The conveyed water by Gualanday Canal has a multiple purpose; it is used for electric generation and 0.8 m³/s are used for human consumption of the Chicoral, Espinal and Coello Municipalities.

The distribution system consists of four main canals, among which can be mentioned the Jaramillo, Zerrezuela and Tolima Canals. The first one is 11.5 km long with a capacity of 6 m³/s and serves 2,000 ha for rice; the second one is 13 km long and has a capacity of 4 m³/s, it serves 6,000 ha; and the last canal is 6.5 km long with a capacity of 2 m³/s and it serves 1,000 ha. The control and division structures have radial gates in combination with a weir with a long crest. Finally the Coello Canal discharges its water by gravity into Coello River.

Table 3.4 Main Characteristics of the Coello and Saldaña Irrigation Districts (Vermillion and Garces-Restrepo, 1996)

Item	Coello District	Saldaña District
Department	Tolima	Tolima
Period built	1949 – 1953	1949 – 1953
Transferred	September, 1976	September, 1976
Design area (ha)	44,100	16,428
Gross irrigated area, 1993 (ha) (1)	25,628	13,975
Net irrigated area (ha) (2)	15,300	12,500
Water users' associations	Usocoello	Usosaldaña
Main soil type	Sandy, loam	Clay, loam
Main crops	Rice, soybean, cotton	Rice
Water source	River Coello	River Saldaña
System type	Run of the river	Run of the river
Intake structure	Radial gates	Radial gates
Irrigation structures	1,666	756
Lowest water measurement point	Secondary canal	Secondary canal
Length of main canal (km)	69	61
Length of canal network (km)	250	162
Area served per km canal (ha)	102	86
Turnout type	Sliding gates	Sliding gates

(1)   Total of the areas irrigated for all crops grown in a year
(2)   Area irrigated for at least one crop in a year

In 1998, the Coello irrigation system was enlarged by the construction of the Cucuana Project that has increased the available irrigation water to 25 m^3/s. The construction of the project started by the Government, but it has been finished by direct participation of the Users Association USOCOELLO.

The Cucuana system captures water from the Cucuana River through a small dam and an intake structure. Next, a sand trap with a capacity of 36,000 m^3 improves the quality of the water that is affected by a high sediment load caused by the deforestation of the Cucuana River basin. The conveyance canal is a lined canal of 37 km and it has lateral gates for the evacuation of the sediment and structures for the water level control, which form the main hydraulic structures. Some reaches of the canal are used for road traffic with a maximum capacity of 20 ton. In the future about 2.5 m^3/s will be used to supply water to the existing GUAMO Irrigation System which covers 1,800 ha and is in operation since 1999. A future enlargement of the Cucuana Canal with 37 km will also be considered.

The Saldaña Irrigation System captures water from Saldaña River through two lateral intakes with a combined capacity of 30 m^3/s. The siltation caused by high sediment load is controlled by a sand trap, which retains only 20% of sediment and which requires dredging by two dredgers. The main irrigation network consists of unlined canals; the conveyance canal is called the Ospina Perez Canal and is 13 km long; its capacity is 25 m^3/s. The distribution network is formed by three canals, namely the Animas Canal of 19 km, which serves 2,600 ha and has a capacity of 4 m^3/s; the North Canal with a length of 19 km and a capacity of 12 m^3/s; and the South Canal with a length of 11 km and a capacity of 9 m^3/s. The total length of the secondary canals is 100 km.

The main drainage system consists of the Saldaña, Magdalena and Chenche Rivers; the internal drainage system is mainly composed by natural streams.

In the early 1960s, the Government of Colombia entrusted the operation and maintenance of the irrigation districts to the Colombian Institute for Agrarian Reform (INCORA). However, the farmers considered the performance of the agency in irrigation management as not satisfactory and at that very moment the percentage of fee collection sharply declined due to the farmers' dissatisfaction with the quality of the system management. The result of the fee collection was harshly contrasted with the early stages of development in the 1950s, where more than 90% of the farmers paid the water fee.

In 1975, the farmers petitioned the government for the right to take over the management of the districts. They based their argument on the fact that during the previous 20 years, they had already repaid their agreed 90% share of the cost of construction. They were also paying water fees to the government and were dissatisfied with the cost and quality of the management that the government was providing. They argued that they could manage the systems more cost-effectively than the government. In 1976, the government agreed with the farmers' demands, expecting that a turnover would save money for the government.

During the transfer process several topics were discussed including the disposition of district staff, ownership status of scheme assets, and the future degree of involvement of the government agency (HIMAT) in the districts. It was finally agreed that most of the existing staff would be retained by the districts and others would be transferred. The ownership of the irrigation structures would remain with the government, although some equipment and facilities were transferred to the farmers' associations. HIMAT would retain a role of oversight for the district management to ensure that the systems were properly maintained meaning that it continued to give its advice and consent for annual budgets, operation and maintenance plans, setting water fee levels, and personnel changes. The farmers' associations obtained direct control over the operation and maintenance of the entire system, including the intakes (Vermillion and Garces-Restrepo, 1996).

During the period 1993 – 1995, Vermillion and Garces-Restrepo (1996) carried out a study to assess the extent to which the turnover of the irrigation management to the farmers in the Coello and Saldaña Irrigation Districts had an impact on the cost of irrigation for the farmers and the government, the sustainability of the irrigation system, and the quality of the water distribution. The sustainability of the irrigation was assessed on both the financial viability of the districts and the physical conditions of the irrigation infrastructure 19 years after turnover. The quality of the water distribution was assessed on the efficiency and equity of the water distribution and on the productivity of water. The main findings of the study were the following:

- Turnover did achieve the government's main objective of ending the governmental financing of operation and maintenance and was accompanied by a significant reduction in government subsidies. However, since the government retained partial control over the irrigation districts after turnover, staff levels declined slowly and the cost of irrigation for the farmers changed little;
- A detailed inventory of the irrigation infrastructure showed that the vast majority of the structures and canals were in good functional condition, 80% of the farmers interviewed stated that the maintenance of the structures was the same after transfer as before and a small percentage said it was better;
- The districts could continue to expand the irrigated area modestly due to better operation and they sustained high levels of production partly due to more flexible

water delivery procedures, a policy to limit rice production and to deliver smaller volumes of water per hectare (conservation of water);

- While the cost of the irrigation did not increase after turnover, the gross economic value of the production per hectare and per unit of water grew dramatically;
- After turnover, irrigation became a relatively small, declining proportion of the total cost.

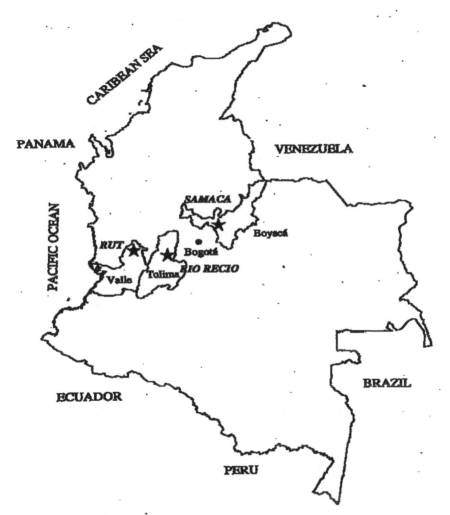

Figure 3.1 Locations of RUT, Rio Recio and Samaca Irrigation Districts (Mora and Garcés-Restrepo, 1998)

This study supports the hypothesis that management turnover programs worldwide will introduce incentives for improvement of the management efficiency and productivity of irrigated agriculture (Vermillion, 1992). Garces-Restrepo and Mora (1997) mention the main characteristics of the following irrigation districts corresponding to the second phase of transfer. Figure 3.1 shows their location.

RUT Irrigation District

The RUT Irrigation District is located in the south west of Colombia, north of the Valle del Cauca Department. The district is within the three municipalities that give origin to its name (Roldanillo – La Unión and Toro). The district lies between the western mountain range and the Cauca River and covers a gross area of 10,000 ha

The area forms the border between the tropical and subtropical climatic zones with an average temperature of 24 °C, fluctuating between 17 °C and 34 °C; precipitation is bimodal with two dry and wet periods. The total yearly average rainfall is about 1,100 mm. Average relative humidity is 72% with an average yearly pan evaporation at 1,080 mm. Average wind velocity is 1.3 m/s and a sunshine duration of 1,936 hours as a yearly average.

Before construction of the district the area was regularly subject to floods from both the Cauca River and the streams descending from the western mountain range.

Feasibility studies for land reclamation and development of what is now the district started in 1958. In 1962 with only 40% of construction finished, lack of funds stopped the work. In 1964 the flood control and land development work continued until 1970 when the construction was concluded. Finally in 1990, the district was transferred to the water users association, ASORUT.

The RUT irrigation system is essentially a flood protection and irrigation/drainage system. The long and narrow, bowl-shaped area is enclosed by a protection dike (from the Cauca River) running along the East and a flood interceptor canal on the western border. A main drain divides the area almost in half, running through the lowest elevation. Parallel to the dike is the main irrigation canal, which is served by three pumping stations with a total discharge capacity of 14.7 m³/s. The interceptor drain also serves as an irrigation canal with users using small centrifugal pumps to serve their individual needs. There is a complementary network of both irrigation canals and drains throughout the area. The main drain discharges freely into the Cauca River during low stages and the drainage water is pumped during the river's high stages. Besides the irrigation and drainage network the RUT has an excellent road network running parallel to the canals. The RUT Irrigation District has 1,200 users, the farm size distribution is of the "minifundio" type; 75% of the farms is smaller than 5 ha, and 11% is between 5 and 10 ha.

The water users association existed at the moment of the transfer, because it had been created at the start of the construction stage and it has performed an advisory role for the management of the government agency. For this reason, training for the association in administrative, operational and maintenance aspects was considered not necessary. Following the government's policy, the users accepted the proposal to give up the subsidies provided by the government for operation and maintenance, which had been 60 to 80% of the total costs before the transfer. However, after transfer the users recognized their acceptance as a mistake given the high pumping costs and they requested the government for a temporary subsidy. On the other hand, given that the system had previously been rehabilitated during the transfer process major improvements for the physical infrastructure were not agreed upon. The management transfer was carried out under the principle of "delegated administration" by which the government through the governmental agency had a shared responsibility with the water users association for the preparation of the budget, operation and maintenance plans, and personnel. Chapter 5 will describe in more detail the current condition of the RUT Irrigation District and the major impacts of the management transfer.

Rio Recio Irrigation District

The Rio Recio Irrigation District is in the Tolima Department in the central part of Colombia. The district has the Magdalena River as the eastern border, the Recio River as its southern border, the Lagunilla River on the North and the central mountain range in the West.

The precipitation pattern is bimodal with two rainy periods, April to June and September to November. The average yearly precipitation is 1,300 mm, the relative humidity is 72%, the average temperature 28 °C, sunshine duration 2,209 hours per year, and the annual pan evaporation is 1,300 mm.

The Rio Recio Irrigation District was initiated by the Ministry of National Economy in 1947 with a double purpose, namely to generate energy and to provide irrigation to the area. It also supplies water (30 l/s) for the drinking water plant of Lerida. The district was constructed during the period 1949 – 1951 and was completed in 1961 through the construction of the Lerida Irrigation Canal. Initially, ELECTROAGUAS managed the system and in the late 1950s, CAJA AGRARIA took over the responsibility for the management of the district and began to expand the irrigation area to other regions. In 1967, the system was handed over to INCORA, which administered the system until 1976. Finally, in 1990 HIMAT transferred the system to the water users association ASORECIO.

Rice is the main crop in the irrigation district followed by pasture, cotton and sorghum, soybean is also grown but on a minor scale. The farm size distribution shows that 25% is smaller than 10 ha, 46% is between 10 and 50 ha, and 29% is larger than 50 ha.

A diversion weir, 32 m long and 7 m high diverts the water from the Recio River. The lateral canal has a capacity of 10.5 m^3/s and is 6 km long. Initially, the system had two power stations, namely Recio 1 and Recio 2. The first one, a low head station, was located 500 m downstream of the headworks with a capacity of 300 KW, but the station is out of service since 1982. The second station is located 5 km from the intake, it has two turbines, which use 3.1 m^3/s by a fall of 100 m and its capacity is 2,500 KW. ELECTROAGUAS was responsible for the administration of the power system, and then the Colombian Institute for Electric Energy (ICEL) took over the management in 1968. Conflicts between INCORA and ICEL arose in view of the water management.

The average discharge available for the irrigation system is 7.4 m^3/s. The irrigation network consists of two main canals (Ambalema and Lerida) and a secondary and tertiary network that is 135.8 km long. The Lerida Irrigation Canal is 3.3 km long and has a capacity of 2.8 m^3/s; the Ambalema Canal is 7 km long and has a capacity of 6 m^3/s. A total of 30 km is lined, most of the lining can be found along the main canal. The entire system distributes water by gravity. Small natural streams from which water is reused by pumping for irrigation constitute the drainage network.

The transfer of the irrigation management took place under the form of a "delegated administration". Like in the case of the transfer of the Coello and Saldaña Irrigation Districts, the users of the Rio Recio District helped with the transfer by the fact that they considered themselves more than sufficient capable to manage the district at lower costs than the government agency. This perception was also supported by the fact that the government agency had included part of the collected water fees in its own budget at central and state level.

A condition for the management transfer included some important improvements of the physical infrastructure to be carried out by the government. The total cost of the improvement was carried by the state without any agreement of repaying the rehabilitation costs by the water users. These improvements included among others the construction of a pumping station with a capacity of 800 l/s with the objective to reuse drainage water that was temporarily stored in natural pools for irrigation, the installation of Ballofet meters at farm level, the rehabilitation of the headworks, improvements of the irrigation canals and vital infrastructure and replacement of machinery. Given the fact that after the transfer part of original state staff was kept, training of the water users association in administrative and operational aspects was not considered to be necessary. Subsidies from the government for the operation and maintenance of the district were fully eliminated after the transfer.

Samaca Irrigation District

The Samaca Irrigation District is located in the eastern part of the Boyaca Department in the central region of Colombia. The ground elevation in the command area varies from 2,600 to 3,000 m+MSL; the area covers approximately 3,000 ha of which about 54% consists of flat land and 46% is hilly. The scheme has deep, loamy soils that don't have salinity problems. Marketing and transport conditions are excellent due to the relative proximity of Bogotá, the capital of the country.

The mean annual rainfall is 690 mm with two pronounced wet periods in October – November and in April – May. The mean daily temperature is 13.8 °C, the relative humidity is 78%, the sunshine duration amounts to 2,117 hours per year, and the pan evaporation is 1,360 mm per year.

The Samaca Irrigation System is supplied by two reservoirs, which are the oldest and largest in Colombia and were built in 1880 by a textile company to generate electricity for its own use. In the late 1930s the reservoir was rehabilitated to provide irrigation water by two hillside contour canals that were finished in 1941. Under INAT, the second reservoir was built and improvements were made in the infrastructure. In 1992, the responsibility for the irrigation management was transferred to the water users association, ASUSA.

At present 2,000 farmers are served by the scheme. The average landholding is 3.5 ha in the valley and 0.9 ha in the hilly area. The farm size distribution is of the "minifundio type" as 95% of the farms is smaller than 5 ha. The main crops are potato, onion and green peas. About 30% of the total command area is currently used for irrigated pasture, mainly on the hillsides. Vegetables, wheat, maize, beans and barley are also grown. In the valley agriculture is predominantly commercial, while on the hillsides subsistence farming is common.

As mentioned above, the Samaca Irrigation District obtains water from two reservoirs at 3,000 m+MSL with a capacity of 4.7 and 1.5 million m^3, respectively. The maximum discharge is 0.9 m^3/s and is controlled by a valve in the lower and smaller reservoir, which forms the origin of a small river that carries water to the valley. In the downstream part, the river serves a dual purpose of irrigation and drainage. On the hillside water is conducted through two lined contour canals of 0.25 and 0.40 m^3/s. These canals feed 21 tanks varying in capacity from 12 to 36 m^3 and they are called irrigation units. The tanks feed a buried pipe system with risers and valves in the individual fields to which sprinklers are connected.

During the transfer process the volumetric water tariff was eliminated and it was agreed upon that the fixed water tariff ($/ha) was raised by 170% in order to achieve financial self-sufficiency. The elimination of the volumetric tariff was supported by the difficulty to measure in a correct way the volume of irrigation water delivered at farm level. As in the previously mentioned cases, the subsidies from the government were fully stopped, any training in view of the administrative and operational aspects was not given, and some improvements of the physical infrastructure were made without any contribution of the farmers to cover these rehabilitation costs. Table 3.5 shows the main characteristics of the RUT, Rio Recio, and Samaca Irrigation Districts. The International Irrigation Management Institute (IIMI) evaluated the impact of the management transfer in the three irrigation districts (Garces-Restrepo and Mora, 1997). Table 3.6 shows some characteristics related to the conditions, arrangements and changes caused by the transfer.

Table 3.5 Characteristics of RUT, Rio Recio and Samaca Irrigation Districts (Vermillion and Garces-Restrepo, 1998)

Characteristics	RUT	Rio Recio	Samaca
Department	Valle	Tolima	Boyaca
Design area (ha)	13,000	23,600	3,000
Irrigated area (ha)	9,700	10,200	2,893
Water source	River lift	River diversion	Reservoir
Intake structure	Pumping station	Gated weir	Vertical gates
Design discharge (m^3/s)	14	11	1
Main canal length (km)	87.7	38.7	29.7
Total canal length (km)	170.7	135.8	58
ha per km of canal	57	74	51
Turnout type	Pump	Sliding gate	Pump and sliding gate
No control structures	16	234	69
Lowest water level measured	Along main canal	Farm inlets	Main intake
Main soil type	Clay, loam	Clay, loam	Clay, loam
Main crops	Cotton, grapes, fruit trees	Rice, sorghum, cotton	onion, potatoes, peas
Built date	1958 – 1970	1949 – 1951	1945
WUA name	ASORUT	ASORECIO	ASUSA

Table 3.6 Characteristics of Transferred Irrigation Districts (Garces-Restrepo and Mora, 1997)

Characteristics	RUT	Rio Recio	Samaca
Transfer date	1990	1990	1992
Transfer type	Full system	Full system	Full system
Transfer mode	By delegation	By delegation	By delegation
WUA pre-established as legal entity	Yes	Yes	Yes
WUA has authority to make O&M plans and budgets	Initially shared with agency	Initially shared with agency	Initially shared with agency
WUA board constituted by major and minor members	Exists but not enforced	No	Yes
WUA can make profit	Under study	Under study	Under study
Water rights are clearly defined	In process	In process	In process
Government subsides exist	Only recently	No	Indirectly
Maximum sanction applied since IMT	Fines	Stop services	Take user to court
Rehabilitation prior to IMT	Pumps only	Major	No
Responsibility for future rehabilitation	Not defined	Not defined	Not defined
Changes in water fees	No	No	Volumetric discharge
Reorganization of personnel structure since IMT	Yes	Yes	Yes
Systems have diversified reserves since IMT	Yes	No	Minor changes

O&M: operation and maintenance WUA: water user association.
IMT: irrigation management transfer

3.2.3   Impacts of Previous Transfers

The impact of the transfer program in Colombia has been a very special issue for national and international institutions. For instance, the International Water Management Institute (IWMI, previously IIMI) has carried out fieldwork and specific studies with the purpose to identify the most important impacts of the transfer program in Colombia. The first study was focused on the impacts of the management transfer for the Coello and Saldaña Irrigation Districts, which have the longest experience with the transfer of irrigation management in 1976. The second study involved the irrigation districts that were part of the second phase of the transfer program (1990) and it included the RUT, the Río Recio, and Samacá Irrigation Districts among others. The studies of IWMI investigated the framework and the basic strategy of the transfer, the powers and functions handed over, and the impacts of the transfer on the irrigation management and irrigated agriculture. The impacts included the costs of irrigation for the Government and the farmers, the financial solvency, the quality of the irrigation operation and maintenance, and the agricultural and economic productivity of the irrigation systems.

In the framework of the present research five irrigation districts were visited in June 2004 with the purpose to identify the current status of the irrigation districts after transfer and to obtain additional information on the impact of the transfer program after a relatively long time under the new management. The irrigation districts visited included the Río Recio, Coello, and Saldaña Districts located in the Tolima Department; the Zulia District in the North Santander Department; and the María La Baja District in the Bolívar Department. A brief description of the Zulia and María La Baja Irrigation Districts will be followed by some reflections concerning the impacts of the transfer program from a qualitative point of view. The Río Recio, Coello, and Saldaña Irrigation Districts were described in the previous sections and the main impacts of the transfer program have been derived from the studies carried out by IWMI. However, some additional qualitative information will be presented concerning the current situation of these irrigation districts and the relationship with the management transfer. Chapter 5 will provide in detail the transfer of the management of the RUT Irrigation District.

Zulia Irrigation District

The Zulia Irrigation District is located in the north-eastern part of Colombia in the North Santander Department, near the frontier with Venezuela. It was built during the years 1962 – 1968 and its operation started in 1969. Initially, the Colombian Institute for Agrarian Reform (INCORA) was in charge of the administration, operation and maintenance of the district; in 1992 the Institute of Hydrology, Meteorology and Land Improvement (HIMAT later INAT) delegated the administration to the water users association which is named ASOZULIA. The potential area of the irrigation district is 38,000 ha, but only 22% (8,500 ha) has an irrigation and drainage infrastructure. Currently, 14,000 ha are under production, which include 8,500 ha of the area with an irrigation and drainage infrastructure. The organization has planned the land improvement of the remainder area in four phases, which include the improvement of 6,000 ha for fruit trees in view of the international markets (Japan).

About 53% of the soils in the district belong to the soil class IV (USDA soil classification); predominately consisting of heavy soils with poor natural drainage, slow internal drainage, and a shallow groundwater table. These physical characteristics are a constraint for the development of crops; it is mainly suitable for the rice crop. In 1998 the land use has been 77% rice, 8% permanent crops, 5% grass, 5% non-permanent crops, and 6% fruit trees.

The irrigation system in the Zulia Irrigation District is a gravity system. The main water source is the Zulia River from which the irrigation water is captured through a lateral intake with a capacity of 17 m^3/s. A sand-trap located at the headwork improves the quality of the irrigation water and allows the siltation of suspended sediment. The growing deforestation process within the Zulia River Basin contributes to a high sediment load in the irrigation water, which affects the maintenance cost. The irrigation water is conveyed and distributed by two main irrigation canals, namely the Astillero and Zulia Canal, from which a secondary irrigation network delivers the water to the farm level. The canals are not lined and the present water losses caused by filtration and leakage are estimated at about 20%. Non-official settlements along the canals and contamination by wastewater form some of the major problems in the irrigation network. The irrigation canals supply irrigation water, but they also provide water for human consumption in several rural communities. They supply on average 25 l/s per community; for this service ASOZULIA does not receive any payment. The municipalities of Zulia and Sardinata, the village of San Cayetano and Ciudadela Atalaya quarter in Cucuta are the communities receiving water from the irrigation district. The water in the network is also used by the Termotasajero thermal-electric plant for the cooling of the turbines.

About 1,200 users benefit from the Zulia Irrigation District. A high percentage of them are small farmers or producers, the average farm size is about 8 ha and most of the farmers are landowners. From an administrative point of view, the irrigation district is divided into 6 zones; five of them belong to the Cucuta Municipality and the remainder to the Zulia Municipality. Each zone has a zonal assembly and one delegate and his substitute is appointed for each 20 users. The delegates together form the Assembly of Delegates, which appoints the Board of Directors. The Board of Directors is formed by two representatives of each zone and by three for the largest zone. Finally, the Board of Directors is formed by 13 members, all of them are considered as main members. At the moment the Government Agency has no representative in the board, although in the beginning, just after the transfer there was one representative.

The Board of Directors appoints the Manager of the organization. The organization has three departments for the management, namely an administrative, operation and conservation department. The administrative department includes the chief of the administrative department and the coordinators for the agro-industry program, conservation, operation, workshop and accounting. The operation department includes the coordinator, one irrigation assistant per each zone, an assistant team for discharge measurements (liquid and solid), and headwork operators. The conservation department is formed by the workshop coordinator and the heavy machinery operators. The Board of Directors is supported by the monitoring and vigilance board, which is in charge of and looks after the compliance guidelines of the organization, the technical, machinery, and credit and training committees. The monitoring and vigilance board was created in 2002 as a result of the newly founded agro-industrial program and consists of six persons of which two are women. The technical committee is formed by three persons; the machinery committee advices the Board of Directors concerning the decisions necessary in view of the purchase of agricultural machinery; the credit committee defines the credit lines to the farmers and advises concerning the credits for production and the purchase of agricultural machinery; and the training committee establishes agreements with public and private institutions in order to train farmers and the technical and administrative staff in agricultural practices and in management and administrative topics. The cooperation agreement with the National Apprenticeship Service (SENA) to develop enterprise capability among the farmers is an example of this type of cooperation.

Water management

In order to obtain water from the irrigation service each user must be registered and supply information concerning the crop to be grown, the cropping area and growing date. The user may not have outstanding bills and his farm must have a well maintained irrigation and drainage canal and control structure. After the fulfilment of these requirements the irrigation plan is approved by the operation department and the water will be available for the next 48 hours.

The traditional crop in the Zulia Irrigation District is rice that has two harvests per year; there is a future expectation for three harvests per year after the improvement of the agricultural practices (land levelling at farm level) and more efficient use of the irrigation water. The organization has established an allocation for rice of 10,000 m^3/ha per harvest, having four supplies of irrigation water per season per user. The water tariff is divided in a fixed tariff and a volumetric one of Col\$ 25,000/ha per semester and Col\$ 6.9/m^3 respectively. Non-efficient use of irrigation water exists at farm level and there is no reliable measurement of the irrigation water delivered to the farmers. Land levelling at farm level is considered an innovation, a modernized practice in order to obtain a more efficient water use, the initial investment is high but after that only maintenance is required.

The distribution of the irrigation water is by rotation, which is considered very appropriate because it allows a good distribution, a rational use of the water and an improved attitude from the farmer side. The maintenance of main and secondary canals is carried out by the organization two times per year and the cost is charged to the users via the water bill. The user is responsible for the maintenance of the tertiary and farm canals; however, the maintenance can be done by the organization and the cost charged to the user.

The most important conflict concerns the fraudulent use of irrigation water during the night; this practice is carried out by farmers living on the farms and it harms the farmers who do not live there. This conflict is stronger during the dry season and is penalized through fines which are charged to the water bill.

Impacts of the management transfer

In 1992 the Colombian Government delegated the administration, operation and maintenance of the Zulia Irrigation District to the Water Users Association ASOZULIA. One of the first consequences of the transfer was the reduction in personnel; it reduced by 70%, namely from 150 to 45. The actual organization mentions that the transfer process took place in a fast way and during and after the transfer process there was not sufficient assistance from the Government side. The economic crisis of the 90s caused by the so-called "Economic Liberalization" affected in a negative way the agricultural sector. Moreover, the action of illegal groups, first the guerrilla and later the paramilitary groups, and the advanced deterioration of the hydraulic infrastructure of the Zulia Irrigation District were factors that led to a critical situation for the organization at the end of the 90s.

The critical situation was apparent by the very poor socio-economic condition of the farmers, which caused them a low level of living, by a serious deterioration of the hydraulic infrastructure and a considerable amount of outstanding bills for the delivered water services. At the end of the 1990s, the outstanding bills became greater than the budget of the user organization; Col$ 1,400 million and Col$ 1,200 respectively for the year 1998.

In the year 2000 the user organization started a reactivation process with a high social content through the development of a pilot project for productive agriculture that initially involved 20 users. After a successful and gradual development the program included 300 users at the middle of 2004 (25% of the total users) with an expectation that the number will grow to 700 users at the end of 2005. The success of the program was based on the establishment of a vision for the future of the organization in which the main purpose was to improve the living conditions of the farmers through an integral and participative approach for the rice-production chain. In this way, the organization changed its traditional role of an administrative entity of the hydraulic infrastructure for the water supply to the farmers by a new role. This new role has a much wider scope and focused on the improvement of the socio-economic condition of the farmers through the introduction of profitable irrigated agriculture practices. In this process the farmers were to lead the future development and are the multiplying agents of the program approach.

The strengthening of the management capability of the organization was a key factor, which has improved the credibility of the organization in the eyes of the users. The organization established a number of criteria for the reactivation process, namely that the agricultural activity has to be a profitable business, that an appropriate organization exists and that all the phases of the production chain are considered; that the inter-institutional relationships with supporting public and private entities are strengthened; that all the actors involved will extensively participate; and that the marketing of the agricultural products will be based on forward contracts.

On behalf of the users the association applied for the so-called "Associated Credit Line" offered by the Governmental Credit Agency, the Agricultural Financial Fund (FINAGRO), for the finance of the production costs of the rice. This credit, equivalent to 70% of the production costs, was transferred to the farmers through a credit supervised by the organization that gave technical support and agricultural inputs to the farmers; in some cases, especially for the small farmers, the organization supplied food products for their subsistence. At the same time, the organization also applied for a credit for the acquisition of agricultural machinery, which allowed for the improvement of the agricultural practices, increased the productivity and reduced the production costs. The savings obtained were transferred to and distributed among the farmers. The marketing of the rice is carried out through forward contracts with the National Agricultural Stock Exchange. In this operation the product price is agreed upon at the beginning of the season and the price can be increased at the moment of selling, but it can never be decreased. In this way the action of the middlemen is controlled.

The efficient use of the financial resources and the transparency of the management actions supported the strengthening of the management capability of the organization and increased the confidence of the governmental agency to that extent that the associated credit line is automatically renewed by the Governmental Credit Agency every year. The successful management of the financial resources lead also to the approval of a credit request for the rehabilitation and modernization of the hydraulic infrastructure, which showed serious deterioration at present and after the management transfer. This credit of Col\$ 10,000 million was approved in 2004 and about 51% was paid for the first phase (2004). The Government contributed 40% through the Incentive for the Rural Capitalization (ICR), and the Departmental Government and the Mayor's Office contributed about 20% for the financing of this credit. From its side, ASOZULIA agreed with the farmers to establish a new and additional fixed tariff of Col\$ 35,000 /ha per user to be paid during 18 semesters.

The credibility of the organization for an efficient and transparent management of the financial resources is reflected in its budget for 2003 and 2004. In 2003, the budget was Col\$ 8,000 million which included three main items: budget for administration, operation and maintenance; credit for agricultural machinery; and associated production credit. On the other hand, the budget for 2004 was Col\$ 1,700 million divided in 65% for conservation activities, 12% for operation and 23% for administration. The budget difference corresponds to the value of the credit given by the Governmental Credit Agency (FINAGRO) to the organization in view of its efficient management of the previous credits. Concerning the financial aspects it is important to mention that from the start of the reactivation process the fee collection is almost 100% and the existing outstanding bills correspond to the period before the reactivation process.

An important fact recognized during the visit to the Zulia Irrigation District is the concept of the management staff and the farmers about the relationship between the transfer program and the successful current condition of the organization. In spite of the fact that the years after the management transfer were very hard and the organization got in a crisis at the end of the 90s, the farmers and leaders believe now that the management transfer brought in time some positive benefits for the organization, because the transfer gave major freedom of action to the organization. This freedom was a very important support to reach the current condition and it was the motor to be more efficient in the productive aspects. The organization was especially able to take full advantage of the financial support from the Government Credit Agency (FINAGRO) and it developed an agro-industry enterprise that improved the socio-economic condition of the farmers.

Nowadays, ASOZULIA shows that they have a successful and large experience in the efficient and transparent management of economic resources that come from the Governmental Credit Agency as they have managed almost Col\$ 9,000 million during the last three years. Therefore, the organization ASOZULIA has become a reference for the management of financial resources for other organizations within the agricultural sector.

The present situation shows an organization that has developed a profitable agriculture focused on rice production; farmers with good socio-economic conditions, are satisfied with the performance of the organization, have a high sense of belonging, and are willing to participate and support the improvement proposals of the organization. Moreover, there are strong inter-institutional relationships with supporting public and private entities; a high management capability and high social development.

In summary can be said that ASOZULIA presents a great organization of the community with a firm leadership from the side of the Board of Directors, great action capability of the Board of Directors to support the productive activities of the farmers: technology innovation for the agricultural practices (land levelling by using laser equipment) to improve the efficient use of irrigation water, the introduction of improved seeds (plant technology), administrative and credit support to the farmers, forward marketing, formulation of programs with a high social content, important role of the women in the organization and a great capability to mobilize financial resources.

Learned lessons

The case of the Zulia Irrigation District offers several lessons to be learned. One of them is the fact that an irrigation system must not be considered as an hydraulic infrastructure that provides irrigation water to the farmers only, but also as an important component of a production system whose final objective is to contribute to the improvement of the living conditions of the farmers through irrigated agriculture under criteria of profitability, equity, efficiency, and an integral and participative management approach. The Zulia case shows that when economic benefits exist, the farmers are more sensitive for proposals to change the situation, are willing to agree to compromises and to participate in the development process of the proposed changes.

The present management approach in the Zulia Irrigation District changed the traditional production system within the agricultural sector. Traditionally, the Colombian Government delivered the land (via the agrarian reform) and financial resources (credit) to the farmers, who lacked training and agricultural experience. As a result, some farmers sold their land and invested the money coming from the credit in other activities than agricultural ones. A similar situation could be observed at the beginning of the transfer process. Farmers without any enterprise experience received also some financial resources from the Government and as a consequence many farmers misused the money and 90% of them got problems with the payment of the credit. After that, the agricultural activities were financed by the middlemen for relatively high interest rates resulting in a low profitability for the farmers and in an increase of outstanding bills for the water service delivered by the organization. Before a total collapse of the district, the organization started a reactivation process that allowed the establishment of a direct relationship FINAGRO-ASOZULIA-Farmer. Through a supervised credit and an integral approach for the rice production chain it has been possible to establish economic sufficiency for the farmers.

The formulation of a future vision by the organization was an important factor for the sustainability of the district. At the moment, the organization considers the creation of funds for education, housing and recreation for the families of the users to strengthen the social welfare in the near future. For several reasons, ASOZULIA considers the present mono-cropping of rice a major risk and therefore it will develop actions to increase the rice productivity (from 5.5 to 7 ton/ha), reduce the production costs and introduce more profitable crops like fruit trees and rubber trees for which good expectations exist in view of international markets.

Maria La Baja Irrigation District

The Maria La Baja Irrigation District is located along the Colombian Atlantic Coast, in the southern part of the Bolivar Department, near Cartagena. The irrigation district was built during 1967 – 1972 for the total cost of US$ 20.1 million, of which 69% was financed by the Colombian Government and the remainder by the Inter-American Development Bank.

The total potential area is about 20,000 ha and the area under irrigation was 12,000 ha in 1995; only 4,000 ha at the end of 2003 and 7,500 ha at mid 2004 just after the intervention of the Government Agency INCODER in October 2003. The main source of irrigation water is formed by two natural reservoirs with a total capacity of 236 million m^3. The total reservoir area is 2,600 ha and the combined discharge is about 20 m^3/s. The irrigation supply is by gravity; the irrigation infrastructure is very simple and consists of an unlined main canal of 58 km and a network of secondary and tertiary canals with a total length of 284.4 km. The control structures are equipped with slide gates. The predominant soils are loam and clay-loam and the groundwater table is 0.50 m below surface in 80% of the area. The average rainfall is 1,890 mm per year and the rainfall has a very irregular distribution.

Rice, sorghum, grass, corn, and cassava are the traditional crops with rice being the most important crop. In the past years African palm was introduced. In the year 2004, about 1,500 ha of rice and 2,900 ha of African palm were cropped. The expectation for the land use for the year 2009 is based on an expansion of the rice area, a target area of 5,000 ha for African palm and the introduction of sugar cane for the production of alcohol as fuel.

Currently, the irrigation district supports about 1,800 users, of which 385 are growers of African palm. About 75% of the farms is smaller than 10 ha, 25% is in the range of 10 – 50 ha, and 3% is larger than 50 ha.

Impact of the Management Transfer

The management transfer was agreed upon in 1994 between the Government Agency (INAT) and the Users Association of the Maria La Baja Irrigation District (ASODIMAR). However, the actual transfer only occurred in 1996 due to the reluctance of the water users and the distrust of the Government Agency in the management capability of the organization. The reluctance of the users was based on the very poor conditions of the hydraulic infrastructure at the moment of transfer and for this reason they demanded rehabilitation of the infrastructure. As a result a joint management was agreed upon for a period of 6 months, which was extended to a full year (1997). During this period training was given to the district staff and it was also agreed upon that a modest rehabilitation of broken-down structures would be done after the transfer and that the users association would have a main say in prioritizing the repair activities.

In the framework of the study carried out by IWMI to evaluate the Impacts of the Colombia's Current Irrigation Management Transfer Program, the functional condition of hydraulic structures and canals was evaluated for Maria La Baja Irrigation District, being this district under rehabilitation. A number of 55 structures out of 250 were inspected, 52% was found not to function, 30% was not functioning properly and only 18% was performing well. On the other hand 13% of the total length of canals has been inspected and about 19% was not functioning, 19% was not functioning properly and 62% was performing well (Garces-Restrepo and Vermillion, 1998). These figures clearly show the deterioration of the hydraulic infrastructure and explain the reluctance of the users to accept the system at the moment of the management transfer.

Once the organization took the full control of the irrigation district, several factors caused a progressive deterioration of the system to that extent that it caused a serious crisis that resulted in an intervention of the Government Agency (INCODER) in October 2003 and the dissolution of the Users Association ASODIMAR. This suspension of the organization seemed to confirm two facts: firstly, the doubt that the organization was competent to manage the district in a good way after the transfer; and secondly, the reluctance of the users to agree with the transfer given the deteriorated condition of the physical infrastructure. During the visit (June 2004) that took place within the framework of this research, the Government Agency was carrying out a strong rehabilitation program for the hydraulic infrastructure as part of the intervention program.

Technical, socio-economic and environmental problems were the main factors which caused the organizational crisis. Some of these factors hindered the full establishment of the organization and the management transfer only accelerated the crisis. In other words, the transfer was not the only cause of the final suspension of the organization, but the risk factors already existed at the creation of the irrigation system.

Serious problems during the design phase, siltation of the reservoirs due to a progressive deforestation, floods, waterlogging and insufficient drainage caused a severe reduction of the irrigation area to the extent that only 20% of the potential area was cropped in 2003. The users expressed that the transfer process took a very short time; that the assistance from the Government side was insufficient, and that the deterioration of the hydraulic infrastructure was a determinant factor. They also mentioned that the "Economic Liberalization" influenced in a negative way the economic capability of the farmers, because before the economic liberalization about 5,000 ha of rice were cropped, but after that the product price decreased and the rice sector went bankrupt. As a consequence the agricultural activity was not stimulated and the land use was changed for cattle.

On the other hand, the organization reached a very low credibility in the eyes of the users due to corruption and misuse of subsidies from the Government, personal interests and influences, which caused the failure of the organization. Other factors that contributed to the critical situation of the organization included a low sense of belonging from the users side, non-official settlements along the canals, increase of outstanding bills, lack of entrepreneurship with social impact, ineffective actions of supporting entities, low management capability, marketing of products via middlemen, low profitability, a low organization culture and an area under the influence of paramilitary groups.

Under these conditions, the Government decided to intervene in October 2003 and created a new organization, called USOMARIA. The Government established a rehabilitation plan for about Col$ 5,500 million with finance from The World Bank; the Government Agency INCODER and the Government of the Bolivar Department were appointed as agents for the intervention. The rehabilitation involved the maintenance of canals, improvement of radial gates, improvement of the road network and civil works. No money was available for community related aspects.

The Government plan was that after the rehabilitation of the irrigation district, it will be given in concession to a private enterprise for its administration. It was said that this enterprise would be constituted by partners, one of which would be the new users association. However, the organizational form is still not clear and the discussions about this topic continue. Therefore, it is expected that a privatization in a more aggressive form will be imposed by the government, which will cause conflicts and will deepen the social differences.

Rio Recio Irrigation District

As was mentioned in the previous paragraphs the Rio Recio Irrigation District was transferred to the Users Association ASORECIO in 1990. During the visit (June 2004) for this research and almost 14 years after the management transfer, the organization showed a stable condition and seemed to have acquired an autonomous management model, which was understood and accepted by the users of the organization.

As a condition for the transfer, the users requested the Government to rehabilitate part of the hydraulic infrastructure, which was in poor condition at that moment. The Government rehabilitated the headwork and the main conveyance canal, repaired the tunnel in the conveyance canal, carried out maintenance works for the canals, built a pumping station for the reuse of waste irrigation water and rehabilitated the road network among others. In general terms and taking into account that the irrigation works are more than 50 year old, it can be said that the current hydraulic infrastructure is functioning well. This means that the organization was able to operate and maintain the system in a good way and that they had sufficient economic resources to carry out the maintenance activities in a timely and sufficient way. It is important to mention that the irrigation system has a multiple purpose; it is used for irrigation, electricity generation and water supply. However, ASORECIO does not receive any economic benefit from the additional services.

The Board of Directors is involved in all activities for administration, operation and maintenance and in the farmers' eyes their credibility is that high that sometimes the manager is requested to solve family conflicts; this means an important acknowledgement of the authority of the organization.

Also a strong tendency to use the water in an efficient way could be identified. The district suffers from water shortage and for that reason an area of 3,200 ha out of the 9,546 ha is on average irrigated per season. A year has three irrigation seasons, namely January – April, May – August, and September – December. In this context the approval of requests for irrigation water and the season closed for irrigation water are the main causes for conflicts with the users.

In view of an efficient water use the discharge in the main canals is controlled by water level gauges; the main canals are lined and their conveyance efficiency is estimated at 90%; the canal maintenance is timely and the measuring structures, type Ballofet, are installed at farm level to measure the volumes delivered. However, an inappropriate installation of the slide gates in the secondary canals was identified. The canal maintenance is by hand due to the fact that the available machinery is inappropriate and will modify the shape of the cross section. The canal maintenance is given by contract to the farmers.

The organization recognizes the fact that the conservation and protection of the water resources is one of the key factors for sustainability and some activities have been developed to secure the water availability. In this way, reforestation, protection and conservation activities within the catchment area of the Rio Recio have been carried out by the ASORECIO. The organization received a positive response from the inhabitants of that area. However, the activities have been hampered by the action of illegal groups.

In the irrigation district rice is still the most important crop followed by sorghum, corn and cotton. The cultivation of cotton is not stimulated given the low market prices. ASORECIO established an irrigation supply of 2 l/s.ha for rice and of 0.7 l/s.ha for other crops. The irrigation volume for rice is 25,000 to 30,000 m³/ha for the traditional cultivation and 11,000 to 15,000 for the cultivation in basins. Irrigation water is only delivered to the farmers when they have no outstanding bills and their irrigation canal is in good condition. Rice cultivation is a profitable business for the farmers and therefore it is an economic incentive to pay in time the water fees and that results in a high fee collection. Improvement of the agricultural and irrigation practices are required at farm level and for this reason the organization gives support and advice to the farmers for farm improvement works, especially for land levelling for rice cultivation.

The irrigation development in the district is limited by several factors, such as water shortage, capacity of the conveyance canals, and the quality of irrigation water. Before the transfer a project to regulate the Rio Recio was considered. However, nowadays and 14 years after the transfer the project is still being considered. The existing tunnels in the main canal limit its capacity and therefore the possibilities to enlarge the irrigation area; technical alternatives to solve this situation have not been considered in an effective way.

From an environmental point of view, ASORECIO faced several problems. The reuse of irrigation water is an alternative to combat water shortage. However, this practice has a risk given the water contaminated by agrochemicals and domestic wastewater. The high sediment load affects the physical quality of the irrigation water. The sediment load is caused by deforestation in the Rio Recio basin and the steep slope of the canals, especially in the Lérida sector. This causes major sedimentation problems in the downstream Ambalema sector. Soil salinization is another risk due to the lack of an appropriate drainage system at farm level and at main level.

As final conclusion for the Rio Recio Irrigation District, it can be said that the organization is consolidated around the supply of irrigation water in an efficient way in spite of some limitations due to water shortage and within the context of a profitable agriculture based on rice cultivation. The organization established rules and regulations, which were accepted by the users and are the basis for decision-making for the water service under water shortage conditions. Other activities concerning the organization of the marketing of products and social programs were not identified.

Coello and Saldaña Irrigation Districts

As mentioned before, the transfer of the management in the Coello and Saldaña Irrigation Districts was requested by the users themselves; this is contrary to the transfer in most of the other irrigation districts where the transfer was considered as imposed by the Government. The experiences of the Coello and Saldaña Irrigation Districts form a reference point for the transfer process as followed later by the Government for the remainder of the irrigation districts. Almost 30 years after the transfer of the management, USOCOELLO and USOSALDAÑA are highly consolidated organizations, which reached autonomy and maturity in spite of the different problems faced by them.

Both irrigation districts maintain the typical organizational structure consisting of the general assembly, Board of Directors and the basic departments for administration, operation, and maintenance; although the Saldaña Irrigation District has besides the technical also a welfare and social security department. About 2,200 users and 2,800 farms are involved in Coello, while 1,600 users and 2,450 farms are involved in Saldaña.

Rice is the predominant crop in the irrigation districts and occupies 80% and 100% of the irrigated area in Coello and Saldaña, respectively. In both areas two harvests per year take place. In Saldaña about 17,000 ha have an irrigation infrastructure, but only 14,200 ha are cropped per season; the expansion of the area for rice is limited by the incorporation of new areas at national level that have increased the national production and reduced the product price. So far the limited experiences with other crops limit the increase of the cultivable area and crop diversification.

In the Coello Irrigation District about 18,000 ha are grown with rice and that corresponds with 80% of the irrigated area, the remainder is grown with sorghum, peanut and cotton, the latter one during the first season. Given the water shortage the rotation of the rice is a compulsory practice for each farmer; this means that a farmer is allowed to grow rice once a year. This action also promoted crop diversification, soil improvement and sanitary control. As compensation for the rotation, a fixed tariff is only charged to the farmers with rice. Before 1998 the water service was closed during one season in view of water shortage. This practice was suspended after more water became available with the opening of the Cucuana Project in 1998, which allowed the expansion of the irrigated area to 6,075 ha.

The Cucuana Project is a good example of the capability of the users association of the Coello Irrigation District (USOCOELLO) for the resources mobilization for the financing of the last phase of the project, which was started by the Government Agency and showed an inefficient performance for the project execution and management of the economic resources. USOCOELLO got the economic resources of US$ 6 million and completed the project in a period of six months.

The inspection carried out for this research showed a good functioning of the hydraulic infrastructure of the two districts, especially the main and secondary irrigation network; the Coello Irrigation District presented the best condition of the two. As mentioned before, the users enlarged the Coello Irrigation District in 1998 with some important hydraulic works: headworks (capacity 26 m^3/s), conveyance canals (32 km) and special works as inverted siphons, viaducts, control structures, etc. In general terms, the irrigation works showed a good maintenance, conservation, and operation condition; therefore it can be said that sufficient economic resources are available for the conservation and maintenance activities.

The water management and technical decisions are highly influenced by the rice cultivation. The procedures for the water supply are defined as a result of the long experiences of the organizations with the agricultural practices of rice. Requests of the farmers for the water service, growing date, and contracts for technical assistance with professional experts, the good condition of the hydraulic infrastructure at farm level, actions and steps at organization's office level, and finally the water delivery by the ditch tender are the most important steps for the irrigation water service.

An inefficient water use was observed at farm level. The control and regulation of the delivered water is based on experience of the irrigation inspector. Nowadays measuring structures; when they have existed in the past they were destroyed by the farmers and other structures were installed in an improper way. Lack of land levelling and the soil quality (light soils) contribute in a high grade to an inefficient water use and it is expected that a larger amount of water is delivered than needed at farm level. Theoretically, the organization established an irrigation modulus of 2 l/s.ha and 1 l/s.ha for rice and other crops, respectively. However, 15,000 m^3/ha per season on average are assumed for the invoices in view of the technical limitation to measure the actual amount of delivered water. In view of a more efficient water use USOSALDAÑA promoted a pilot project in 2,500 ha by using laser technology for land levelling and by the transplanting of the rice plants. The training for the irrigation practice has been another activity developed by USOSALDAÑA and a license as trained irrigator must be obtained to perform that activity.

The rice crop is a profitable business compared to other crops; the profitability for 2004 was Col$ 2.5 million/ha in case the user is landowner and Col$ 1 million when he is tenant; this means that rice growing is a good business when he grows directly or let his farm. In the Coello Irrigation District most users are tenants, while in Saldaña 64% are tenants and 34% are owners.

On the other hand, rice is an important factor to increase the land value to the extent that the value of a hectare of rice land (Col$ 25 million) is 5 times more than the one for other crops. Therefore, some users use their political influences within the organization to register their lands as suitable for rice and to get water allocation in spite of the fact that their lands are not suitable for rice given the light soils. This aspect contributes to inefficient water use and also promotes corruption. It does not stimulate crop diversification and when the Cucuana Project started to operate some farmers stopped the growth of fruit trees and improved the lands for the rice crop.

Also a tendency of land fragmentation has been identified, which helped the concentration of small and middle producers (minifundio). In Coello about 50% of the farms is smaller than 5 ha and they cover 10% of the irrigated rice area, while in Saldaña about 65% is smaller than 5 ha and cover 18% of the area. The land size ranges from 5 to 20 ha in 33% of the farms that cover about 32% of the irrigated area in Coello, while in Saldaña about 26% cover 43% of the area and range from 5 to 20 ha. The rice growth demands intensive agricultural practices, which can be more easily attended by the small producers than by the larger ones given the small area size.

The rice crop is a strong employment generator. Torres (2003) mentions that according to the Federation of Rice Growers the national average of day wages for rice cultivation is 32/ha and about 12,928,000 day wages were generated for the year 2002. He mentions that in the Saldaña Irrigation District 55 day wages per ha were employed and that the transplanting demanded 71 day wages. These figures were based on the small producers with permanent rice production.

Torres (2003) gives some additional information, which explains the intensive practice of rice cultivation in the irrigation districts in the Tolima Department. For the year 2002, 43% of the rice area in Colombia was in the Tolima and Huila Departments with 6.4 and 7.2 ton/ha for the average national and regional yield, respectively; a comparable figure for the USA gives a yield of 6.3 ton/ha. The irrigation costs for 2003 were 56.8 and 66.5 US$/ha for Coello and Saldaña, respectively, while in the USA the costs were 74 US$/ha. Moreover, the yield was 0.45 kg/m^3 irrigation water for a water allocation of 16,000; these values are 0.62 and 10,000; and 0.25 and 20,000 for USA and Asia respectively.

As mentioned before rice growing is a profitable activity, which allows the farmer to fulfil his financial obligations such as the water fees. For this reason, the outstanding bills are very low in the two organizations. The financial self-sufficiency of the farmers is based on a timely payment of the water fees, which has an additional incentive, namely the guarantee of a permanent water allocation, especially for the rice crop, which is also a good mechanism to give a high value to the land or to get a high value for its renting.

In general terms, the two irrigation districts show financial self-sufficiency, and in some cases, like the Coello Irrigation District, their functional budget is larger than that for small municipalities. This situation has evoked the interest of political sectors and some user associations have become politicized. For this reason, corruption, conflicts from personal interests and individual influences are some of the problems faced by the users associations. Lastly, Coello kept some internal disputes that were generated by the open disapproval of corruption and private interests, which needed specific Governmental intervention.

The dynamics and development of the rice growth caused that the main activity of the user's organizations is focused on the supply of irrigation water, while additional activities like marketing are developed by other interested parties. The rice agro-industry is well developed and the enterprises for rice processing act as intermediary agents within the production system; they give agricultural inputs and financing to the farmers and then they receive the production and establish the product value.

No cooperative forms for marketing exist in view of past, negative experiences, which created distrust amongst the farmers and helped individual actions and particular interest. The user's organization complained about the minimal and inefficient presence of the government agencies that provide security and supporting services. The relationship with the Environmental Authority is poor in spite of the fact that Col$ 300 million were paid by USOSALDAÑA for water concession. The money for financing given by the Governmental agency is insufficient to attend the financial requests of the farmers; and the security problems created by the action of illegal groups also affect the user's organizations to the extent that USOSALDAÑA mentioned the payment of an insurance against damage to and attack on the hydraulic infrastructure.

From an environmental point of view the user's organization faces serious threats caused by the potential danger of soil salinization. The user's organizations have not yet considered detailed salinization studies, but it is expected that salinization problems will develop given the inefficient water use and the lack of a drainage system at farm and at scheme level. The excessive use of agrochemicals and the inefficient management and disposal of solid waste are also risks for the health of the farmers and for the quality of the natural resources, especially the water resources.

The two irrigation districts have serious problems with the water quality that is affected by the high sediment load as a result of the serious deforestation of the surface water basins. Saldaña and Coello have heavy machinery at the headworks for the extraction of sediment that increase the maintenance costs and generate an environmental problem due to the disposal of the sediment. In Saldaña two dredgers extract 200 m^3 sediment per day during the dry season and 1,500 to 2,000 m^3 per day in the wet season. As a consequence, USOSALDAÑA included in the water bills for 2004 a cost of Col$ 15,000/ha per season for the sediment control. In Saldaña the change of the river course required the construction of a new, additional intake. In Coello the water is now captured by a diversion canal due to the siltation of the original structure. Cucuana is a recent project and the control of sediment, including a sand-trap with an approximate capacity of 36,000 m^3.

From a social point of view it can be said that the two irrigation districts have been influenced by the action of groups who act on the fringes of the law; in this case, paramilitary groups, which have displaced the guerilla during the last years. However, the administrations of the districts and the users have reached some coexistence agreements with these groups.

An important concern is the apathy of the users to participate in the assemblies. As explication for this situation the users mention that most of them are tenants and the landowners live outside the irrigation district; the tenants are only interested in the economic profitability, while the landowners are interested to receive the money from renting their lands. Also the exchange of interest inside the organization, conflicts among small, middle and larger producers, corruption at the level of the Board of Directors are some factors contributing to the apathy of the users.

3.2.4   Conclusions

One of main objectives achieved by the Transfer Program is related to the relationship between Government and the land improvement sector. From financial point of view, it can be said that the financial burden of the Government was significantly reduced since the total costs for administration, operation, and maintenance were transferred to the water users associations of the irrigation systems. The partial and in some cases the total elimination of the subsidies after the transfer of the systems contributed also to the reduction of the irrigation costs for the Government.

The Transfer Program helped also the further development of the new institutional structure for the land improvement sector and contributed even more to the costs reduction for the Government and facilitated the implementation of the policy of decentralization and delegation for the land improvement sector. The removal of power of institutions as HIMAT first and then INAT, and the fusion with others institutions in the current INCODER is a sample of that policy.

The impact of the delegation of functions of the land improvement sector to the regional and local institutions cannot yet be evaluated as these institutions only just started to perform the delegation. At present the Government of the Departments of Atlantico, Bolivar and Tolima started with the delegation of functions. In 2004, the Atlantico Government delegated the contract for the rehabilitation works for the Repelón and Santa Lucia Irrigation Districts, and the Tolima Government delegated the contract for the initial works of a new project called "Triangulo del Tolima". Halfway 2005, the Central Government transferred money to the Bolivar Government for the rehabilitation of the Maria La Baja Irrigation District. The Department in turn will delegate some of their functions to regional and local entities, such as the Departmental Agricultural Secretary and the Municipality Units for Technical Assistance (UMATA). Scepticism exists among the leaders of the water users organizations concerning this type of delegation, because they believe that the regional and local delegated institutions historically did not have an effective presence in the area of the irrigation districts, are influenced by the political sectors, and maintain a low credibility due to a high corruption level and low transparency for their activities.

The development of the transfer program showed a clear difference between the first stage of the transfer program and the next phases. It was clear that the management transfer of the Coello and Saldaña Irrigation Districts was promoted by the users themselves, who believed that they were able to operate and manage the financial sources of the system more efficiently than the Government Agency. In this case the Government Agency was against the transfer, because they believed that they will lose the political management of the land improvement sector.

The motivation for the following phases of the transfer Program came especially from the Government side, while the users only accepted the Governmental transfer proposal under certain conditions. From a detailed reflection of the role of the Government in the following phases of the transfer program can be concluded that the transfer program was a kind of strategy of the government to face the fiscal crisis of the 1990s and to obtain the necessary compromises with the international banking, which demanded for the implementation of a policy of decentralization and delegation of functions. This seemed to be confirmed by visits to the organizations during this research; the latter said that during the process of management transfer the Government Agency was more interested to develop the process as quickly as possible and for that reason the support and training from the Government side were insufficient.

According to the transfer program all the irrigation districts would have been transferred before 1997. At present about 67% of the systems have been transferred, which cover about 64% of the total area. An important fact is that 90% of the total area with irrigation and drainage infrastructure has already been transferred, but 87% with only drainage infrastructure has not yet been transferred. It seems that there exists no incentive for the management transfer of the systems with only a drainage infrastructure.

About 30 years after of the start of the transfer program the researched users' organizations presented a varied grade of development. Some of them, as in the case of the Coello, and Saldaña Irrigation Districts (transferred 30 years ago) and in a lesser grade, in the Rio Recio (transferred 15 years ago), showed a significant organizational stability that is based on a financial self-sufficiency resulting from an adequate provision of the irrigation services and on a focus of their main activities on operational aspects within the practice of a profitable agriculture around an only crop: rice.

This situation was helped by the presence of supporting entities for the agro-industry and marketing, research and technology transfer activities, specially put in motion by private parties with particular interests. Therefore, it can be said that the transfer program contributed in a high grade to the consolidation of the production system, where the different phases of the production chain were developed by special parties and that allowed them to satisfy their wishes and interests and to devote their efforts to specific activities; in the case of the water users association to the provision of irrigation water services.

However, Maria La Baja Irrigation District presented the other face of the coin. Today the organization has been shut down by the Government and is in a reorganization process. After its transfer, less than 10 years ago, the organization did not find an organizational model that allowed them to adapt it to the new self-sufficiency conditions established during the transfer and therefore it arrived very quickly at a crisis situation. It seemed that the existing traditional culture of subsistence agriculture supported by a paternalistic governmental tradition and the lack of a clear vision of the agricultural enterprise contributed to the negative impacts after the management transfer.

On the other hand, Zulia Irrigation District that was transferred 13 years ago, showed five years ago great dynamic developments from an organizational point of view, after it reached a very critical condition almost directly after the management transfer. Contrary to the present conditions as previously mentioned for the other irrigation districts, the Zulia Irrigation District has an organization model that is based on an integral and participatory approach around a profitable agricultural activity that is focused on rice production and on a clear vision for the future development of the organization and the district.

Consequently, it can be said that in the post-transfer process and from an organizational point of view three crucial phases seem to take place, namely adaptation, maturity or full development, and consolidation. Adaptation is a critical phase where the organization with the best capabilities will adapt itself in a good condition to the responsibilities demanded by the management transfer. The quality of the training during the pre-transfer and transfer process and the effective assistance given through the adaptation phase prove to be key factors for a successful performance and future sustainability of the organization. The maturity or full development phase is the period in which the organization has developed its full capabilities and looks after the integral development of its beneficiaries based on a profitable irrigated agriculture practice under equity, efficiency, competitiveness and sustainability criteria. In this phase the leadership and an integral and participatory management approach are the most needed requirements. The consolidation phase is the situation in which the organization as a whole has achieved its maximum development condition to the extent that some specific activities of the production chain can be carried out by other parties while the organization maintains the integral management approach. Under this vision it can be said that the Coello, Saldaña and Rio Recio Irrigation Districts are now in the consolidation phase, but they lack an integral management vision; the Zulia District is in the maturity phase and Maria La Baja District failed during the adaptation phase.

The management transfer must not be seen as a simple transfer of responsibilities to the water users associations with the purpose to free the government of financial burden and to contribute to a lighter fiscal deficit. The management transfer must take place in those cases that have favourable conditions which allow for the full development of the management capability of an organization in order to exploit the potential of the existing socio-economic, natural and human resources in view of the improvement of the living standard of the users. If no favourable conditions exist, the organization managing the transferred system will show a relatively low performance and will have to carry the financial burden and fiscal deficit that before the transfer were suffered by the government.

Increasing the water use efficiency is another clear objective of the management transfer to the water users associations. However, the researched irrigation districts plainly showed that the use of irrigation water is a critical issue hampering the environmental sustainability of the irrigation systems. In view of an efficient use of the irrigation water, most of the operation departments carry out discharge measurements and control of the irrigation water in the main and secondary distribution system, but there exists a series of difficulties to control the irrigation delivery to the users given the non-existence of measuring structures. Moreover, it is expected that there exist an inefficient use of irrigation water at farm level and the danger of soil salinization in view of the lack of an appropriate drainage system.

The water shortage in the Rio Recio Irrigation District called for a more efficient use of the available water and consequently Parshall flumes were installed at farm level. The installation of this type of flumes in the Coello and Saldaña districts was not a success, because the farmers destroyed them all in a short time. In these two districts the water shortage is not so critical. From these observations, it seems that there is a close relationship between the availability of irrigation water and a tendency to use the water in an inefficient way. Moreover, the availability of irrigation water as a key element to increase the value of the land and the social status is also a factor affecting the efficient use.

The financial self-sufficiency of the organization as a result of the water selling, especially for crops that demand a large amount of water like rice, does not seem to be a strong factor that could contribute to the efficient use of water. From this perspective it could be thought that the administration of the organization would not have an incentive to promote the efficient use, because the more water is sold, the more revenues are received and the more the financial self-sufficiency. However, the promotion of efficient water use by the administration could generate additional revenues through the enlargement of the irrigated area taking into account that most of the irrigation districts have not yet developed all their potential land for irrigation.

From environmental point of view it should be mentioned that the deforestation processes are still growing and that they affect in a negative way the watersheds of the main surface waters as visits to most of the irrigation districts have shown during this research. These processes affect amongst others the quality and costs of the maintenance and operation activities and the availability of water. Although, most of the river basins are outside the influence area of the irrigation districts, the organizations are conscious to develop the needed actions for re-forestation, protection and soil conservation in order to assure water security for the future. Some actions have been affected by the presence of illegal groups and ineffective action of the environmental authority.

Lastly of all, the experience of the management transfer in Colombia opened the opportunity to consider a wider perception of the role of an irrigation system. The systems should not only be considered as a simple technical infrastructure that only provides irrigation water to the farmers, but it should be recognized as an important component of a production system, whose final objective is to contribute to the improvement of the living conditions of the farmers through irrigated agriculture under criteria of profitability, equity, efficiency, and an integral and participatory management approach. Under this approach, the management of the irrigation systems by the communities will increase the management capability of the organizations and when economic benefits exist, the farmers are more sensitive for proposals to change the situation, are willing to agree with compromises and to participate in the development process of the proposed changes.

3.3   Management Transfer in other Countries

Schultz et al. (2005) mention that in relation to the improvement and expansion of irrigation and drainage systems there are certain specific issues that deserve everyone's attention. In the case of the emerging countries, several issues are far from being resolved and significant efforts will be required from the parties concerned to achieve sustainable solutions. One of the issues includes the institutional reforms in the direction of stakeholder-controlled management and government support for modernization and reclamation. In many countries institutional reforms in irrigation and drainage system management towards stakeholder-controlled management are ongoing (Japanese National Committee of ICID, 2000; Czech Committee of ICID, 2001; Ukraine National Committee of ICID, 2002). Transfer of systems or of responsibilities is especially taking place in the emerging countries (Asia, Central and South America), and in countries with a transition economy (Central and Eastern Europe).

These transfers are generally desirable, since government-controlled organizations in several countries and over the years have not really been able to improve the management significantly. However, the transfer process is still in its inception stage in many emerging countries and therefore, it will take several years to show the benefits of such transfers. The transfers may require quite different approaches (Schultz, 2002). In the emerging countries, there is generally a population of farmers and here the transfer concerns the transfer of responsibility and maybe of ownership of parts of the systems from the government to the farmers. In these countries a significant part of the systems is more than 30 years old. So the transfers will have to go hand in hand with modernization. However, in countries with a transition economy, there are several specific problems, such as: unsuitable layout of the systems, which is mostly based on the former large-scale type of agricultural production; the uncertain future of the agricultural sector; required funding for modernization and resulting operation and maintenance costs; lack of good governance, unaffordable pumping systems and environmental degradation (ICID Yalta Declaration, 2002).

In some of the countries, there is not even a clearly identified farmers' group. These issues make the transfer process quite complicated. Several economic and financial aspects of the transfer are of special importance and they may include the government policy on agriculture and rural development; required area of agricultural lands and the part of these lands where irrigation and/or drainage can be applied under the new conditions; farmers' position in the local economy; and to find solutions that result in sustainable irrigation and drainage system management after transfer. With respect to the first two issues it has to be realized that in several of the countries a complete agricultural reform will be required before a transfer of irrigation or drainage system management can be successfully planned and implemented, while in these countries the farmers are completely uncertain about their future and therefore not in a position to commit themselves to responsibilities that they cannot afford. With respect to the farmers' position in the local economy, it is important for them to know whether their standard of living from agriculture will remain acceptable to them, while more feasible alternatives through employment in industry and urban activities may arise. The economic and financial questions that arise with respect to sustainable irrigation and drainage system management after transfer, concern especially the determination of the best modernization options; the cost of modernization; the resulting cost and efforts for operation and maintenance of the modernized systems; the full cost recovery or sustainable cost recovery (Tardieu and Préfol, 2002; Tardieu, 2003); the cost sharing and the capacity to pay. The issues of cost sharing and capacity to pay are the more important in the light of the sustainability of the modernization and transfer activities.

In the next sections, a general description of the transfer process and its main results will be presented for some countries with conditions similar to the Colombian case. The countries are Mexico, Indonesia, Philippines and Turkey.

3.3.1 Mexico

The economic crisis of Mexico in the 1980s led to radical and extensive reforms in the agricultural sector. Among the most significant institutional reforms was the program to transfer the irrigation management responsibilities for large-scale irrigation districts from the sole control of the public sector irrigation agency to a joint management arrangement with newly created water user organizations (Kloezen et al., 1997).

In 1989, in an attempt to recognize the problems in the irrigation sub-sector, as part of the National Development Plan (1989 – 1994), the Government of Mexico created the National Water Commission (CNA). The commission was created with an explicit mandate to define a new policy for the management of the waters of the country. This led to the development of the National Program for Decentralization of the Irrigation Districts (or Transfer Program), which was designed to establish a system of co-responsibility between CNA and the water users and where the 80 public irrigation districts would become financially self-sufficient (Johnson, 1997).

In the past, the Government built, operated and managed 3 million ha of large surface irrigation systems organized into 80 irrigation districts. With the transfer program, the management of these systems has been handed over to water users organizations (WUOs), known as Asociaciones Civiles. These water users' organizations manage the irrigation sub-systems or "modulos", varying in command area from 5,000 to 20,000 ha; these organizations are responsible for the operation and maintenance of the secondary irrigation and drainage systems. At the end of 1994, full or partial management responsibility for 55 irrigation districts (69% of the total) with a command area of about 2.5 million ha had been transferred to 316 WUO's and 319,451 farmers, and by the end of 1996, 87% of the area with medium and large scale irrigation had been transferred to water users associations (Palacios, 2001).

The work involved countless meetings at various levels, from discussions with leaders of producers and marketing associations to one-on-one discussions with users. The transfer program was initially focused on the most productive irrigation districts, with the most commercially oriented farmers. The most important criterion for selecting districts was the potential of the water user organization to become financially self-sufficient, with users paying the fees to cover the costs of operation, maintenance, and administration.

Groenfeldt and Sun (2001) suggested that the Mexican Government offered the farmers an incentive to accept the higher costs for their irrigation. In fact, there was a carrot as well as a stick. The carrot was the management autonomy and the transfer of mechanized equipment from the agency to the farmers association. The farmers would become the owners of this equipment, and would be free to set their own rules for when to clean the canals, how to distribute the water, and which technical staff to employ for this purpose. The canal would be theirs on a 20- year concession, which is in practice a transfer of ownership. But there was also a "stick". If farmers refused to take over the management, the government could not offer any assurance that the canal network would be kept in a good shape. The government in effect threatened to nullify the conventional and standard understanding with the farmers regarding the subsidy levels in the irrigation sector. Many farmers, and particularly the commercially oriented ones, could not accept the risk that the irrigation infrastructure might collapse. They preferred to take over the management, and with a few exceptions, they haven't looked back. They are paying much more for their water without the government subsidy, but the reliability and responsiveness of their new management structure is well worth the price.

CNA continues to play a strong role, first of all in the direct management of the headworks and the main system, and in facilitating the associations' management of their section. The Government did set up the legal framework that defined the new relationship between the water user associations and the state. The details of that relationship could then be tailored for each association when its memorandum of understanding with the government agency was negotiated. The Government monitors the performance of the association, both financially and technically and it provides technical assistance, e.g., in monitoring the soil quality in poorly drained areas, and installing measuring devices. The Government continues to be a source of indirect financing for the associations through subsidized loans for equipment. Major rehabilitation and physical improvements will continue to be handled by the Government, although the associations can determine what is done within their area (Groenfeldt, 2001)

Gorriz et al. (1995) also mention that the implementation of a reform program requires a strong government commitment of top political leadership and policy managers; the establishment of sound legal and institutional frameworks to enable and simplify the management transfer of irrigation districts to WUO's; and the adjustments to the new roles by both farmers and the government irrigation agency.

Johnson (1998) mentions that as the irrigated districts contain some of the most productive agricultural areas in the country, the overall thrust of the program was to ensure that these areas continued to be highly productive by keeping the irrigation systems in operating condition. Over time it was expected that agricultural production would continue to increase, but that was not the primary objective. The Mexican Transfer Program was designed to ensure sustainability of the irrigation districts; to reduce the financial burden on the Government; to pass responsibility for operation and maintenance to the users; to increase efficiency of the use of water, and improve and sustain system performance; and to reduce the number of public employees in the irrigation districts.

As results of the Transfer Program, Johnson also mentions that:

- WUOs have proven capable of operating and maintaining the "modulos", even up to 50,000 ha or more. Water fees collected by the users have not only supported the "modulos" operation and maintenance activities but also have funded most of the operation and maintenance activities by CNA staff at the main canal and water source level. This is in sharp contrast to the situation that existed, when the systems were heavily dependent upon government subsidies and consequently was deteriorating rapidly due to lack of stable funding;
- The number of CNA staff has been reduced and, in most districts, the systems are being operated with less staff, although often the "modulos" have recruited staff with higher levels of education. The ability to hire and fire their own staff has improved the responsiveness of the operational staff to the needs of the users;
- With increased operation and maintenance budgets, including more funds for maintenance and a more responsive staff, the transfer program has created a situation that is much more sustainable than the situation in the irrigation sector before transfer.

According to Johnson, additional changes are required to ensure that the program is sustainable over time. The system of water fees needs to be changed so that the districts develop a reserve fund for emergencies, for future replacement, and modernization. They also need to shift to a system where the "modulo" collects a fixed amount to pay the costs of the staff and other facilities of the "modulo" and a volumetric fee to cover the variable costs of delivering water. In addition, it is necessary to clarify the water laws to protect agricultural water rights. Mexico's legal system does not clearly specify the rights that exist or not for irrigated agriculture and how those rights can be protected against demands for water from municipal and industrial users.

Since the districts, divided into 'modulos', were transferred to their water users associations, a new group of problems have appeared. These are termed second-generation problems, which include conflicts over water, often with municipalities, due to the poorly specified rights; insufficient revenue to support a proper operation and maintenance; poor accounting and bookkeeping practices; widespread firing and hiring of staff when directors change; nepotism in staff appointments; and the use of director positions as political springboard. Other problems stem from the poor condition of the irrigation infrastructure and the failure of the government to fulfil modernization commitments and duplication of efforts and poor coordination between associations and CNA. After its downsizing, CNA is top-heavy with managers and directors and requires a thorough reorganization to adjust its structure to its new role (Palacios, 2001).

The new National Water Law provides a suitable framework to solve most of the problems related to the use, exploitation and management of the hydraulic resources. Water rights are generally defined, a registry for water rights is being established, and the outlines of a market for water rights are given. However, although implementing regulations for this bill already exist, there are many legal gaps. Needed is guidance on ways how to solve specific problems at the level of the basin, irrigation district, and water users' organization, as the Law itself requires.

3.3.2 Indonesia

Vermillion (2001) carried out a study with the purpose to evaluate the outputs and impact of the Small-Scale Irrigation Turnover Program (SSITP) in Indonesia that was initiated in 1987, and subsequently to examine to what extent the new Irrigation Sector Reform Program (ISRP) presently in progress is likely to address shortcomings identified in the Small-Scale Irrigation Turnover Program (SSITP). A summary of the main characteristics and outputs of Vermillion's study are presented as follows.

Agriculture employs 44% of the Indonesian labour force and contributes for 16.5% to the gross domestic product. About 71% of Indonesia's more than 9 million ha of rice is irrigated and other major irrigated crops are sugarcane, maize, soybeans, tomatoes, chilli, onion and vegetables. In Java most farms are only 0.25 – 0.35 ha in size; outside Java the average farm size is about 0.5 to 1.0 ha. The total area under irrigation is approximately 7.1 million ha, of which about 5.5 million ha are listed officially as public irrigation systems and 1.6 million ha are managed solely by farmers (Soeparmono and Sutardi, 1998, cited in Vermillion, 2001). More than 2.1 million ha of irrigated land (30%) are served by small-scale systems with command areas of 500 ha or less, of which about 900,000 ha were considered public systems in 1987. The large majority of small-scale irrigation systems in Indonesia are river diversions with free or adjustable intakes, unlined canals and small fixed proportioning division structures along canals and at farm inlets. Small systems sometimes have overflow structures; silt flushing gates; division boxes; flumes and culverts that were built by the government.

The period from the 1960s to the early 1990s was the construction era for the irrigation sector. Between 1969 and 1994 Indonesia constructed new irrigation systems and upgraded or rehabilitated existing ones. As a result, the total irrigated area expanded by 1.4 million ha and 3.4 million ha of existing systems received physical improvements. During this period of rapid development of the irrigation infrastructure, projects were implemented by the government through contractors with little if any participation of farmers in the design, construction and management of the new or improved systems. This approach to irrigation development frequently resulted in farmers' dissatisfaction with the appropriateness or quality of the irrigation system design and construction. The positive results of the construction era were the large and rapid expansion of the irrigated area. Average rice yields rose from 2.4 tons/ha in 1968 to about 4.5 tons/ha by the early 1980s. One study attributed approximately 16% of the increase in average yields over this period to improvements in the irrigation infrastructure (Varley, 1997 cited in Vermillion, 2001).

However, the construction era left a number of less positive legacies. These included: a narrow focus by the government on infrastructure development, lack of creation of viable water users associations, faulty or inappropriate design or construction, over-dependency on foreign financing and a weakening of the farmers' sense of responsibility for irrigation management. These legacies did not create an organizational capacity, especially at local level, to cope with the rising problems of competition for water and the physical and financial non-sustainability of irrigation systems.

By the late 1980s the Government of Indonesia recognized that it needed to make adjustments in its management of the irrigation systems to address the growing insolvency of the irrigation sector. The following factors could be attributed to this:

- Provincial governments were unable to allocate sufficient funds for routine operation and maintenance;
- The central government was unable to provide more than about 50% of the estimated subsidy required to add to the amount the provincial governments were providing, hence there was a need to reduce costs to the government for the irrigation sector;
- The Land and Building Tax and the old irrigation fee were not producing significant amounts of revenue for irrigation;
- Numerous small-scale systems, which were traditionally farmer-developed and managed, had been automatically re-classified as government systems after rehabilitation projects. This unnecessarily increased the government's burden in the irrigation sector;
- A recognition that irrigated rice yields and cropping intensities had levelled off and that future increases in productivity would depend on more demand-responsive water management;
- The previous major rehabilitation program for small-scale irrigation and many other systems were completed more than ten years ago, were in often dire need of repair due to deferred maintenance. A new "reform" program was seen as a means to obtain additional funds from external donors for more rehabilitation;
- External donors, especially the World Bank and the Asian Development Bank, promoted policy reforms as condition to new irrigation investment.

As a consequence, the Indonesia Government issued the Government Policy Statement on the Operation and Maintenance of Irrigation Systems. The three key components in the new policy included the transfer of the management of all public irrigation systems of 500 ha or less to water users associations; a new irrigation service fee to be adopted in all public irrigation systems, which are not transferred to farmers management (those larger than 500 ha); and the introduction of more efficient operation and maintenance procedures in public irrigation systems.

The turnover of public irrigation systems of 500 ha or less was implemented through the Small-Scale Irrigation Turnover Program. The policy stated that after turnover the operation and maintenance and the financing for operation and maintenance would become the responsibility of the water users associations. The provincial irrigation services were to organize the water users associations (at least one per system), supervise the minor rehabilitation where needed, and change their roles from direct management to provision of support services and regulation. Approximately 900,000 ha or 21% of the total area of public irrigation systems in Indonesia was served by small-scale government irrigation systems. This constituted about 70% of all public irrigation systems in the country. The policy stated that turnover would be implemented gradually over a 15-year period, ending in the year 2003.

The immediate objectives of the turnover program were to increase the farmer participation in operation and maintenance of small-scale irrigation systems; to decrease or eliminate the need for the government to finance operation and maintenance by promoting self-reliant water users associations; and to improve the operation and maintenance performance of small-scale irrigation systems.

The second component of the new irrigation policy was implemented through the Irrigation Service Fee Program, which started in 1988. As farmer payments could not be sustained, the program was discontinued by the late 1990s.

The government formed an inter-departmental working group at national level to develop a strategy for implementing the turnover program. A pilot stage in 1987 and 1988 was implemented in West Java and West Sumatra using community organizers to guide farmers in the formation and training of water users associations. Based on the results, the Ministry of Public Works issued a regulation on procedures for implementing the Small-Scale Irrigation Turnover Program. The regulation also described the powers of the water users associations.

The first step of the implementation at district level was an inventory of all systems below 500 ha. The inventory collected information about the technical level of the systems, the functional condition of the infrastructure and whether a water users association already existed or not. Results from the inventory were used to classify systems into one of three categories:

A. systems without any agency staff that had an water users association and had been wrongly classified as a public system;
B. systems which had an agency staff assigned and which did not need repair or improvement of the infrastructure;
C. systems that had an agency staff and that needed repair or improvement of the infrastructure.

Following the inventory and classification, the provincial irrigation service at the district level prioritized and scheduled the systems for transfer. When a system was selected for turnover, the community organizer and a staff of the sub-district office conducted a "socio-technical profile" of the system. The profile was a detailed examination of the system infrastructure and its functional condition, socio-economic conditions, administrative context, who was currently performing what management tasks and identification by the farmers of what physical improvements and institutional strengthening was needed. Farmer representatives, working with the community organizer, prepared a "farmers version" of the design of needed repairs and improvements. These were simple sketches and descriptions.

Local engineering firms, or in some cases staff of the provincial irrigation service, produced a "technical design" for improvements, based on the farmers' version. This was followed by construction. Preference was given to hiring farmers of the system as labourers. Parallel to the physical improvements, the community organizer and local provincial irrigation service staff facilitated the creation or strengthening of the water users association.

The first systems were turned over to full farmer management in 1990, in West Java and West Sumatra. The Small-Scale Irrigation Program rapidly spread to most of the provinces of Indonesia. Early observations indicated that the results of turnover were relatively positive, without any important negative consequences (Mott-MacDonald, 1993; Bruns, et al., 1994).

Over time, the government increasingly dealt with more problematic systems. Field research identified the following main problems that arose during implementation of the Small Scale Irrigation Program:
- Farmers motivation to take over the full management and financing of systems was hampered by regulations imposed by the local authorities to dictate official cropping patterns, which were often at odds with farmer preferences;
- The irrigation agency was often preoccupied with physical repair works, often to the neglect of sufficient efforts in organizing water users associations and capacity building;
- Farmers inputs to the design of repairs were overlooked or ignored during construction;
- Design and construction was financed entirely from government funds, with only occasional voluntary additional contributions by the farmers;
- Long delays in construction and legalization of water users associations sapped the enthusiasm of the farmers for the program;
- Water users associations were legally and politically weak, had no water rights and were not included as member of the district-level irrigation committee. Hence, water users associations had no voice to deal with the rising issues of water allocation along river basins and little influence to enforce its own rules, collect water charges, settle water disputes among members and develop agri-business ventures;
- The Small Scale Irrigation Turnover Program was structured and implemented according to top-down administrative instructions, quotas and standardized training materials. Little room was included for meaningful negotiations and joint planning with farmers;
- Some reluctance and resistance by the Provincial Irrigation Services to implement the Small Scale Irrigation Turnover. Also, the staff of the provincial irrigation service was often not transferred from the small-scale systems after turnover;
- The provincial irrigation service provided little if any technical or managerial guidance after turnover and did not significantly change its own role from implementer to provider of support services.

The program slowed down and by 1997 only 420,000 ha had been turned over. This was 47% of the 900,000 ha area targeted for turnover by the year 2003. Delays were caused by the decision that all systems should be repaired or improved before transfer, delays and irregularities in construction contracts, reluctance about the Small Scale Irrigation Turnover by provincial and district authorities, disagreements between farmers and Provincial Irrigation Services and funding constraints.

The Small-Scale Irrigation Turnover Program was a modest reform that affected only a small part of the irrigation sector. Only a limited amount of authority was devolved to farmers and the provincial irrigation services still maintained control over the intake, and in many cases, over the main canals of the small-scale systems. The basic co-dependency between farmers and the provincial irrigation services for deferring maintenance and obtaining externally financed rehabilitation projects was not changed; hence, the Turnover Program did not overcome the problems of financial and physical sustainability of irrigation. Water users associations were still relatively weak organizations in the rural institutional landscape of Indonesia. They lack water rights, do not own the infrastructure, have difficulty in obtaining credit from banks and do not have an influential link to river basin management forums. Irrigation intensities are already relatively high in Java and have very limited potential for increasing agricultural productivity.

Clearly water users were mostly on the sidelines of the turnover process, watching as the government, once again, provided free repair and upgrading of irrigation infrastructure. The results were often unsatisfactory designs and construction and a reinforcement of the perceptions among farmers that the government was responsible for repairing and upgrading irrigation systems. Farmers could rely on the government to return again. It made sense to defer maintenance.

It has become apparent in recent years that more far-reaching reforms are needed in order to achieve better benefits for farmers and more economies for government through devolution. In order to create more meaningful incentives for farmers to invest in long-term sustainability of irrigation systems, there is a need to expand the economic niche of farmers, link them to river basin management and have a clear policy requiring joint investment in repair and rehabilitation of the irrigation infrastructure. The water users associations need stronger legal power, help in developing agri-businesses and more clear water use rights.

Vermillion (2001) states the following conclusions:

- The Small-Scale Irrigation Turnover Program, initiated in 1987 was a first attempt to reform the irrigation sub-sector in Indonesia and to address the insolvency of the irrigation sub-sector, while enabling farmers to sustain and improve the productivity of irrigated agriculture. The study carried out to assess the results of the Small Scale Irrigation Turnover Program did not show evidence of any significant impact. Although improvements in water distribution and prevention of water disputes in some systems were reported, the turnover program did not lead to a significant reduction in the cost of the irrigation sub-sector for the government, which was the principal objective of the program;

- The design of the reforms was also narrow and incomplete. It did not include a new system of water use rights, legal empowerment of water users associations, restructuring the irrigation bureaucracy or strategic planning for provision of new support services. Moreover, the reforms did not include devolution of authority for medium and large-scale irrigation systems;

- By the mid 1990s it was apparent that progress in all three components of the 1987 policy on irrigation operation and maintenance had slowed down to a halt. The Irrigation Service Fee was not being collected, needs-based budgeting was not being implemented, there had been no basic change in the method of allocating funds for operation and maintenance, under-investment in maintenance continued, rapid deterioration was still the norm and farmers still had little voice or choice in irrigation investments in small-scale systems and in operation and maintenance in medium and large-scale systems;

- The Irrigation Sector Reform Program can be considered as a comprehensive reform program in that it deals with the fundamental under-lying problems in the sector: accountability, incentives, transparency, and choice of service provision and empowerment of water users. The next great test is whether or not the initial wave of political commitment for reform will be sustained through to the adoption of consistent practices and procedures at the local level. This will require not only sustained commitment but also willingness of the government to shift its role to regulation and provision of support services.

3.3.3 Philippines

The National Irrigation Administration (NIA) is responsible for the overall irrigation development in the Philippines. The irrigation systems can be categorized into national (NIS) and communal irrigation systems (CIS); 42% corresponds to the national systems while 50% of the total irrigated area belongs to the communal systems. The NIS systems cover areas of more than 1,000 ha and are jointly managed by an irrigator association (IA) and the NIA. The farmers pay an irrigation service fee. The CIS are systems with areas between 30 and 1,000 ha, which were constructed by the farmers. After construction, the systems were transferred to and managed by the farmers, who amortize the capital cost without interest for not more than 50 years. Being successful with the transfer of the operation and maintenance to the farmers of the CIS systems, the NIA started to transfer the NIS systems to the IA's in 1989 (Hernandez, 1995).

Oorthuizen and Kloezen (1995) carried out a case study to assess the impact of the removal of operation subsidies to the NIA on the management performance of a small canal irrigation system in Southern Luzon. Their study showed that the creation of financial autonomy of the agency did not fully produce the often-assumed positive effects regarding both accountability relationships and irrigation management performance.

NIA, founded in 1964 as a public corporation, was required to become a financially autonomous irrigation agency for its operation budget. To increase its revenues, the NIA management designed policies to increase the payments of the water users. In the smaller, farmers managed communal systems the farmer–beneficiaries were to amortize a substantial part of the NIA's capital investment. For the larger, government managed national systems the irrigation service fees to be paid by the water users were increased.

The offices of the national systems and the regional and provincial offices were also forced to become financially viable, which meant that the operating costs of an office in a given year needed at least to equal its revenues from fee collections and other sources. The NIA management also took steps to reduce its operating expenses; one mechanism used was the reduction of the staff. Large staff reductions were made possible through the devolution (management transfer) of the responsibility for several management tasks to farmers, organized in associations (IA). In this way, NIA designed three stages of devolution:

Stage 1:   an IA is contracted to maintain the canals in a specific water master section and to help the water-master in the operation;

Stage 2:   besides maintenance, the IA draws contracts to collect irrigation service fees from its members;

Stage 3:   IA enters into a turnover contract. Under the overall supervision of NIA, the IA becomes fully responsible for all management tasks.

By 1989, 518 IA's had gone into stage 1 or 2, covering about 140,000 ha, while for stage 3 about 9,000 ha were turned over to 35 IA's. The 149,000 ha under the three stages equals one-quarter of the total irrigated area under the national systems. Depending on the specific type of devolution, the reduction in NIA staff in a sample of involved systems ranged from 13 to 75%.

The most important findings of the study by Oorthuizen and Kloezen (1995) on the pilot project were the following:

-    NIA's policies implemented after the removal of its operation subsidies led to a much more financially viable organization, both at national and system level;
-    About the impact of turnover on the management performance:

♦ the assumed positive impact of NIA's financial autonomy did not fully materialize as far as system operation is concerned. NIA's policy reforms for financial autonomy did not bring the operational performance to a higher level. Though the costs of the management reduced after turnover, the association was not able to solve the major operation problems of the sub-systems;

♦ financial autonomy did not improve the quality of maintenance. Before and after turnover, farmers took care of regular cleaning and weeding of the farm ditches next to their fields, without any involvement of either the NIA or the board of the association;

♦ NIA's financial autonomy did not bring clear improvements in the management of the sub-system's physical deficiencies;

♦ the impact of NIA's policy reforms for financial autonomy on the financial performance was mixed. The positive point was that the fee collection rate under IA management increased to 81% in 1989, which was much higher than the system-wide average of 20% by NIA prior to turnover. On the other hand, IA failed to tackle major operation problems, while the lack of canal maintenance will threaten system performance in the long rum;

- At national and local level the agency designed and implemented policies to reduce costs and to increase income. But unlike the theory, the balance between the interests of the agency to make ends meet and the need to meet local demands was not easily found;

- The quality of some of NIA's services was negatively influenced by the financial needs of the agency. The system office attained financial viability at the cost of institutional support towards the association, while the board was pressed to meet amortization due at the cost of canal maintenance and infrastructural improvements.

3.3.4   Turkey

Turkey is a rapidly modernizing country of about 66 million people with a population growth rate of 1.5% per year. At the beginning of the 1980s, national economic strategy switched from a policy of industrialization based on import substitution to a policy aimed at allowing a greater role for markets.

The irrigation potential in Turkey is 8.5 million ha of which 93% (7.9 million ha) is irrigable from surface water sources. About 53% of this area has been developed (4.2 million ha) and the government expects to expand the gross irrigated area by about 50,000 ha per year. The gross area currently irrigated by large-scale surface systems is under different managing entities: 1 million ha under private systems, 1,094,000 ha operated by the local government or cooperatives, 356,481 ha managed by DSI (Directorate General of State Hydraulic Works), and 1,706,019 ha are transferred to organizations. The transfer process has been developed in two stages. The first one, called The Irrigation Group (IG) Program, was addressed by the Irrigation Groups which are less formal organizations and aimed at operating and maintaining the lower level facilities in an irrigation system. It is considered as the pioneering program to that to be developed during the second stage and is named Accelerated Transfer Program (ATP).

The Irrigation Group Program dated from the 1960s when the General Directorate of State Hydraulic Works developed a program to transfer the operation and maintenance responsibilities for secondary and tertiary distribution networks to the Irrigation Groups. Under this program village headmen or majors of municipalities entered into a contractual arrangement with the General Directorate of State Hydraulic Works to take over the administrative responsibility for tasks such as collecting and submitting farmer water demand application forms to the General Directorate of State Hydraulic Works, managing water distribution below the secondary canals, and cleaning and minor repair of hydraulic structures. They are also responsible for hiring labourers for maintenance work and ditch riders for water distribution. In 1994, approximately 600,000 ha, or about 40% of the General Directorate of State Hydraulic Works developed area, were partly managed by Irrigation Groups.

In comparison with the General Directorate of State Hydraulic Works this program has been considered more cost efficient in implementing operation and maintenance tasks, with local administration seen to use less labour at lower cost. The effectiveness of water use was also improved with this program. Some Irrigation Groups reached irrigation ratio values of 77% compared with 60% on the systems run entirely by the General Directorate of State Hydraulic Works.

However, the Irrigation Group Program has faced some problems, which have caused a limited impact on overall scheme operation and maintenance costs, cost recovery rates, and cost for the General Directorate of State Hydraulic Works and other personnel. Among these problems can be mentioned the lack of direct farmer involvement in the Irrigation Group establishment and operation, reluctance by some Irrigation Groups to take management responsibility, and changes in the commitment of the staff of the General Directorate of State Hydraulic Works to the Irrigation Group Program. Irrigation Groups have also faced difficulties associated with the fact that the Irrigation Group boundaries are based on settlement areas rather than on irrigation boundaries, complicating the operation and maintenance of canals that cut across village boundaries. The absence of any mechanism for articulating local units into a structure, which could manage entire systems or hydrologic units, has been the most serious constraint. Without it, the potential impact on public operation and maintenance expenditures, staff levels, and cost recovery rates is extremely limited, even if the Irrigation Groups work effectively and efficiently.

The Accelerated Transfer Program started in 1993. It was the result of the discussions carried out in 1990 in order to face the crisis of the irrigation sector caused by a combined effect of a national budgetary crisis and rapid growth in the wage costs of labour, especially of members of the labour union. In this point the World Bank played an important facilitating role in initiating and sustaining the program in its early stages. The program was initiated in four pilot regions with initial intention to start with smaller systems (less than 3,000 ha) and to gradually expand the area covered. The initial plans called for transferring about 150,000 ha per year, achieving 1.35 million ha by the year 2000. In 1995 the area transferred was more than three times the area planned for transfer in that year, and new higher targets for subsequent years were established. The high pace of transfer in 1995 resulted in 61% of the General Directorate of State Hydraulic Works developed irrigated area in the country having been shifted to local control by the end of that year. The pace declined in subsequent years, but still averaged about 120,000 ha per year through the end of the decade. Presently just over 80% of the irrigated area developed by the General Directorate of State Hydraulic Works has been transferred to local entities to manage.

The most important organization forms which act as recipients of operation and maintenance responsibility in the transfer program are the village management, municipal management, and irrigation association management. In the first one, the irrigation system serves only a single village, in the second one only a single municipality, and the third one the irrigation system covers more than one local administrative unit (village or municipality). The transfer process takes 6 to 9 months from initiation to implementation and covers the stages of Initiation (creating the interest or willingness of the receiving group), Transfer Agreement (document sets out the rights and responsibilities of the parties), Transfer Protocol (catalogues and describes all of the characteristics and facilities of the irrigation unit being transferred; the ownership of facilities is not transferred and remains with the state), Preparation of operation and maintenance plans, and finally Implementation by the organization with assistance from the General Directorate of State Hydraulic Works as needed. Svendsen and Nott (2001) concluded about the Transfer Program as follows:

- Staffing intensity on systems managed by Irrigation Associations (IA) is only about 56% of that prevailing when the General Directorate of State Hydraulic Works was the sole managing entity, showing strong gains in operational efficiency from the transfer program;
- For the Directorate General of State Hydraulic Works, the transfer program has resulted in significant declines in its own operation and maintenance staff levels. However, staffing intensity for the remaining systems under the General Directorate of State Hydraulic Works control and the overall General Directorate of State Hydraulic Works staffing levels remain extremely high;
- The General Directorate of State Hydraulic Works operating costs have also fallen sharply. However, the operating costs per ha on the 20% of the systems still managed by the General Directorate of State Hydraulic Works are roughly double those on Irrigation Group managed systems;
- The Irrigation Groups are currently charging about US$ 78 per ha in irrigation fees. The General Directorate of State Hydraulic Works charges about 13% more than this, in nominal terms, but collects far less.

Svendsen and Nott (2001), mention that some outcomes of the transfer program are evident at this early stage, while others will not be assessable for several years. Public costs of operation and maintenance begun to fall and will very likely continue to do so over the next few years. Private costs have increased and will likely continue to increase as more and more responsibility is transferred to local agencies. Cost recovery has improved dramatically. The General Directorate of State Hydraulic Works operation and maintenance staff levels have fallen marginally, though more dramatic declines will depend on resolving issues of transfer and termination with the powerful unions representing the support staff of the General Directorate of State Hydraulic Works. Associations have gained control over many operational decisions and secured the opportunity to stabilize and improve system performance.

The impacts of transfer on the quality of irrigation services are not yet assessable, though early evidence in three systems suggests that irrigation associations may, in some cases, be able to expand the irrigated area they manage beyond previous averages. Moreover, some important issues of future sustainability remain. Still, in comparison with efforts in other countries, the early achievements of the Accelerated Transfer Program in Turkey show considerable promises for achieving the objectives held both by the government and by local associations.

Svendsen and Nott (2001) mention also that second generation problems and challenges are already emerging in the wake of the early successes of this initiative. These can be categorized in terms of the party on which they have their primary effect.

Challenges for the General Directorate of State Hydraulic Works include among others: the difficulty in reducing overall staff levels in general, and operation and maintenance levels in particular, in the wake of transfer; the absence of a charging mechanism for bulk water supply to irrigation associations, and the consequent absence of an economic restraint on demands for water; and the indistinct vision of a new role for the agency in supporting existing irrigation in the post-transfer era.

Emerging problems for irrigation associations include: the undefined nature of the water rights in Turkey, and the consequent insecurity of their claims on irrigation water; restricted options for obtaining heavy maintenance equipment; the lack of a legal basis for forming federations of irrigation associations for joint purchasing and supplying "combined" services such as equipment maintenance; the lack of a clear policy on capital cost sharing for rehabilitation and/or new system construction; the need to increase direct farmers participation in irrigation associations governance and to reduce the dependency on village and municipal leaders in filling the leadership roles in the irrigation associations; and the weak support services for irrigation associations in some areas and regions.

3.3.5   Conclusions

From an analysis of the previously described experiences can be concluded that the governmental transfer policy for the irrigation sector has been highly influenced by the fiscal crisis of the governments during the 1980s and 1990s. During that fiscal crisis the governments found in the irrigation sector a way to reduce their financial burdens by transferring the responsibility for the administration, operation and maintenance and the corresponding costs of the existing irrigation systems to the water users' organizations. The effective reduction of the governmental staff for the management of the irrigation systems and the reorganization of the institutional infrastructure were other facets of the contribution of the irrigation sector to the alleviation of the financial crises of the governments. In the case of Mexico the reduction of the governmental staff was about 71%, while in the Turkey the cutback was about 54%. This is clear in the cases of Mexico and Turkey; especially in the case of Turkey for the second phase of the transfer program named Accelerated Transfer Program, for which the Mexican experience was the reference for the implementation.

In the cases of Indonesia and Philippines the direct crisis of the irrigation sector caused by its financial insolvency was the main driving force for the management transfer. In Indonesia, after the peak of the construction era and an inappropriate management approach of the Government, the irrigation systems fell in an organizational, physical and financial deterioration and under these circumstances the external donors enforced a policy of reform (management transfer) as a condition for new irrigation investments (rehabilitation). In the Philippines, the management transfer was promoted as a mechanism through which the public irrigation agencies became self-sufficiency from financial point of view.

The previous experiences showed a variety of differences for the development of the management transfer program. The pace of the implementation of the program varied from an intensive way, as in Mexico to a gradual one in the Philippines; and in Turkey only after the implementation of some pilot projects. There is an increasing consensus about the unsatisfactory performance with a gradual transfer or with pilot projects. Some cases, such as in the Philippines, are taking many years without a real transfer of responsibilities and authority to the users.

The target areas to be transferred showed also significant differences. In Mexico, the irrigation area to be transferred was about 3 million of ha, of which 87% had already been transferred after 6 year that the program has been effective. In Turkey the target area to be transferred before the year 2000 was 1.35 million ha at an expected rate of 150,000 ha per year. Indonesia and the Philippines presented some smaller areas to be transferred. In Indonesia the program was focused on the small-scale irrigation systems with a target area of only 900,000 ha, which is just 21% of the public irrigation systems; while the Philippines focused its program on 149,000 ha, which was about 25% of the total irrigated area under national systems.

The potential productivity and the perspective to develop, to sustain or increase the profitability of the existing production systems to be transferred could be considered as an incentive for the users to accept and participate in a more than active way in the development process and implementation of a program for the management transfer. This was perceived for the Mexican model, which has been considered as a very successful experience. The transfer program in Mexico was focused on the most productive irrigation districts with a majority of commercially oriented farmers. Maybe, this was the case of the irrigation districts located in the North of the country with an irrigated area of more than 200,000 ha, which have an intensive production system that is mainly oriented on the export and the national market.

The transfer process has allowed a change in the management approach of the irrigation systems, passing from the governmental approach that is based on a supply oriented management to the demand management approach in which the users determine who will be their managers and who will have the authority and the responsibility to operate the system in the way as they desire. In the governmental approach the government agency supplies not only the water, but also the system managers and the operation rules that resulted in many cases in irrigation systems that are poorly maintained and that show a clearly and steadily deteriorating infrastructure.

4

Framework for Sustainable Management of Transferred Irrigation Systems

4.1   Basic Concepts and Approaches

This thesis will formulate a framework for the sustainable management of the RUT Irrigation District after transfer by the Colombian Government to the Water Users Association ASORUT. In this chapter the formulation of a conceptual framework will be presented, but that section will be preceded by a description of some general concepts and approaches of sustainable water management and especially the sustainable use of land and water resources and the management of land and water systems. Concepts and approaches concerning sustainable water management are presented by several authors and from different points of view. Some of them will be used in the formulation of the framework for sustainable management to be developed.

Sustainable development is defined as the approach towards meeting the needs and aspirations of present and future generations. It describes the means by which economic and social progress could be achieved without compromising the integrity of the environment (Brundtland, 1987). According to the International Commission on Irrigation and Drainage (ICID) and based on current trends, the population growth may exceed in the near future increases in food production. During the past, irrigation and drainage have significantly contributed to the increase in food production and nowadays one-sixth of the cropped lands that are irrigated produces one-third of the world's harvest of food crops and drainage has enhanced the productivity of another tenth of cropped lands. These contributions have lead to the fact that irrigated agriculture consumes 70 to 80% of the fresh water used in developing countries. Moreover, drainage of agricultural lands impacts downstream water quality and about 40% of the irrigated lands are at risk from waterlogging or salinization, or are already affected. Drainage, apart from improving food production, can also help in making agriculture more diversified and competitive, promoting rural well being, improving public health, etc. The performance of many irrigation systems is below their potential levels and the future growth in irrigated agriculture will be limited by the scarcity of water and land, by increasing competition for water, by the degradation of the environment, by the rising cost of development, by the deterioration of existing systems, and, finally, by the inadequacies of management.

Humanity has the ability to make sustainable development to ensure that it meets the needs of the present without compromising the ability of future generations to meet their own needs (Brundtland, 1987). Development and environment need not to be in conflict, provided that humanity adjusts to the requirements imposed by sustainable development. Firstly, development must not damage or destroy the basic life support system: the air, water, soil, and biological systems. Secondly, development must be economically sustainable to provide a continuous flow of goods and services derived from the earth's natural resources. And thirdly, it requires sustainable social systems, at international, national, local and family levels, to ensure the equitable distribution of the benefits of the goods and services produced, and of sustained life supporting systems.

The concept of sustainability implies limits: not absolute limits, but restrictions imposed by the present state of technology and the social organization of environmental resources and by the ability of the biosphere to absorb the effect of human activities (Brundtland, 1987).

A main consequence of the sustainability concept is that not a sectoral approach, which sees and solves the different water problems sectorally, but a more integrated approach should be followed to attain sustainability. In this view, land and water problems are intertwined with social problems at different levels and in which scientific problems arise in various significances. International awareness for the importance of water management issues is increasing and there is a growing consensus on the need for integrated approaches. There is a clear understanding that effective water management has a key function in sustainable development.

The International Conference on Water and Environment (ICWE, 1992) in Dublin stated in the Dublin Principles that:
-   Fresh water is a finite, essential and vulnerable resource having an economic value with social, economic and environmental implications, which should be managed in an integrated manner;
-   Development and management of water resources should be based on a participatory approach, involving all stakeholders;
-   Women play a central role in the provision, management and safeguarding of water;
-   Water has an economic value and should be recognized as an economic good, taking into account affordability and equity criteria.

Associated with these principles is integrated water management, which implies:
-   An inter-sectoral approach;
-   Representation of all stakeholders;
-   All physical aspects of the water resources;
-   Sustainability and environmental considerations;
-   Sustainable development, reliable socio-economic development that safeguards the resources for the future;
-   Emphasis on demand driven and demand oriented approaches;
-   Decision-making at the lowest appropriate level (subsidiarity).

4.1.1   Needs in Land and Water Development

Some interference of man with the hydrological cycle is inevitable: the changes of land use and the use of water for irrigation, industry and human consumption affect the hydrological cycle and consequently the availability of water.

Sustainable development implies not only economic and resource policies, but also ethical aspects. The value systems of societies everywhere are involved, and water management is only one aspect of development. From only the water point of view, land and water development projects are sustainable, if water of sufficient quantity and quality and at acceptable prices is available to meet demands now and in the future without causing the environment to deteriorate.

The translation of sustainability into actions depends on spatial (or area) and time scales. In physical planning different levels can be distinguished, for instance national, regional and local planning. In planning the following stages can be mentioned, namely proposal, institutional consultation, public consultation, and decision (Schultz, 2005).

Spatial scales consider the rural and urban area. The rural area scale comprises a different set of development aspects at field, system, local, regional, country and global level. At field level the following items may considered to be of importance: basic unit in the layout of an area, other influencing factors (soil type, field irrigation and drainage systems, farm size, and flexibility), and determination of optimal field sizes (field dimensions; construction cost for irrigation and drainage facilities, hydraulic structures and roads; and operation and maintenance costs). At system level it will make a substantial difference if a project concerns land reclamation or the improvement of a cultivated area. In the latter case attention has to be paid to aspects like history, social status, and land property. At local level the attention will include the influence of the municipality, land use plans and settlement types. At regional level the plan is generally based on an area or on an integrated regional development plan. National planning, strong and weak points, and physical conditions are important aspects to be considered at country level.

Time scales take into consideration one season, one year, and lifetime of elements, a generation, and a century. One season is the smallest level to be considered. It determines the functions of the system components, and the required operation and maintenance versus the capacity to get this implemented. The time scale of one year considers aspects like accessibility of an area and the functions that would have to be fulfilled by the water system; here the requirements of the system during different parts of the year are important. The lifetime scale considers each element with its specific lifetime, after which it has to be renovated or replaced. Generation level is a long-term process in which the impact of a land and water development project is expected to influence the living conditions. It means that the users would have to accept and appreciate the results of the project in order to be able to improve their living conditions. This scale also implies that gradual improvements generally have a better overall result compared to rapid large-scale improvements. The last step in the time scale concerns the century level. Here long-term perspectives concerning, for example population growth and environmental sustainability, have to be considered. Various development approaches can be followed for land and water development projects, for instance large scale rapid development, small scale gradual development, directly based to the final stage, and step wise development. The different approaches have to take into account that a project will have to follow various stages, and should include the socio-economic and environmental consequences of the proposed development.

There is a great need for land and water development, aiming at the improvement of the living and production conditions in the rural areas, land reclamation, and the development of urban and industrial areas. The living conditions of the users will be determined by considering the need for the development, level of service required, role of the Government, and side effects of the development (Schultz, 2005). A successful project will on the one hand realize the objectives and on the other hand maintain the environmental impact at an acceptable level.

The need for development in rural areas is generally determined by the necessity to increase and/or to rationalize food production and to promote sustainable rural development. In other words, there is a direct link between the investments to be made and the benefits to be expected. This direct link enables planners to identify which investments may be justified. Meeting the needs is the objective of a project and only under favourable conditions can the objectives be met. Usually, a development project must be planned under restrictions imposed by nature or by other users. Obviously, no development is possible without the financial resources, especially for operation and maintenance and many projects fail due to insufficient funds. Moreover, the success of a project is strongly determined by the creation of an attractive environment for the users to initiate and continue the proposed activities. This means that the project has to be attractive and implies that it can be maintained adequately. However, the determination of the required level of service is very complicated as the interaction between water management and crop yield is not easy to quantify. In most land and water development projects the Government plays an important role, as it initiates developments that fit in its policy, and by preventing unwanted developments.

In the case of improvement of existing areas, the following aspects play a role:
- *Role of the Government.* It generally plays a guiding role during the whole process;
- *Determination of improvement options.* Various options or combination of them arise, like improvement of physical infrastructure (water management system, road system, or water transport system), improvement of agricultural production conditions (land reallocation, agricultural extension, crop diversification), and improvement of market conditions (credit facilities, storage facilities, market development);
- *Consultation with users.* The objective of the projects is to improve their living, or production conditions. This requires consultation and approval procedures with the parties involved in order to find out which improvement options may be most successful;
- *Institutional reforms and cost recovery.* In many instances the Government is the owner of the water management system and has the main responsibility for operation and maintenance. Related to improvements this position cannot be maintained and the involvement of the stakeholders will increase. This will require institutional reforms and cost recovery;
- *Land ownership.* It has often resulted in significant problems and delays in land improvement projects. Therefore this aspect deserves a lot of attention from the early stages.

Sustainability in Land and Water Development

Scarcity and misuse of freshwater pose a serious and growing threat to sustainable development and protection of the environment. Human health and welfare, food security, industrial development and ecosystems are at risk, unless water and land resources are managed more effectively at present and beyond than they have been in the past (ICWE, 1992). Sustainable development of land and water is essential for a stable development and it implies extended criteria for planning, design, operation and maintenance of the developed systems.

Prerequisites and criteria for sustainable development are:
- Political will to plan, and operate the project in a sustainable way;
- Finances and time to collect the necessary data for planning and management;
- Finances to continuously operate and maintain the project.
A development plan should be based on:
- A view of the project as an integral part of the society and considering all interactions with society and environment;
- Optimally adapted structures and considerations of non-structural solutions;
- Robust structures, which after failure can be replaced or repaired with minimal disruption of services;

- Considerations to alleviate water quality problems during operation;
- Full assessment of beneficial and adverse environmental impacts, and of means to alleviate or mitigate the adverse ones;
- Assessment of risks and safety measures to prevent disasters; like possible failure of structures, as well as of the system;
- Considerations of uncertainties of both supply and demand;
- Considerations of social impacts e.g. caused by dislocations of people or system failure;
- People affected by the project activities should be encouraged to become optimally adapted to the new living conditions and environment;
- Provisions to cope with changes in demand or land use;
- Elimination of all potentials for future water conflicts.

Society should demand that these criteria are met and that the legal and political frameworks required are established. A well-planned project is no guarantee of a sustainable system as sustainability also implies that the system is operated to perform as planned for the foreseeable future. A prerequisite for appropriate management is that it will include:

- A competent staff of managers, technicians, and workers for operation and maintenance;
- The assurance of a continuous supply of water at all times for all needs and provisions have to be made for times of scarcity;
- Continuously improvement of the systems performance to achieve maximum efficiency;
- Cost consciousness by considering means of covering the cost of operation and maintenance in a socially accepted and efficient way;
- Monitoring of the performance of the system.

The management has to consider a large number of often conflicting demands on the available water, and has to operate the system under various constraints. The different interests require a decision process involving multiple objectives and decision makers, multiple users and constituencies. Water management strategies for sustained growth and development have to be optimized by strategic planning. As water becomes a limiting resource, competitions, among water use sectors and disputes between upstream and downstream users in sharing water will increase. Relevant laws and policies are needed to ensure such sharing which should be equitable and efficient. Competitions in water use and deterioration of water quality have resulted in less water being made available for agriculture. Water saving irrigation techniques and water quality management measures will have to be promoted in order to cope with the decreasing water availability and the pollution of water. The demands on water for different uses might be settled by a trade-off among the water users.

While the irrigation sector is the largest user of water, it is necessary to optimize water use, promote conservation, and improve irrigation efficiencies. Optimization requires modern and viable management systems for planning, water distribution, operation and maintenance including engineering, agronomic, social, economic and financial aspects. The adoption can help to raise productivity and water use efficiency. Socio-economic development and environmental conservation are also closely linked; the economics may be based on the concept of internalized environmental cost in setting the prices. Most important is that sustainable development should be based on adequate participatory approaches for planning and management as well as mechanisms for accountability and democratic control (Savenije, 1997).

4.1.2 Management Requirements

The following components can be considered for the concept of sustainability:
- Institutional sustainability (capacity to plan, manage and operate the system);
- Technical sustainability (balanced demand and supply, no mining);
- Financial sustainability (cost recovery);
- Social sustainability (stability of population, stability of demand, willingness to pay);
- Economic sustainability (sustaining economic development or welfare and production);
- Environmental sustainability (no long-term negative or irreversible effects).

Once a water management system exists, management has to translate the demands into an operational and maintained system that will provide a certain flow of water to meet the needs of the users. For providing water a staff of technicians and managers ranging from plan supervisor to revenue collectors is generally needed. Conservation and efficient use are essential to reach a balance between growing demands and finite supplies. This does not require new techniques, but only more effective use of existing technology, better management, and effective transfers of knowledge and experience. According to the Copenhagen Statement the management of water resources for rural communities should follow the principles that:
- Water and land resources should be managed at the lowest appropriate cost;
- Water should be considered as an economic good, with a value reflecting its potential use.

Among the management tools to control the water consumption in a region, water pricing is one of the most effective. Water may be provided free to all, as part of a communal service, on one end of the scale, or it may be charged at a price that covers all costs for collection and distribution. Water pricing is also an effective way for saving water. Economists stress the role of water prices in resource protection and water saving, and politicians are challenged to use their results in plans to preserve the water resources.

The environmental, technical, financial, economic and social aspects of sustainability of irrigation and drainage systems require close attention, continuous monitoring and evaluation, for which comprehensive information systems will have to be developed. They allow an analysis of performance aimed at determining the extent to which the objectives of a system are being achieved. Performance analyses are indispensable to assess the effectiveness of the management and may help to prevent mistakes. The increased demand for irrigation service coupled with a reduction in new investments and increased water scarcity requires an increase of the productivity and efficiency of existing irrigation systems in order to achieve higher production. However, the low performance of most of the existing irrigation systems is a serious constraint to reach higher efficiency and productivity. In reality, the irrigation sector in many countries shows a low performance in which a lack of maintenance produces poor irrigation and drainage services leading to farmer dissatisfaction. Institutional reform is needed to improve this performance. The aim of benchmarking is to improve by a systematic and continuing process of (a) measuring performance, either within the same irrigation scheme against its previous performance and desired future targets, or externally against similar systems, (b) analyzing the results, identifying best practices and (c) implementing change. Hence, benchmarking is a process for achieving improvements in an irrigation and drainage system through comparison of its performance against other systems and its own objectives and targets. Real improvement is achieved only when problems in management, operation and maintenance are identified and best practices with respect to specific problems are adopted. Benchmarking differs from conventional performance evaluation in that it is a continuous and systematic process that leads to improved performance by going through a chain of activities that cover three aspects: measurement, analysis, and making change (Malano and Burton, 2001).

Sustainable development is mainly based on the application of well-developed principles and concepts of engineering and management. Unfortunately, most of the systems face financial constraints and investments mainly follow from the needs for replacement of deteriorated parts and for expansion of existing facilities. Often new investments are difficult to obtain and then by giving high priorities to basic needs like disaster reduction and disease prevention may result in the required investments. Shortage of funds affects not only the implementation activities, but also the required training and education of the management and the operating staff.

4.2   Formulation of the Conceptual Framework

Taking into account the time elapsed since the start of the first transfer of the irrigation management to the water users associations and the diverse problems faced by them for the management, operation and maintenance of the systems, a conceptual framework will be proposed for the sustainable management of the irrigation and drainage systems after transfer. This framework is supported by certain general principles, approaches and definitions, some of which were stated in numerous events and conferences concerning improvement and preservation of the natural resources with special emphasis on land and water resources.

There is a clearly identified need for institutional reform in irrigation as a precondition for the modernization of the existing irrigation systems. In order to enable the implementation of sustainable transfers all the legal, economical and organizational aspects should be identified and formulated:

-   *Legislation and regulation*: the development of national water legislation should take into consideration environmental, economic and social conditions of the country as well as technological features of the irrigation and drainage systems;
-   *Institutions and organization after institutional reforms*: this process needs a clear view of responsibilities and accountabilities; water users should participate in water management decisions at all levels;
-   *Financing and economy*: the transformation should result in a viable financial basis for irrigation, based on a substantial Government funding and incentives for modernization and declining support for operation and maintenance, and an increasing farmers share for operation and maintenance up to the level of full cost recovery within a time frame of a number of five years.

The proposed conceptual framework will mainly follow the Principles of the Dublin Conference (ICWE, 1992) and the concept of sustainable development as explained before. In addition, the following approaches have been taken into account:

-   The hypothesis that: "fully autonomous organizations accountable to their customers and managing a single irrigation system will exhibit the highest performance, will prove most adaptive to changing conditions and therefore will prove to be most sustainable" (Merrey, 1996). The hypothesis means self-

sufficiency of the management by the water users associations without major dependency on external agents;
-   Service-oriented management, which includes the ability of an irrigation and drainage organization to respond to the needs of its customers. This type of management is measured by the effectiveness of the management processes and the way by which they have achieved the overall Government's social and economic objectives, and have met the expectations of the farmers.

In the context of the irrigation and drainage sector, the management by water users organizations is defined as the process by which the irrigation and drainage service is provided (output) by using the allocated resources (input) in a sustainable and cost-effective manner. To attain the organizational goals and objectives specific activities are needed, such as planning, organizing, managing and controlling the land and water resources. According to Uphoff (1991 cited in Malano and Hofwegen, 1999) three activity levels in irrigation management can be distinguished; and they are focused on: Water (acquisition, allocation, distribution and drainage); Structures (planning and design, construction, operation, maintenance); and Organization (decision-making, resource mobilization, communication, conflict resolution).

4.3   Development of the Conceptual Framework

The sustainability of an irrigation and drainage system refers to the physical and organizational infrastructures for management, administration, operation and maintenance during the lifetime of the system according to established specifications and giving the users the expected benefits without causing damage to the environment.

4.3.1   Criteria

The concept of sustainability involves three elements that play an important role in the definition of the conceptual framework for sustainable management of an irrigation system. The three components are Community (C), Environment (E), and Science & Technology (S&T) and they are interconnected within the existing institutional and legal contexts. This interrelationship determines the sustainability of an irrigation system. Figure 4.1 illustrates the mentioned interrelationships.

Community (C) represents the different interest groups in the irrigation and drainage sector; in which the farmers are the main beneficiaries. The Irrigation Authority and the Government are also part of this component. Each group may have different objectives and expectations concerning the irrigation and drainage service. The main objective for the *farmers* is to maximize benefits and to minimize costs for irrigated agriculture as a production system. Taking into account that water is the main input, the characteristics demanded by farmers for the service provision include the operational performance of the irrigation authority (how well irrigation water is supplied) and the flexibility (frequency, rate and duration). The community and public have an important role in the water and land resources management. The participation of the stakeholder is essential for sustainable development of the land and water resources and there is a need to enhance the role of the water users' organizations. Management transfer of irrigation systems will only result in sustained development and management of the resources when the community organization has received the necessary economic and legal powers.

The *Irrigation Authority* is the organization that manages the system with the main objective to deliver adequate water to users in an equitable and reliable way to allow for optimal productivity at farm as well as at system level. Other objectives are to maximize the return per unit of capital invested, maximize use of labour intensive methods, to realize an efficient and equitable water distribution at least cost, and to increase the irrigated area. The irrigation authority together with the users must formulate clear rules for operation and water delivery through service specifications. On the other hand, the objectives of the *Government* are based on macro-economic considerations like increasing food production, employment, and foreign exchange, and helping to alleviate rural poverty. In this way, the Government uses irrigation and drainage as a tool for social and economic development insuring that negative effects do not harm the interest of the society and creating mechanisms for an effective operation of irrigation and drainage within the context of integrated water management.

Finally, the community with its specific socio-economic-cultural characteristics and organizational forms interconnects with the Environment and Science & Technology components in order to contribute to the sustainable management of the systems.

Environment (E) provides the natural resources for irrigated agriculture as production system. Climatic conditions, soils, ecosystems, and water resources are the main inputs provided by the environment component. Factors that affect the relationship between Environment and Community in two directions form risk factors. The climatic conditions include rainfall, temperature, evaporation, seasonality, etc. From soils point of view their physical and chemical characteristics may form the risk factors; the most important are salinization, fertility, waterlogging, erosion, topography, natural flora and fauna, etc. Water is an important input for the agricultural production and its risk factor may follow from its reliability of availability, quantity and quality, temporal and spatial variation, lowering the groundwater table and long-term deterioration of the aquifers.

In turn, some practices from the Community affect the availability of natural resources among which can be mentioned deforestation, over-exploitation of natural resources, inefficient water use, pollution by using fertilizers, herbicides, and wastewater, inappropriate cropping patterns, environmental negative impacts by construction of the physical infrastructure for the natural resources management, and finally competition by other sectors.

Science and Technology (S&T) represents the scientific and technological knowledge, methods, tools, and experiences that are available and can be used for the community with the purpose to intervene the natural environment in order to satisfy its needs and desires. The Science & Technology constitutes the mechanism by which the risk factors mentioned before for the relation between Environment and Community can be controlled or overcome. In this way, farming under controlled climatic conditions (greenhouse), and establishment of resistant crops to drought are some of the practices to control the risk factors depending on climate. Appropriate land use, conservation and protection of lands, land improvement, drainage, control of erosion and flood protection are some of the technical measures to control the risk factors concerning soils. Measures to ensure a reliable water availability as quantity, quality and timeliness (water storage), methods to determine in a realistic way demands and availability of water, appropriate physical infrastructure for acquisition, conveyance, distribution, and delivery of water to users, and irrigation methods constitute the mechanism to control the risk factors related to water resources.

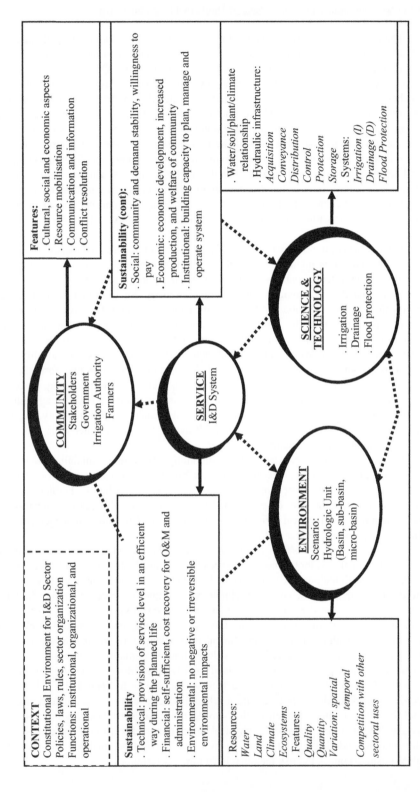

Figure 4.1   Schematization of components of the developed conceptual framework

However, a key point is the relationship between Science & Technology and Community. The technological alternatives for the provision of an irrigation and drainage service must be in line with the cultural, socio-economic, technical, and financial conditions of the Community. Here, the concept of Level of Service is important; the level is defined as "a set of operational standards established by the irrigation and drainage organization in consultation with irrigators and the Government and other affected parties to manage an irrigation and drainage system" (Hofwegen and Malano, 1999). Thus, level of service specifications, cost of the service, willingness to pay for the service provision, organizational forms, technical capacity, participation for operation and maintenance, decision-making, conflict resolution, and resource mobilization, play an important role in the relation Science and Technology and Community. In this way the adoption of technological alternatives from the Community side would have to be in such a way that it will result in a sustainable autonomous management without any major intervention of external actors.

4.3.2 Required Components

For the development of the conceptual framework of the transferred irrigation systems to communities, four components or actors have been identified namely the Government, the Water User Associations, the Farmers, and Supporting Entities. Each one of them plays an important and specific role, which interaction contributes within a particular context, to the sustainable management of the transferred system

Role of the Government

During the last 40 years the role of the Colombian Government for the development of the land improvement sector has been changed from very active role to a one with very different characteristics.

With the new institutional arrangement the political strategy of the Colombian Government can be described as a strengthening of the regional entities and their communities and in offering the private sector the main role in the provision of most of the support services for the irrigation districts. On the other hand, the Government will increasingly restrict its role to regulate the sector and will provide subsidies in a more modest and discriminating way than in the past, with an emphasis on offering subsidies in such a way that they will stimulate corresponding investments by the local water users.

Within this new institutional context the decentralization is understood as a gradual process for more regional autonomy, where the identification, formulation, negotiation, execution, monitoring and social control of programs and projects for the rural development are the result of the interaction between local and regional actors.

In this way several institutions at national, regional and local level play an important role within the institutional arrangement for the land improvement sector, namely the Ministry of Agriculture and the Colombian Institute for Rural Development (INCODER) at national level; the Regional Autonomous Corporation, the Agriculture Secretary, and the Departmental Associations of Water Supply Systems and Wastewater Disposal at regional level; and the Municipality Unit for Agricultural Technical Assistance (UMATA), and the Major's Office at local level.

The Ministry of Agriculture is in charge of the policy to promote the competitive, equitable, and sustainable development of the agricultural, forest, and fishing processes and rural development by using criteria for decentralization, consensus and participation contributing to the improvement of the welfare of the Colombian population. The Colombian Institute for Rural development, INCODER, is charged with the execution of the policy for rural development by facilitating the access to productive factors, by strengthening the regional entities and their communities and by promoting the voicing of the institutional actions in order to contribute to the socio-economic development of the country.

The Regional Autonomous Corporations have as objective the execution of the policy, plans, programs and projects concerning environment and renewable natural resources and to enforce the current legal rules for the administration, management, mobilization, and use according to the regulations, and guidelines issued by the Ministry for the Environment. The Agriculture Secretary is in charge of the planning and sustainable development of the agricultural and agro-industrial sector at department level according to the national and department policy. The Departmental Association of Water Supply Systems is a public departmental enterprise and is in charge of the provision of the services for drinking water supply and the disposal and treatment of wastewater for the municipalities conform the rules and regulations of a department.

The Major's Office is a body for the municipality administration and is responsible for the efficient resources management and for the achievement of the institutional objectives. The Municipality Unit for Agricultural Technical Assistance (UMATA) is responsible for the provision of the technical assistance to small farmers on agricultural, livestock, fishery, environmental, social, and gender issues.

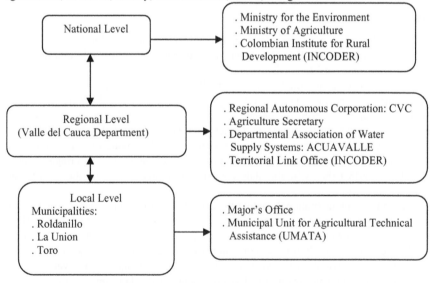

Figure 4.2 Governmental institutional arrangements within the RUT Irrigation District

From an environmental point of view, the departments and municipalities are also responsible for the promotion, co-funding, and execution of projects and works for irrigation, drainage, land improvement, flood control and use and conservation of basins. These activities have to be developed in coordination with the national body, which is responsible for the national system for land improvement.

With this institutional arrangement the Government will continue to give support to the land improvement sector and within this framework the water user associations will find the governmental inputs for the management of the transferred irrigation systems in view of their sustainability. Figure 4.2 illustrates the governmental institutional arrangement with an emphasis on the RUT Irrigation District. In this context, the contribution of the Government to the sustainability of the irrigation systems after they are transferred to the water user associations should put emphasis in the following aspects.

Protection of the Agricultural Sector

An important incentive for the farmers is to improve their living conditions trough profitable agriculture; in this way, factors such as production and productivity, production costs, production value and market conditions play an essential role.

At the beginning of the 1990s the Colombian Government tried to respond to the globalization wave in the world trough the so-called "Economic Liberalization". However, this process of economic liberalization did not stimulate the agricultural activities in the country due to the import of products with much lower prices and the increase in cost for the agricultural inputs. As a consequence, the agricultural sector did not grow, the total agricultural area was reduced and many jobs were lost.

Nowadays Colombia plans to be involved in new international market agreements; for example in the case of the Free Market Area for the Americas (ALCA) and the Free Market Agreement with the United States (TLC). The ALCA agreement involves all economic sectors, amongst them the agricultural sector and it has as main objectives the generation of economic growth, marketing of goods and services, the improvement of the conditions for the accessibility of the market, the elimination of obstacles and constraints, and the establishment of mechanisms in view of this major gateway to technology trough economic and technical assistance.

In this way, the country has to enter these negotiations by taking into account the particular characteristics of its agricultural sector and by establishing clear, stable, and transparent rules concerning marketing practices, duties, subsidies, etc.; resulting in a strengthening of the agricultural sector and improvement of the well- being of the farmers.

Hydraulic Infrastructure Ownership

The development of an appropriate "sense of belonging" at the water users-side has been considered as a major factor to influence and develop the sustainability of the land improvement systems. The Law 41 of 1993 for the land improvement sector establishes that the ownership of the physical infrastructure of the irrigation systems built after the enacting of the law will be of the users after all the investment costs have been fully recovered. However, most of the irrigation districts in Colombia were built before the endorsement of Law 41.

This has caused that the water users do not have any feeling that they can be identified as the owners of the systems, because the Government has kept the actual ownership of the infrastructure. As long as the infrastructure is seen as the ownership of the Government, the users will not identify themselves with it, and they will expect that the Government will be responsible for all the investments for maintenance, rehabilitation, and modernization.

Negotiating with the water users associations concerning a form of "Common Property Right" might contribute to the feeling that they have "a sense of ownership", so that they will take care of the infrastructure. The property rights will give the users the incentives for a major and efficient involvement in the management, because the rights provide them with sufficient confidence that they as holder and users of the rights will reap the future benefits of investment and careful management, and that they will bear the losses incurred by misuse. By giving them authority and control over the infrastructure will stimulate the mobilization of the community. On the other hand, the assignment of the property rights to the water users by the state can be interpreted and explained as a demonstration of the political will of the Government and its commitment to the sustainability of the irrigation systems after their transfer.

However, interviews carried out during the fieldwork for this research showed that the users' organizations considered the property rights of the hydraulic infrastructure as a less important issue. At the contrary, they believe that when the Government keeps the property rights of the infrastructure, it will be a very effective mechanism to obtain the participation of the Government in future programs for rehabilitation and modernization.

Water Rights

The National Code for Renewable Natural Resources and Protection of the Environment has been formulated in 1974 through the decree No. 2811. In 1978, the decree 1541 established the necessary by-laws for non-marine waters. This decree addresses amongst others its domain, how to obtain water rights, the use, protection, conservation and re-cycling of water, the regulation of the groundwater, water administration, water users' associations and sanctions.

The law stipulates that to obtain a water right within the context of agriculture a concession or permit is needed. The concession is defined as the "title by which the environmental authority provides a person a right to extract, use and benefit from the water resource for the purpose of irrigation in an irrigation district". The law also establishes a procedure to transfer or withdraw concessions in case of misuse. In the case of irrigation districts, the concession is provided to the water users' association as a whole and not to individual users.

After its establishment in 1993, the Ministry of the Environment began to charge a fee to the districts for water diverted from river basins. The Regional Autonomous Corporations became responsible for the collection of the water use fees for the Ministry for Environment. The fee is charged on a volumetric basis per semester and 90% of the takings are allocated for the conservation, protection and control of the watershed in each district. This fee is in addition to the fee for the operation and maintenance services collected by the districts to pay for their own expenses.

There has been some resistance to the fee from water users but it is now generally accepted as a permanent institution. The Government made an effort to make the fee more acceptable for the water users by allowing the districts to replace the cash payments for the fee by in-kind contributions, namely in the form of implementing activities, which are directly related to the conservation of the watershed.

However, in view of the sustainable management of the irrigation systems after transfer the Government trough the environmental authority (Ministry for the Environment, Regional Autonomous Corporation) should update and revise the criteria used to allocate water concessions to water user associations. The water concession should no only be based on the criterion of the water quantity of water needed, but it should also be based on minimum standards for the water quality to be allocated to the users.

The deterioration of the watersheds caused by deforestation processes and the contamination of the watercourses have a negative impact on the water quality in the RUT region and have changed the water quality into a risk factor or constraint for the agricultural development. The deforestation processes cause a high sediment load in the surface water that results in a silting-up of the irrigation network, increase the cost for operation and maintenance and impact the financial sustainability of the irrigation system. The contamination by wastewater disposal and industrial waste affects the water quality from a chemical point of view and has a negative effect on the agricultural productivity and the production quality. Especially the low quality of the irrigation water limits the present possibilities to bring the agricultural produce to international markets, as the produce does not meet the high quality standards, especially the ones in view of health risks.

In other words, the task of the environmental authority is to provide and assure an allocation of water to the water user association according to the general criteria of a good physical-chemical water quality as mentioned in the water concession. The supply of water from the sources should be available for the users with acceptable minimum standards for the quality. This is possible through the design, execution and evaluation of a monitoring process of the water quality of the sources from which irrigation water is provided to the irrigation district. The environmental authority should also enforce the existing regulations for solid waste and wastewater disposal and their required treatment before the water is discharged into the watercourses; and should formulate and implement a program for the preservation of watersheds in order to reduce or even stop the present deforestation processes.

For the most critical cases, the environmental authority should compensate the water user association affected by low water quality through financial contributions to cover the additional costs for operation and maintenance in view of the low water quality. In the case of contamination by solid waste and wastewater disposal the companies, industries or municipalities identified as the polluters might be pressed to compensate the production losses and extra costs for operation and maintenance; meanwhile, in the case of deforestation the compensation for loss and extra costs should be provided by the Regional Autonomous Corporation and the municipal administration.

In the specific case of the RUT Irrigation District, the Regional Autonomous Corporation of the Valle del Cauca (CVC), the Departmental Association of Water Supply Systems and Wastewater Disposal of the Department of Valle del Cauca (ACUAVALLE), and the Municipality Units for Agricultural Technical Assistance (UMATA), and the Major's Offices of the municipalities of Roldanillo, La Union, and Toro play an important role in the preservation and protection of the watersheds.

The Regional Autonomous Corporation of the Valle del Cauca (CVC) is responsible for the protection and preservation of the Cauca River basin, which shows a high sediment load in the river and a contamination from wastewater disposal and chemical waste from the municipalities and industrial areas upstream of the RUT

Irrigation District. Enforcement of the regulations to treat the wastewater and to dispose the solid waste of the municipalities would contribute enormously to a better water quality for the RUT district as the water is pumped by the Tierra Blanca and Candelaria pumping stations to the irrigation system.

The Departmental Association of Water Supply Systems and Wastewater Disposal of the Valle del Cauca Department (ACUAVALLE) is responsible for the treatment of the wastewater from the municipalities: Roldanillo, La Union and Toro that are located in the area that influences the RUT district. ACUAVALLE needs to improve the performance of all the treatment plants, which would contribute to a better water quality in the interceptor canal.

The Municipality Units for Agricultural Technical Assistance (UMATA), and the Major's Offices of the municipalities of Roldanillo, La Union, and Toro with the support of the Regional Autonomous Corporation of the Valle del Cauca (CVC) need to formulate and implement plans for the improvement, conservation and protection of the watersheds of the streams coming from the western mountain range and that discharge into the Interceptor Canal. The expected reduction of sediment load towards the Interceptor Canal would contribute on the middle term to an improved operation, reduction of costs for preservation and maintenance, and will contribute in this way to the financial sustainability of the RUT Irrigation District.

Rehabilitation and modernization

According to a declaration of the International Commission on Irrigation and Drainage (ICID Granada Declaration, 1999) rehabilitation and modernization of systems will have to be brought about by encouraging the formation of water users' organizations, by effective participation of users in the initial assessment of needs and by evaluating alternatives. Rehabilitation and modernization must result in additional benefits to farmers and be financially viable in that operation and maintenance costs should be at an acceptable level.

The majority of irrigation districts in Colombia are more than 40 years old and they have already been totally or partially under rehabilitation programs financed by the Government through international loans. The results of the rehabilitation were far below the expectations in terms of changes in area irrigated, cropping intensity, and productivity. The rehabilitation program probably served the Government to reinforce the farmer perceptions that the district and the infrastructure belonged to the Government (Garces-Restrepo, 2001).

Rehabilitation has been an important issue in the second phase of the transfer process, which occurred during the 1990s. The issue did not only include the need to rehabilitate, but also who would be responsible for the payment of the rehabilitation costs. Finally, in those cases where the rehabilitation was included in the negotiations for the transfer, the Government paid the greatest majority of the costs, except for minor contributions by water users associations for very specific activities.

Within the new institutional context for the land improvement sector the implementation of future rehabilitation and modernization programs would have to be the shared responsibility of the Government and the water users associations with a more predominant role for the water users associations. The Government would have to adapt a policy with defined criteria and transparent procedures that reflect the new relationship between Government and water users associations concerning the rehabilitation and modernization of the irrigation districts. This policy should be very different from the one followed during the past.

The initial idea for rehabilitation or modernization should arise from the water user associations as a result of a consensus amongst the farmers that implies that the improvement of the hydraulic infrastructure allows them to maintain or even increase their living conditions trough profitable agriculture. This approach includes the essential requirement that the farmers recognize that agriculture is an important means for their livelihood and welfare and that a program of improvement would even boost this condition more. In response to this approach, the Government would give the organizational support in order to address and formulate the improvement process by using governmental organizers or contracting experts from other entities. The Government also needs to provide financial support to pay the cost of the feasibility studies for the improvement, to establish credit lines to the water users associations with soft interest rate, and to be sponsor for the water user associations when they request support from national and international moneylender entities.

Experiences with this type of relationships between the Government and (water) users associations existed already in Colombia. One example is the relationship between the Government and the Users Association of the Coello Irrigation District, USOCOELLO, before the end of the Cucuana Project in 1998 in which the users organization played a very important role in view of the mobilization of financial resources. Another successful experience has been the Zulia Irrigation District, where the users' organization took advantage of the conditions offered by the Government for the rehabilitation and modernization of the irrigation system through a credit-line that was established by the Agricultural Financing Fund (FINAGRO). Moreover, the organization got also a financial subsidy from the Government that covered about 40% of the total project cost, next to the mentioned FINAGRO finances for rehabilitation and modernization. This subsidy was named Incentive for the Rural Capitalization (ICR). The users' organization got also a very essential contribution from the local and departmental Government, namely the funds that covered the costs to secure the credit of FIANAGRO.

It can be said that FINAGRO is the contribution of the Government to promote and develop production activities through the execution of projects that are feasible from technical, financial, and environmental point of view. Users' organizations showing a sound organization and a high management capability will have access to these financial resources for their productive and profitable projects.

As general criteria the Government needs to take into account various factors, such as the interest of the community, judged by the amount of money that they can contribute from their own sources; the districts' long-term agricultural and economic productivity; the number of hectares and beneficiaries involved; and an environmental impact assessment of the proposed works.

On the other hand, the Government also needs to promote amongst the water user associations the culture for improvement programs by the creation of a (reserve) capital fund with seed capital provided by the Government and with periodic contributions agreed by the parties. The returns and the capital contributions will constitute an important reserve fund as basis for the undertaking of a rehabilitation and modernization program.

This new culture also would have to facilitate a change in approach. This means that to wait until the infrastructure is in a full state of total collapse and then to undertake a rehabilitation program would have to be replaced by an ongoing, incremental, pro-active and smaller-scale approach of repair and restoration. In this way, the works will be done as the needs arise, before they become serious and require large investments with substantial external funds.

The water users associations with a more active role in their responsibility of the rehabilitation programs could also strengthen their position in bargains with the Government and other public and private entities when they have all the property rights of the hydraulic infrastructure. It has been shown that those associations with more assets (especially property rights) are treated better and are better able to negotiate than those without any asset (Agrawal, 1997; Haddad, Hoddinott, and Alderman, 1997; Quisumbing, 1994 cited in Meinzen and Knox, 1998). Users with recognized property rights are more likely to have a say in the decision-making process.

Role of the Water Users Associations

In the conceptual framework for the sustainable management of irrigation systems after transfer, the organizational form of the community through the water users association will play a central and crucial role. Some forms of community organization existed before the transfer of the irrigation management, but these organizations played a secondary role and were subject to the decision-making by the Government Agency. That was one of the main elements that caused a very low sense of belonging at the side of the farmers.

The transfer of the irrigation management was based on the fact that the users' organizations would become fully responsible for the administration, operation and maintenance activities, including all the associated costs and without any help or participation of the Government through subsidies. After the transfer the users organizations would have to play a more critical and active role than the one performed before the transfer. However, an analysis of the role and the performance of most water users' organizations after the transfer showed that the management of the irrigation districts followed the same management style as the one previously developed by the Government agency. This means that the management was only focused on the service to provide irrigation water, from which the organization obtained its revenues (water tariffs) for the sustenance and financing of the organization. In other words, the users organizations acquired and inherited the management style as followed by the Government agency before the transfer process, but now with the additional difficulty that the organization did not have any governmental subsidy. In the eyes of the farmers the transfer of the management brought only a change of actors in the management and therefore the users' organization was still seen as an extension of the Government agency.

A few years after the transfer, the financial resources of the users organizations became insufficient and were even depleted; they were negatively affected by a low paying capability of the farmers and by agricultural activities with small profits. This situation caused a serious crisis for the organizations at different levels, which resulted in a state close to full collapse. This fact may lead to the conclusion that the delegation of management from the Government to the users organizations had a limited scope and that the organizations failed to reach the initial objective of the delegation, namely the full responsibility for the administration, operation, and maintenance of the irrigation systems.

In view of a sustainable management of the irrigation systems after the transfer the users organizations will have to achieve and develop sufficient managerial capabilities to pay full attention to new developments, which are required by the management transfer, including the enlargement of its activities to fields beyond the mere operational aspects, which consist of the provision of irrigation water services. In this way, the final objective of the users organization will be to contribute to the improvement of the socio-economic conditions of the farmers as first beneficiaries through an integral and profitable approach of irrigated agriculture and the management of the irrigation district under principles of equity, competitiveness, sustainability and multi-functionality in line with the legal framework established by the Government. Within this context, the irrigation system is considered as a very important element of the production system and one of the tasks of the users' organization is the efficient management of the system in order to provide irrigation water in an effective and efficient way to the farmers.

In view of the sustainability of the systems after the transfer the users' organization would have to develop an integral and participatory management approach. This means the full participation of all the actors involved; the farmers as the first and most important beneficiaries; next the technical and administrative staff; the local, regional and national authorities for the land improvement sector; and representatives of the public and private supporting entities. An integral approach needs to be based on the knowledge of all the physical aspects of the natural resources and their conservation, on a reliable socio-economic development, with emphasis on a demand-driven and demand-oriented approach; moreover, the decision-making process will have to take place at the lowest appropriate level. A bottom-up approach based on decision-making at the lowest level (subsidiary principle) is recommended as it will incorporate the active and wide participation of all, directly and indirectly involved actors, allowing for the formulation of actions and interventions in a mutual way. This approach will have to form a sound base for the implementation, creation and sustainable management with a high degree of success.

The implementation of this conceptual framework for the role of the users organizations will have as starting point a clear and complete statement from the organization about its future mission and the establishment of objectives and goals for the short, middle, and long term that are in line with the interests and expectations of the farmers. In view of these goals, the users organization will need to develop sufficient capability to formulate, plan and implement development-coherent programs, which include social-economic, technical, and environmental components and have to be based on an integral, participatory and communal approach involving all the actors that are directly engaged in the management (farmers, technical and administrative staff) and the ones indirectly engaged (stakeholders, supporting private and public entities, and the society at large). These integral development programs will have as main pivot the promotion and practice of irrigated agriculture and will include all the stages of the agricultural production chain, namely production, harvest and post-harvest improvement, added value, agricultural industry, agro-business and marketing.

The implementation of this conceptual framework for the role of the users organization will result in an autonomous and self-sufficient organization that is less dependent on external agents for the efficient management of the transferred irrigation system and that will focus its main activities not only on the efficient provision of irrigation water, but also on the integral and participatory development of programs with favourable socio-economic impacts for its beneficiaries.

Under these circumstances the organization will have to show a high capability to monitor, interpret, realize and satisfy the interests and wishes of its direct (farmers) and indirect beneficiaries by using approaches for local initiatives and incorporating participation and subsidiary principles. If the organization is capable to satisfy the interests of the farmers, the latter will be willing to participate and support the development process proposed by the organization, and it will result in the recovery and strengthening of the sense of belonging and self-respect of the farmers.

Transparency, accountability, clear actions, administrative efficiency, and equity will be the main management characteristics to be exhibited by the organization and requirements demanded by the farmers to create confidence and credibility. Leadership and a high capability to establish inter-institutional relationships will be necessary characteristics for the organization as contribution to a sustainable management. Leadership is needed to propose and implement development projects based on profitable irrigated agriculture and in view of the improvement of the socio-economic conditions of the farmers. Leadership is also needed to promote parallel economic activities in order to recover and maintain the traditional domestic practices as a mechanism that contributes to food security and improvement of the domestic economy which will facilitate the participation of the women and young people and will assure a new farmer generation.

On the other hand, inter-institutional relationships are needed to prevent that the organization will become isolated from local, regional and national developments and a very strong linkage with supporting public and private institutions will help to get the required approval for and to obtain the resources to develop the necessary, commercial programs for its beneficiaries. In view of the strengthening of the inter-institutional relationship the users' organization might consider the participation of representatives of supporting public and private entities, local administration, and the land improvement sector. They can participate in the deliberative meetings of the organization with a right to speak, but without a right to vote, warranting the transparency of the decision-making process and assuring a more effective and direct commitment from the supporting entities.

Role of the Farmers

The farmer along with the Government and the users organizations is one the main actors responsible for the sustainable management of the irrigation systems. Within the proposed conceptual framework for sustainable management of the irrigation systems after transfer, the role of the farmers will change from a passive role, as was described in the previous sections to another, more active, positive and receptive one. In this scenario the farmers will have as incentive the development of profitable irrigated agriculture (in contrast with a subsistence one), which will allow them to increase their income and therefore to raise their living conditions. The income might even more increase when the farmers follow an integral approach for their agricultural practice; this approach will have to cover all the phases of the production chain and includes production, post-harvest management, agro-industry, and marketing.

This new role will demand from the farmers, new attitudes and aptitudes to move from a condition of subsistence farming to a condition of farming with profits. On one hand, the farmer will have to check, consider, and optimize the quality of the operations at farm level with the purpose to be more efficient in view of the management of the invested economic, human, and natural resources; to increase the production and productivity and to keep in mind the principles for the protection and conservation of the natural resources. This means that the farmers will have as central objective a decrease of the production costs (inputs) and an increase of the production (outputs) and its associated value in view of attaining the largest possible benefits. In this way, modernization in land preparation, appropriate cropping patterns, crop rotation, sowing dates, selection of improved crop varieties (resistant to diseases or drought, major yields), introduction of cash crops, improvement and optimization of agricultural practices (application of fertilizer, herbicides, fungicides, etc.), irrigation management at field level in a more efficient way (irrigation methods, irrigation scheduling, drainage, storage, reuse of wastewater, and other aspects), conservation, protection and reclamation of soils, management of the post-harvest activities and added value for the agricultural production are some of the main aspects to be taken into consideration by the farmers in view of the modernization and optimization of the agricultural activity. Moreover, with a broader vision of the agricultural activities at farm level the farmer will develop complementary activities, which will bring him amongst other benefits, the improvement of his economics. As examples of these complementary activities can be mentioned the production of organic fertilizer from solid waste (harvest waste, muck, etc.), the establishment of market gardens and animal husbandry in view of food security, nursery and small agro-business (crafts). These parallel activities will not only give the possibility to increase the income, but they will also allow the involvement of the whole family, especially the women and the young people.

On the other hand, the new role of the farmer will demand the further development of his capability to establish new and to strengthen existing relationships with other farmers and the users' organization. The relationship with other framers will allow him to become familiar with successful experiences developed by progressive farmers through pilot and innovative projects, to carry out development projects in a shared way and to facilitate the acquisition of new knowledge and the transfer of technology. The relationship with the users organization will allow him to enjoy the benefits that follow from the by the organization established objectives, which include the promotion of the socio-economic welfare of the farmers and their families and to strengthen the social contacts and mutual help through a profitable irrigated agriculture. In this scenario the farmers would have to accept and follow the rules and regulations introduced by the organization for the development of the basic activities, to participate in the formulation, implementation and development of projects and to contribute to the decision-making process in a positive and active way. The relationship with the organization will allow the farmer to face and overcome crucial constraints within the production chain of the agricultural system. Added value for agricultural outputs (agro-industry processing), marketing under favourable conditions, access to credit lines and modernization and optimization projects are important issues that can be better tackled by the farmers in a collective manner than in an individual way.

Role of the Supporting Entities

The supporting entities are not directly responsible for the management of the irrigation systems; however, they can play an important role in the sustainable management of the systems, because their actions will provide the basic services for the system and might help to attain the final objectives of the land improvement sector policy, which include the increase of the production, improvement of the living conditions of the farmers and an integral development in the command area of the irrigation districts under criteria of sustainability, equity and competitiveness. In general terms, it can be said that the objectives of the supporting entities are to contribute to the improvement of the efficiency of the administrative, operation and maintenance activities; to increase the profitability of the economic activities of the irrigation districts and the income of the farmers; to strengthen the organization and management capability of the users; and to guarantee the environmental sustainability at long term.

The services provided by the supporting entities show clearly multi-purpose characteristics like interdependency, concentration and relationship with the regional development and structural factors. Thus, the provision of supporting services concerns the organization, the irrigation management, the production and the environment aspects (multi-purpose); their demand depends on the development phase of the irrigation system (design, construction, implementation, full maturity), users and type of production system, while the actual activities are related to the level of development within the regional context. The supporting services are inter-dependent and the efficiency and effectiveness of one service determine the performance of other services. The provision of supporting services tends to be concentrated in urban areas and from there they are dispersed to middle and small cities and rural areas depending on the level of development of the economic, regional activities and the demand for services; the main urban and regional economic developments, the accessibility of the supporting services and vice versa. The provision of supporting services is mainly determined by policy factors like economic globalization, decentralization, privatization, fiscal crises and the policy for the land improvement sector, which will determine the conditions for the market of products and services.

The responsibility to provide the supporting services has changed in a gradual manner, namely passing from governmental agencies to a more intensive role of private agencies according to the policy of decentralization as followed by the Government for the land improvement sector. Whoever is responsible for the provision of the supporting services, the availability of these services will require an organizational structure that satisfies the needs of the irrigation districts; an accessibility for the users in view of affordable costs and facilities; opportunity, efficiency and quality of the services, maximum coverage and effectiveness that will satisfy the demands and needs of the users.

The conceptual framework for the sustainable management of the irrigation systems after transfer assumes that the actions of the supporting entities are concentrated on four basic fields, namely community organization; administration, operation and maintenance of the irrigation systems; production system; and environment. For each field the information, research, technology transfer, financing, and training will be cross-cutting aspects.

The community organization is closely related with all the other organizations to carry out the basic and complementary activities in the irrigation districts. An organized, motivated, and trained community is a fundamental requirement to reach the main objectives of the users' organization and to access the supporting services through its management capability. The technological improvements in view of the production process will not result in the expected profitability and improved standard of living when the community is not organized in view of marketing and supporting services. The management capability of a users' organization is related to its organizational strength and the scope of objectives. Thus, an organization that has as its single activity the distribution of irrigation water will develop a much more limited management capability than one that develops additional activities related to an integral production system, protection of the environment, improvement of the living conditions, and regional development. The organizational strength is influenced by factors like the participation level of its members, the leadership of the organization and the relationship with other organizations. The participation level depends on the sense of belonging of the farmers and their motivation that are influenced by the quality of the services provided and their related costs, the existing mechanism for conflict resolution, and the existence or lack of organizational and participatory structures. The leadership is a key factor for the strength or weakness of the organization; major aspects of a good leadership are its capability to organize and manage the organization. The relationship with other organizations will give social strength to local organizations improving their management capability in view of their contacts with larger institutions.

From an organizational point of view, the supporting entities will need the development of research and training in the field of social aspects that will help to solve the problems faced by the users' organizations after the transfer of the systems and to strengthen the organization in order to achieve an efficient management of the irrigation districts.

For the administration, operation and maintenance of the irrigation districts the supporting entities will have to focus their efforts on the evaluation of the effectiveness of the administrative and operational activities and on a detailed uncovering of the maintenance needed for the machinery and the physical infrastructure. They also have to define a realistic tariff structure, which will result in sufficient funds to cover the present expenses and future investments taking into consideration the Government policy of eliminating its subsidies. Research is required in water management at system level, especially research that includes aspects concerning the irrigation water requirements according to crops and soils, the total efficiency of the system (conveyance, distribution and application), water distribution and flow control. The activities of the supporting entities will have to contribute to satisfy the training needs at managerial, operational and users level. At managerial level training is required in aspects related to strategic planning, financial management and team work. At operational level training is required in aspects related to an efficient use of irrigation water, water management at farm level, operation of hydraulic infrastructure, zonification of the water distribution, irrigation scheduling, soils management, crops diversification, etc.

The production system is considered as a key issue for the sustainable management of the transferred irrigation systems, because it involves an integral vision for a profitable irrigated agriculture that will allow the farmers to pass from a situation of subsistence farming to a situation of farming with profits. To get this effect, the agricultural practice will have to consider all the stages of the production chain namely production, harvest management, added value through chemical and physical processing (agro-industry), and market strategies in view of local, regional, national or international markets; this means that the production system will have to be managed by the farmers under enterprise criteria. In this way, the supporting entities will have to supply research, training and extension services to the farmers in order to optimize the quality of their operations at farm level pursuing an efficient use of the invested resources and seeking the maximum profitability for their agricultural activities. The supporting entities will also have to promote and facilitate the formation of the other types of organizations (cooperatives, associations) of the farmers in order to manage post-harvest of the agricultural production (storage centres) or to develop parallel activities (agro-business); they will also have to facilitate and promote inter-institutional relationships with other organizations or institutions for the development of agro-industrial processing projects, marketing in favourable conditions, and access to the credit lines for productive development, modernization and optimization.

Concerning the environmental aspects the contribution of the supporting entities is especially required in order to secure the water availability for the irrigation districts. Most of the river basins suffer at present under deforestation, which will cause unreliable availability of irrigation water and will affect its physical quality due to the high sediment load. This situation influences the operation activities at system level in a negative way and affects the financial basis of the users' organization due to the increased costs for operation and maintenance. The water quality is also affected by the contamination of the water sources with wastewater and solid waste coming from industry and urban areas. At irrigation system level the activities of the supporting entities will have to be focussed on the prevention of negative environmental impacts by the excessive use of agrochemical products, disposal of solid waste, degradation of soils by agricultural machinery, waterlogging and soil salinization.

The choice of supporting entities from the governmental side is very wide. The most important entities include:

- Colombian Institute for Rural Development (INCODER);
- Departmental Agriculture Secretaries and Municipality Units for Agricultural Technical Assistance (UMATA), which by delegation are in charge of the implementation of the governmental policy for the land improvement sector at departmental and municipal level. These organizations at regional and local level play an important role nowadays as they are the link between INCODER and the community organizations and they will deal with the available resources for investment in an efficient way;
- Colombian Agricultural Institute (ICA) states that it contributes to the sustainable development of the agricultural sector through research, technology transfer, and prevention of sanitary, biological and chemical risks for vegetable and animal species;
- Colombian Corporation for the Agricultural Research (CORPOICA) supports the development, research, and technology transfer for the agricultural sector and the promotion of the processes for technological innovations;
- National Apprenticeship Service (SENA) carries out all the required activities that are essential for the investments in the social and technical development of the Colombian workers, offering and executing an integral professional approach for the incorporation of individuals in all kind of productive activities that contribute to the country's social, economic and technological growth. It also offers services for the continuous development of human resources that are linked to the companies; it provides information; orientation and employment creation; support to the entrepreneurial development; technological services to the productive sector, and support to innovation projects, technological development and competitiveness;
- Agricultural Financial Fund (FINAGRO);
- Autonomous Regional Corporations;
- Academic Sector consists of the universities and the centres for technology education. This human resource has a great potential for the support to the irrigation districts in developing research, extension, training, and technology transfer activities. Some irrigation districts like ZULIA and RUT have established

cooperation agreements with regional universities for the development of various supporting activities.

4.3.3 Analysis and Conclusions

The management of the irrigation systems that was transferred to the users associations in the framework of the policy of delegation and decentralization implemented by the Governments at the end of the 1980s has been of crucial concern for policy makers, planners, and stakeholders of the land improvement sector. An evaluation of the performance of the users organizations in view of their management of the systems has shown wide-ranging results passing from successful experiences as in the Mexican case to less successful ones as in Indonesia and the Philippines; in fact the experiences have been also mixed within the own context of each country. However, researchers in this field recognize that even nowadays there is not sufficient evidence concerning the impact of the management transfer and that more comprehensive methodologies are needed to assess the impact of the delegation policy. The delegation process has only been considered very recently and it needs more time to collect more evidence to know more about the process and its subsequent stages.

The knowledge of the existing experiences with the transfer of the management of irrigation systems to users' organizations has led to the fact that the international community of the land improvement sector will focus its efforts on the search of management approach models in order to attain the fundamental conditions for sustainable management. At the first place, a general consensus exists that the sustainability concept has to be considered in a very wide view involving institutional, technical, financial, economic, social and environmental criteria. Previous approaches for sustainable management of the irrigation systems after transfer only emphasized the technical and financial aspects and lacked an integrated approach for the formulation and implementation of the management model and bringing the organizations to critical conditions.

In the particular case of Colombia, the management experiences of the transferred irrigation systems showed some successful experiences like the cases of the Coello and Saldaña Irrigation Districts, some poor ones like the Maria La Baja Irrigation District, and some promising ones like the case of the Zulia Irrigation District. It can be said, that the management of the systems after transfer was mainly influenced by some external factors, like the governmental paternalistic policy before the irrigation management transfer, the conceptualization and negotiation conditions before and during the transfer process, and the real potential for productive development of the command area of the systems.

The paternalistic policy of the Government before the transfer process caused a very low sense of belonging of the farmers, a tendency to follow the practice of subsistence agriculture and the full responsibility of the Government for the management of the irrigation systems and its associated costs. For most of the cases, the transfer process was seen by the Government as a simple transfer of the responsibility for the administration, operation and maintenance activities to the users' organizations. The process was characterised by the lack of any integral training for the development of the systems under enterprise criteria. On the other hand, a close relationship was identified between the real potential for productive development of the command area and the performance of the organization in view of the management of the system. This aspect can be explained by comparing the environment of the regional, socio-economic development of the Coello and Saldaña Irrigation Districts with the environment of the Maria La Baja Irrigation District. In the first case, an active and dynamic socio-economic development is found around the rice crop, this in contrasts with a poor development in Maria La Baja Irrigation District.

The proposed conceptual framework establishes as starting point that the users' organizations will have to display some clear characteristics of autonomy and self-sufficiency in view of a sustainable management of the irrigation systems after transfer. The autonomy is related to the capability of the organization to establish in a participative and concerted way its mission and future views; plans and programs to be reached at short, middle and long term; rules, norms and regulations for the development of its essential functions; appropriate organizational and governance forms; and mechanisms for monitoring and control. All this within the juridical- legal and institutional framework established by the Government for the land improvement sector and taking also into consideration the socio-economic and cultural environment of the community that will benefit from the transfer. The self-sufficiency is related to the capability of the organization to attend all the aspects of management without any major intervention of external agents.

The sustainability of a users' organization is closely related with its capability to interpret and satisfy the needs and wishes of the farmers as main beneficiaries and its capability to adapt to changing conditions caused by external agents to the organization. The farmers have as main objective to increase their income with the wish to improve the living conditions of their families. Thus, an integral approach of an irrigated agriculture under equity, efficiency, competitiveness, and profitability criteria will contribute to the fulfilment of the needs and wishes of the farmers. The integral approach means the consideration of all stages of the productive process, namely production, post-harvest improvement, added value through agro-industrial processes, and marketing. In this context and in view of the sustainable management of the transferred irrigation systems, the management scope given by the users organizations will have to go beyond the administration, operation and maintenance of a physical and organizational infrastructure for the provision of the irrigation water service and will have to be enlarged to the development of plans and programs with high socio-economic content pursuing the welfare of the farmers.

The sustainable management will be helped when an apparent convergence of interests exists between the users' organization and the farmers. This is possible when the main objectives of the organization are in line with the interests and expectations of the farmers. If the organization is capable to satisfy the interests of the farmers, it might be expected that they will be willing to participate and support the development processes.

5

STUDY SITE: THE RUT IRRIGATION DISTRICT

The RUT Irrigation District is located in the South-west of Colombia, in the northern part of the Valle del Cauca Department between 4° 25' and 4° 40' N and 2° W. The district lies between the Western Mountain Range and the Cauca River and covers a gross area of 10,000 ha. The RUT Irrigation District lies within the three municipalities that give their name to it, namely Roldanillo, La Union, and Toro. The altitude of the district ranges between 915 and 980 m+MSL. Figure 5.1 shows the location of the RUT Irrigation District within the Valle del Cauca Department.

Figure 5.1 Location of the RUT Irrigation District within the Valle del Cauca Department (Burbano and Forero, 1999)

5.1   Background

Before the construction of the RUT Irrigation District the area (11,500 ha) was regularly subject to floods from both the Cauca River and the streams descending from the Western Mountain Range. As a consequence an area of 1,500 and 3,500 ha was under permanent and periodic floods respectively. As a result of the high water an area of about 2,500 ha had a high moisture content that only allowed the growth of grass, and an area of 4,000 ha that could only be used for extensive cattle breeding.

At the start the RUT Irrigation District was foreseen for small farmers according to the legal framework given by the Agrarian Reform Law. Each family would receive 5 ha and moreover they would receive technical, financial and organizational support from INCORA for the development of the land. At the same time, an area of 50 ha was established as a maximum for the land property.

Feasibility studies for land reclamation and development of the area what is now the district started in 1958. In 1962 with only 40% of the construction works finished, lack of funds stopped the work. In 1964 the flood control and land development works continued until 1970 when the construction works were concluded.

In general terms, the RUT irrigation system is primarily a flood protection and irrigation/drainage project. The long and narrow bowl-shaped area is surrounded by a protection dike, running along the East border with the Cauca River and a flood Interceptor Canal on the West side. A main drain divides the area almost in half, running through the lowest elevations. Parallel to the river dike is the main irrigation canal, which is served by three pumping stations with a total capacity of 13.8 m^3/s. The flood Interceptor Canal also acts as an irrigation canal for the water users who use small centrifugal pumps to serve their individual needs.

There is a complementary network of both irrigation and drainage canals throughout the area. The main drain discharges freely into the Cauca River during low stages and the drainage water is pumped during the high stages of the river. Besides the irrigation and drainage network the RUT has an excellent road network running parallel to the canals. Figures 5.2 and 5.3 show the cross section and plan view of the RUT Irrigation District.

The development of the RUT Irrigation District can be described in the following chronology (Blanco, 1996):
- 1958: Regional Autonomous Corporation (CVC) started the construction, financed by the Agrarian Bank (Caja Agraria);
- 1962: construction works were suspended by lack of funds;
- 1964: Incora renewed the construction works with appropriate economic resources under the supervision of CVC and in two stages:
- First stage: development of 10,000 ha;
- Second stage: development of 3,300 ha. During this stage the area was reduced to 1,500 ha and finally the whole area was cancelled. The RUT area was to be irrigated by 3 pumping stations that are located in the Interceptor Canal near the locations of the Roldanillo River, Union Stream, and Toro River, respectively. Nowadays, an area of 1,200 ha is irrigated by using water from the Interceptor Canal, which is supplied by Tierrablanca pumping station.
- 1965 - 1966: The Inter-American Development Bank (BID) approved a loan to finish the construction works;
- 1968: Incora received from CVC the management of the RUT Irrigation District;
- 1969: With BID (Inter-American Development Bank) resources the pumping stations at Tierrablanca and Cayetana were build; also the main irrigation or marginal canal, the protection dike, the main and secondary irrigation and drainage systems, administrative office, and road network were developed;
- 1973: The Candelaria pumping station was built at km 8+100 of the marginal canal with the aim to improve the irrigation in the areas between the left bank of the

Main Irrigation Canal and the right bank of the Main Drainage Canal (sectors 2 and 4);

- 1976: The management of RUT Irrigation District was handed over to HIMAT (Institute for Hydrology and Meteorology and Land Improvement);
- 1982: Rehabilitation works were carried out with economic resources of the World Bank (Loan 1996 – CO). The rehabilitation works included protection works and the raising of the protection dike in some reaches, silt cleaning of the Interceptor Canal, repair of the pumping stations, improvement of the main drainage canal, lining of irrigation canals, rehabilitation of the road network, and repair and replacement of heavy machinery and vehicles. These rehabilitation activities were part of the compromise obtained from the Government as an essential requirement for the future transfer process;
- 1989: The management of the RUT Irrigation District is transferred to the Water User Association, ASORUT by delegation.

5.2   General Aspects of the RUT Area

The RUT area forms the border between the tropical and subtropical zones with an average temperature of 24 °C, fluctuating between 17 °C and 34 °C; the precipitation is bimodal with two dry and wet periods. The average yearly rainfall is about 1,100 mm, the average relative humidity is 72% and the average yearly evaporation is 1,080 mm. The average wind velocity is 1.30 m/s and the sunshine duration is 1936 hours as a yearly average.

The topography of the district is nearly flat with a light longitudinal slope (0.13 m/km) in the South – North direction. The cross-section of the area has a concave form with a slope of 1.2 m/km in the direction of the main drainage canal, as shown in Figure 5.2.

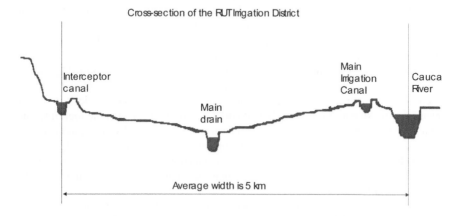

Figure 5.2 Cross section of the RUT Irrigation District area (ASORUT, 2000)

The RUT area presents as most recent geological formation alluvial and colluvial sediments from the Pleistocene period conforming the filler of the Cauca River plain.

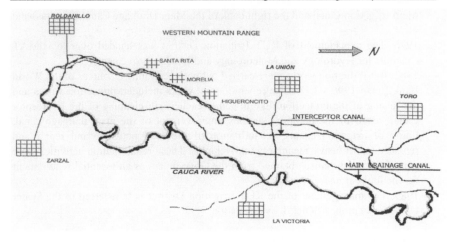

Figure 5.3 General plan of the RUT Irrigation District (Burbano and Forero, 1999)

The geological structure of the basins of the RUT area was caused by the Andean orogenesis ending tertiary period. Intense folds are not noted within the area; however, quite some fractures and micro folds are observed. The geological faults belong to a system whose general orientation is North–Northeast or Northeast with some associated secondary faults. In general, the geological structures of the RUT area follow an Andean direction (Muñoz, 1973).

The landscape of the area presents 3 basic forms: namely the foothill alluvial plain, fluvial-lacustrine plain, and flood plain. Figure 5.4 shows these forms. The depositions have been caused by the streams coming from the West Mountain Range, which flow to the Cauca River. The main depositions resulted in the colluvial sediment basin, alluvium combined with colluvium, alluvial marsh, alluvial beds, natural fluvial dikes, and alluvial fans. The alluvial plain is the most important deposition formed by a series of alluvial cones, which comprises gravel, sand, clay, and pebble brought by the above-mentioned streams. The colluvial sediment basin is found as hillside deposit with its characteristic sloping form. The alluvial cones or fans are located at the foot of the hill in a downward direction. The alluvial beds consist of gravel and sand deposits, and are formed by the water flows during the rainy season. The natural fluvial dikes are relatively permeable deposits formed by sand and clay, and coarse particles next to the river. The alluvial marshes are formed by deposits mainly with clay; including muddy areas artificially dried up.

About 63% of the area of the RUT Irrigation District has clay loam soils, characterized by little limitations for agriculture and a low salinity; 14% presents problems with salts, waterlogging and a heavy texture, and 21% has a high content of gravel, sand, and no deep soils. The salinity problem is intensified by inappropriate irrigation methods, failure of the drainage system, and the use of the drainage water for irrigation. From a geological point of view the salinity is caused by:
- Release of the water contented in rocks due to weathering processes;
- Hydrolysis of mineral material which produces calcium and sodium salts that are carried in soluble form towards the plain;

- Formation of alluvial marsh that contributes to the concentration of soluble substances coming from the mountain areas. These deposits cover large areas of the alluvial plain.

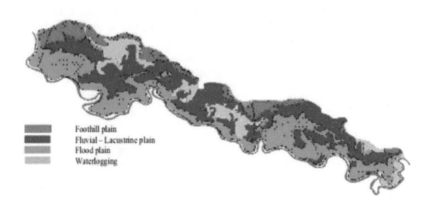

Foothill plain
Fluvial – Lacustrine plain
Flood plain
Waterlogging

Figure 5.4 Map of the plains in the RUT Irrigation District (Burbano and Forero, 1999)

Moderate-slow permeability, good - fairly good available moisture, infiltration rate 0.2 - 1 cm/h, content of organic matter 1.7% - 2.2%, pH 6.8 - 7.7, medium - low fertility, and soil depth 0.90 m are some general characteristics of the soils in the RUT Irrigation District. Table 5.1 shows the soils classification in the RUT Irrigation District.

Table 5.1 Soil Classification in the RUT Irrigation District (ASORUT, 2000)

Soil Class	Area		Remarks
	ha	%	
I	1,170	10.8	Good quality. Suitable for several crops. Appropriate management is needed for fertility conservation
II	6,945	64.7	Fairly good - good quality for crops. Very good quality for grass
III	2,235	20.8	Poor - fairly good for crops. Fairly good - good for grass
Sub – total	10,350	96.3	
	400	3.7	Area without classification. It is located between the Cauca River and the main irrigation canal
Total	10750	100	

Figure 5.5 shows the soil classification according to salinity and Appendix I presents the detailed classification and the physical and chemical characteristics of the soils in the RUT Irrigation District.

5.3   Description of the Irrigation and Drainage System

The RUT irrigation system is essentially a flood protection and irrigation/drainage project. The physical aspects of the system comprise the protection works, the irrigation, drainage, and road networks. The information on the physical aspects was obtained from technical documents and reports; interviews with the administrative and technical staff of ASORUT, with field operators, farmers and retired workers; and especially from a thorough field survey in order to identify in details the hydraulic infrastructure, its physical condition and management.

Table 5.2 summarizes the general characteristics of the hydraulic infrastructure and figure 5.6 shows the location of the main infrastructure of the RUT Irrigation District

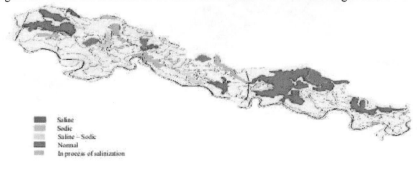

■ Saline
　Sodic
　Saline – Sodic
■ Normal
▥ In process of salinization

Figure 5.5 Salinity in the RUT Irrigation District (Burbano and Forero, 1999)

Table 5.2 Hydraulic Infrastructure of the RUT Irrigation District (ASORUT, 2000)

System	Work	Characteristic	Capacity (m³/s)
Irrigation	- Pumping Stations:		
	Tierrablanca	4 units	6.7
	Candelaria	2 units	2.8
	Cayetana	3 units	5.1
	- Canals		
	Conveyance	1.4 km	6.0 – 10.0
	Interceptor	31.0 km	6.0 – 10.0
	Main	44.0 km	0.7 – 4.8
	Conveyance 1.0	12.0 km	0.8 – 1.8
	Secondary	35.3 km (total)	0.2 – 0.6
	Tertiary	32.5 km (total)	0.05 – 0.1
Drainage	- **Pumping Stations**		
	Candelaria	2 units	2.8
	Cayetana	3 units	5.1
	San Luis	2 units	0.7
	Portachuelo	2 units	0.7
	- **Canals**		
	Main	26.0 km	26.0
	Secondary	76.6 km (total)	1.5 – 5.0
	Tertiary	41.6 km (total)	0.4 – 0.8
Flood Control	- Interceptor Canal	31.0 km	6.0 – 10
	- Protection Dike	44.0 km	Return period: 1:10 years
Roads	- Roads	350 km	
	- Bridges	40 units	Load maximum: 15 Ton

Figure 5.6 Hydraulic infrastructure of the RUT Irrigation District (Burbano and Forero, 1999)

5.3.1   Protection Infrastructure

The protection works safeguard the RUT area against floods and include the protection marginal dike and the Interceptor Canal respectively.

Protection Dike

The protection dike is 44 km long and runs along the left bank of the Cauca River. It begins in the southern part of the district, next to the Tierrablanca Pumping Station and ends in the northern part, about 10 km downstream of the Cayetana Pumping Station. The dike protects the area against floods caused by high water levels of the Cauca River. Along its total length, the protection dike has some highly vulnerable sectors as a result of the erosive action of the Cauca River. Los Millanes, La Bodega, and Remolinos are sectors in which protection works like sheet piles, rock lining, and groynes have been built on the left bank of the river in order to guarantee the stability of the dike.

On the other hand, the height of the dike forms a serious risk factor. Several points have been identified in which it is necessary to increase the height in order to prevent that it overflows during high water levels in the Cauca River. A critical situation occurred in 1999 when due to a heavy rainy season the marginal dike was very close to overflowing; which would have affected the economy of the region in a negative way. During that occasion the users have put sand bags on the top of the dike in the critical reaches with the purpose to increase its height as a temporal protection measure. Another risk factor is formed by the unofficial settlings on the strip of land between the left bank of the dike and the right bank of the main irrigation canal (marginal irrigation canal). There, small farmers have established their housing and crops causing serious risks to the stability of as well the main irrigation canal as the protection dike. The number of these settlings has been increased during the past and a solution to this problem belongs more to the local authority than to the district administration.

Interceptor Canal

The Interceptor Canal is 31 km long and runs along the West side of the area, starting in the Roldanillo Municipality and discharges by gravity to the Cauca River in the North of the RUT area. Its functions are to protect the area against floods from the small rivers coming from the Western Mountain Range, as well as, to supply irrigation water for some areas. Among the streams intercepted by the Interceptor Canal are the Roldanillo and Toro rivers; Rey, San Luis, La Union, and San Antonio streams.

The Interceptor Canal is affected in a negative way by these streams, which deposit a large amount of sediment that has a negative impact on the hydraulic performance of the canal as a protection work, on its maintenance and safeguarding. The high sediment load deposited on the canal bottom decreases the water carrying capacity and increments the water levels for the peak discharges and put the area under risk by a possibly overflowing of the canal. The streams coming from the Western Mountain Range transport high sediment loads due to deforestation and erosion processes taking place in their basins. In this respect, the basin of Rio Toro presents the most critical situation, which is even worsened by the geological conditions, being the presence of the Toro´s fault. Moreover, the presence of sediments in the canal play an important role in the growth of the aquatic floating vegetation, which has a similar effect on the hydraulic performance as that described before. The sediment has also an impact on the water quality. Among the aquatic vegetation can be mentioned aquatic iris (eichhornia crassipes), aquatic lettuce (pistia stratidtes), tifas (typha sp), and najas (najas sp). The operation of the canal and hydraulic structures is also affected by the presence of sediments and aquatic vegetation. The impact of the wastewater discharges to the Interceptor Canal will be discussed later on.

Another point of concern is the removal of the sediments and aquatic vegetation that involves high costs for the RUT administration. This activity is almost permanent around the year taking into account the Interceptor Canal length and the large amount of deposited sediments by the streams. Moreover, the removal of the sediments with equipment has changed the shape of the cross sections.

On the other hand, the sediments removal also has some environmental impacts, namely a unpleasant odour, and changes in the landscape. Once the sediments are removed, they are placed inside an excavation pit located parallel to and with the same length as the canal reach to be cleaned. There the sediments dry in a natural way during several days, and next they are transported to other areas on request of the farmers. Photo 5.1 shows the treatment of the removed sediments in the Interceptor Canal.

Photo 5.1 Excavation for sediments drying in Interceptor Canal

5.3.2   Irrigation Infrastructure

As shown in Table 5.2, the irrigation infrastructure comprises three pumping stations; and the primary, secondary and tertiary canals network with their correspondent hydraulic structures.

Pumping Stations

There are three pumping stations, namely at Tierrablanca, Candelaria, and Cayetana. They are located at km 00+000, km 18+199, and km 34+000 along the main irrigation canal. Tierrablanca is the main pumping station and has a total capacity 6.8 m^3/s. Its function is to pump water from the Cauca River to the main irrigation canal and/or conveyance canal. Each unit discharges in an individual way to an outlet box by a pipe with a flap valve.

A limitation of the pumping station is the difficult operation when the water levels in the Cauca River are low, especially during the dry season when irrigation water is required. The water users consider this situation as a risk and for that reason some alternative solutions have been considered. Moreover, the pumping units do not have discharge measuring devices, which allow a monitoring of their performance. However, several years ago a calibration curve was obtained by which a relationship between the water level in the Cauca River and the discharge of the pumping units was established. Nowadays, according to the experience of the field operators, this relationship does not work any longer in a good way. Photo 2 shows the Tierrablanca Pumping Station.

Table 5.2.2 Pumping Stations of the RUT Irrigation District

Pumping Station	Location	Purpose	Pump units	Q m^3/s	Dynamic head (m)	Power hp	Velocity rpm
Tierrablanca	00+000	Irrigation	4	1.7	7.45	247	705
Candelaria	18+199	Irrigation/ drainage	2	1.4	6.80	175	585
Cayetana	34+000	Irrigation/ drainage	3	1.7	6.80	175	580

Photo 5.2 Tierrablanca pumping station

The Candelaria Pumping Station was built 5 years after the construction of the Tierrablanca and the Cayetana Pumping Stations. It has 2 units with a total capacity of 2.8 m³/s. The station has a double function, namely pumping water for irrigation from the river to the main irrigation canal; and pumping drainage water from the main drainage canal to the Cauca River. The first function will improve the water availability for irrigation of the surrounding area of this station; reduce the travel time in relation to the time needed when the water was pumped from the Tierrablanca Pumping Station. The water can flow in both directions of the irrigation canal. The second function will help to evacuate water from the main drainage canal, especially when high rainfalls occur in the southern part of the RUT Irrigation District. In this way, it supports the Cayetana station in its specific drainage function. Another important characteristic of the Candelaria station is the possibility to introduce water to the main drainage canal from the Cauca River by gravity. It is very important to decrease the salts concentration of the main drainage canal in view of the fact that the farmers use water from the drain for irrigation. Also the pumping units at Candelaria do not have discharge measuring devices that allow for a monitoring of their performance.

The Cayetana Pumping Station was built in a simultaneously way as the Tierrablanca Station. It has a double function, namely for irrigation and drainage. For the first function, it pumps water from the main drainage canal to the main irrigation canal, from which the water flows in both directions (downstream and upstream direction). For the second function, it pumps water from the main drainage canal to the Cauca River. It has three pumping units with a total capacity of 5.1 m³/s. The water quality pumped for irrigation from the main drainage canal is considered as a risk factor; the water contributes to the salinization process of the area.

There is no doubt that the three mentioned pumping stations play an important role in the operation of the RUT Irrigation District. Appendix II shows the monthly pumped volume (m³) of water by each one of the pumping stations during the period 1989 – 1995.

An analysis of the information for this period shows the following characteristics:
-   The importance of the Tierrablanca Pumping Station follows from the fact that about 60% of the water pumped during that period comes from this pumping station, about 15% from Candelaria, and 25% from Cayetana. See figure 5.7.

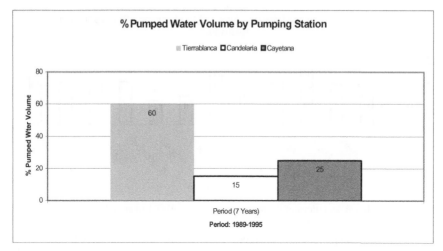

Figure 5.7 Contribution of the pumping stations to the operation of RUT Irrigation District

-   Also at monthly level, the Tierrablanca station is very important. The percentage
    of pumped water ranges from 49% in April to 75% in July; whereas the values for
    the Candelaria and Cayetana stations range from 12% in April to 20% in June and
    August, and from 11% in July to 39% in April, respectively. See Figure 5.8.

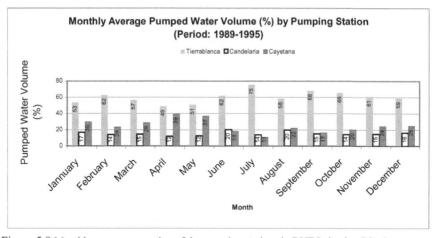

Figure 5.8 Monthly average operation of the pumping stations in RUT Irrigation District

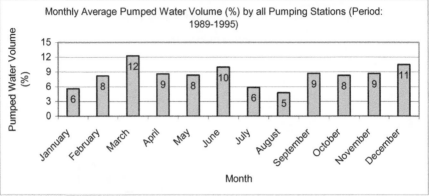

Figure 5.9 Combined operations of the pumping stations in RUT Irrigation District

- The combined operation of the three stations during an average year for the period 1989 – 1995 shows that the highest percentage of pumped water is 12% in March, followed by 11% in December; while 5% and 6% are the lowest values in August and January, respectively. See figure 5.9:
- The average monthly discharge of each station and their combined operation during an average year (1989 –1995) shows the highest values for the Candelaria station with a maximum value of 1.54 m^3/s in May; only in August by the Cayetana station has a higher discharge (2.7 m^3/s). This station has the lowest values during the remaining months ranging between 0.5 to 0.7 m^3/s. On the other hand, the Tierrablanca station gives intermediate values, ranging between 0.8 and 1 m^3/s. The discharges for the combined operation of the three pumping stations range from a minimum of 2.5 in October to a maximum of 4.7 m^3/s in August. See figure 5.10.

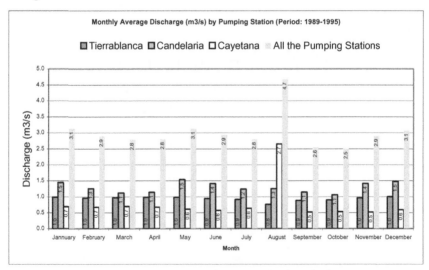

Figure 5.10 Discharge of pumping stations in RUT Irrigation District

Appendix III presents the values for the irrigation and drainage volumes pumped during the period 1991–2002 by all three pumping stations and it shows the water volumes charged and the energy consumption. An analysis of this information for that period gives the following characteristics:

- The importance of the pumps for irrigation in comparison with the drainage. On an average, 84% (38,774,146 m^3) of the total pumped volume (46,142,126 m^3) corresponds with irrigation, whereas 16% (7,367,980 m^3) with drainage. However, this pattern shows an important exception for the year 1999 year when the drainage volume (53.8%) was larger than the irrigation volume (46.2%). This year was a historic one for the RUT area because of very high rainfalls and high water levels in the Cauca River, which almost caused flooding in the District. The pumping pattern for irrigation and drainage is given in figure 5.11;

- The analysis for an average year during 1991 – 2002 shows that the period from June to September has the main requirement for irrigation pumping, whereas the period from March to May is the major one for drainage. In general terms, this pattern corresponds to the wet and dry seasons in the RUT Irrigation District. See figure 5.12.

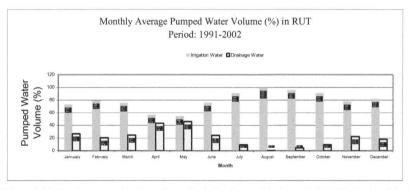

Figure 5.11 Yearly pumped water volume for irrigation and drainage in RUT Irrigation District

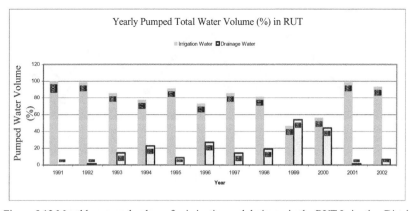

Figure 5.12 Monthly pumped volume for irrigation and drainage in the RUT Irrigation District

The irrigation and drainage distribution pattern for three years (1993, 1998, and 2002) within the period 1991 – 2002 is as follows (see figure 5.13.1, 5.13.2 and 5.13.3):

- The distribution pattern for the drainage presents similar characteristics with small differences. Small values for the first three months of the year, an increase for April and May with an exception for May 2002, a decrease for June; and low values again for the period from July to October with an exception for October 1998; and finally an increment during November and December with an exception for December 2002;

- The irrigation distribution pattern is more irregular than the drainage pattern. The values increment for the first three months with an exception for March 2002; they decrease for April and May with an exception for May 2002 which shows a dramatic increase; the increase for June is small in 1998, dramatic in 1993, and moderate in 2002; July presents a decrease in 1993 and an increase in 1998 (light) and 2002 (moderate); In August the irrigation requirement decreases in 1993 and increases in 1998 and 2002. September presents an increase in 1993 and a decrease in 1998 and 2002; October shows an increase in 1993 and 1998 and a decrease in 2002; and finally December shows an increase in1993 and a light decrease in 1998. Appendix IV shows records for the pumping hours and energy consumption for the three pumping stations during the period 1990 – 2002. The average values for this period indicate once more the importance of the Tierrablanca Pumping Station. About 49% of the operation hours correspond with the Tierrablanca station, whereas 12.5 and 38.5% with the Candelaria and Cayetana stations, respectively. In relation to the energy consumption about 68.9% corresponds with the Tierrablanca station, while 12.7 and 18.4% with the Candelaria and Cayetana stations, respectively. See figures 5.14.1, 5.14.2 and 5.15.1 and 5.15.2.

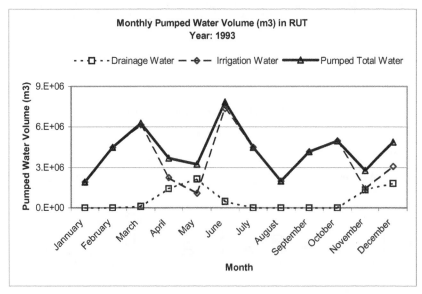

Figure 5.13.1   Irrigation and drainage pattern in RUT Irrigation District – 1993

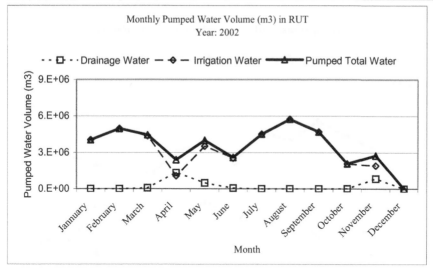

Figure 5.13.2 Irrigation and drainage pattern in RUT Irrigation District – 1998

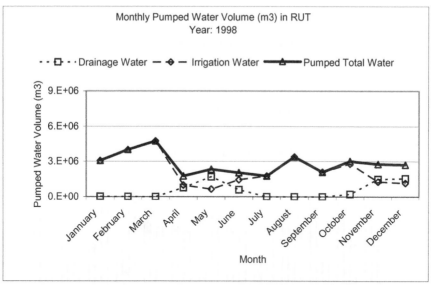

Figure 5.13.3 Irrigation and drainage pattern in Rut Irrigation District – 2002Figure 5.14.1 Yearly pumping hours in RUT Irrigation District

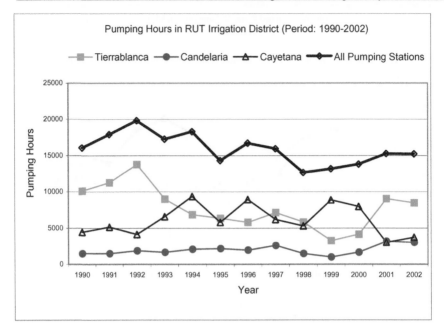

Figure 5.14.1 Yearly pumping hours in RUT Irrigation District

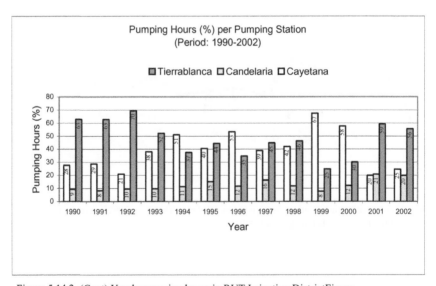

Figure 5.14.2 (Cont) Yearly pumping hours in RUT Irrigation DistrictFigure

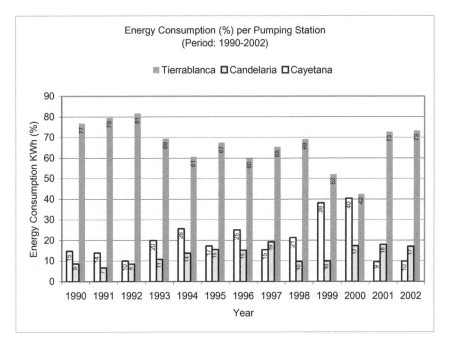

5.15.1 Energy consumption in RUT Irrigation District

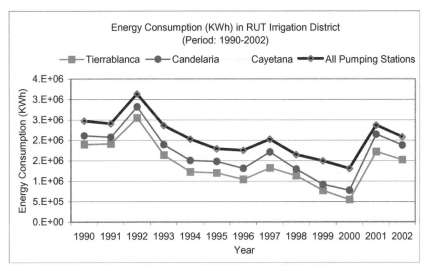

Figure 5.15.2 (Cont.) Energy consumption in RUT Irrigation Distric

Irrigation Network

The irrigation network comprises the main system and a system at farm level. The first one consists of a primary and secondary canal network and is managed by the ASORUT (Users Association of the RUT Irrigation District), which plays the role of an Irrigation Authority. At farm level, the irrigation network is very simple and in some cases does not exist.

The Main Irrigation Canal (M.I.C) or also called Marginal Irrigation Canal, the Conveyance Canal (C.C), the Conveyance 1.0 (C.1.0), and Interceptor Canal (I.C) constitute the main canals of the irrigation network. In some places of the RUT area the Main Drainage Canal (M.D.C) functions also as an irrigation canal.

The Main Irrigation Canal M.I.C. is 44 km long and runs parallel to the Protection Dike, the cross section has a trapezoidal shape, the bottom width is 4 m and the total depth is 2.5 m, it is unlined for the major part of its alignment, and has a total capacity of 4.8 m^3/s. It starts at the Tierrablanca Station and ends in the northern part of the RUT area, about 10 km after the Cayetana Station and is connected with the M.D.C through an outlet structure (km 43+188) and a canal (417 m long). As mentioned before, the canal is supplied by the three pumping stations: Tierrablanca (km 00+000), Candelaria (km 18+199), and Cayetana (km 34+00); the first two stations take water from the Cauca River, and the other one from the M.D.C.

Along the canal are five control structures, four of them with downstream control and located at km 6+021 (number 1), km 11+220 (number 2), km 19+850 (number 3), and km 37+860 (number 5). The structure number 4 has a large sliding gate and is located at km 25+450 (La Victoria Bridge). From a hydraulic point of view the downstream control structures are seriously deteriorated and they cannot be operated in the way as they have been designed. Nowadays the structures are modified in view of their mechanical operation by the "irrigation inspectors" who uses chains, flywheel and operates according to their experience and good judgment. See photo 5.3.

M.I.C has several direct offtakes (lateral gates) and primary canals; these in turn supply other canals at secondary and tertiary level. The Tierrablanca Pumping Station supplies water to the M.I.C along the first 11 km; the Candelaria Station to the following 14 km, and finally the Cayetana Station to the last 17.8 km. Table 5.3 illustrates the list of offtakes and canals in M.I.C. Table 5.3 shows that only 15 out of 41 offtakes are being operated and that 26 structures are out service.

On the other hand, the offtake structures and the branch canals have some characteristics, which deserve to be mentioned. In the first case, the offtakes were designed with a device to measure the discharge under the sliding gate. Each offtake has two small wells and through a connection with piezometers the upstream and downstream static head of the gate are registered. In this way, the discharge could be measured and regulated according to the needs. This provision has been made for all the offtakes built during the initial construction works. The new ones, especially those established for the sugar cane sector, from 1995 onwards, have a single sliding gate. This difference can be attributed in the first place to the traditional irrigation culture of the sugar cane sector, and in the second place due to the lack of knowledge of the hydraulic concepts of the district.

For the branch canals two types of structures exist. One is similar to the direct offtakes with a device for the measurement and regulation of the discharge; normally this type is used for relatively small branching canals. For the major canals the structures have two vertical sliding gates and work under the hydraulic principle of the

Constant Head Orifice (CHO). It is a combination of a regulating and measuring structure (Bos 1989, Kraatz 1975). The upstream gate (orifice gate) is an adjustable submerged orifice for measuring the flow by using small wells as mentioned before; whereas the downstream gate is operated to get the required constant head (normally 0.06 m). See photo 5.4.

Table 5.3 Direct Intakes and Derivation Canals in M.I.C

Supply Station	Direct Intakes (Number)	Branch Canals Marginal	Location at km	Dimensions		Capacity (m^3/s)	Length (m)
				b(m)	h(m)		
Tierrablanca	14 (*)						
		1.3	1+300	0.90	0-65	0.21	93
		3.78	3+780	1.50	0.75	0.50	1,702
		5.02	5+020	2.00	0.70	0.50	1,983
		6.38	6+380	2.00	0.90	0.60	2,484
		8.64	8+640	0.90	0.64	0.14	
		9.34	9+340	1.20	0.69	0.21	1,008
Subtotal	14	6					
Candelaria	1						
		13.23	13+230	1.50	0.93	0.60	3,750
		14.14	14+140	1.20	0.65	0.21	873
		16.93	16+930	1.50	0.75	0.70	330
		17.38	17+380	1.20	0.70	0.35	1,539
		18.52	18+520	1.50	0.85	0.50	1,920
		20.75	20+750	1.50	0.65	0.35	1,288
		22.10	22+100	1.20	0.65	0.21	1,542
		23.22	23+220	1.20	0.70	0.35	1,455
		25.13	25+130	1.50	0.86	0.60	2,320
Subtotal	1(**)	9					
Cayetana		27.84	27+840	1.20	0.70	0.14	420
		29.65	29+650	1.50	0.90	0.50	1,053
		30.96	30+960	1.50	0.80	0.60	2,250
		33.58	33+580	1.20	0.70	0.28	1,158
		33.84	33+840	1.20	0.65	0.28	1,073
		35.83	35+830	1.20	0.70	0.21	206
		37.59	37+590	1.50	0.85	0.50	1,640
		42.92	42+920	1.20	0.65	0.28	807
Subtotal	0(***)	8					
TOTAL	15(****)	23					

Remarks:

(*):   Located between km 0+300 – km 11+000

(**):   Located in km 12+200. Outlet 14 more out off service between km 11 – km 25

(***):   Outlet 12 out off service between km 25 – km 43+880

(****): Total out off service: 26

Photo 5.3 Control structure modified with chain (left) and flywheel (right)

Photo 5.4 Direct intake (left) and constant head orifice (right) in M.I.C

The field survey revealed the followings findings in relation to these structures:
- No knowledge is available on the hydraulic functioning of the Constant Head Orifice by a part of the field operators (irrigation inspectors and field operation engineer) and as consequence the operation is not in line with the one intended by the designer. Nowadays, the irrigation inspector operates only the upstream gate, as the most downstream gate has no sliding sheet, flywheel and screw;
- No knowledge about the function and existence of the small wells for the discharge measurements. Most of them are filled with soil, covered with vegetation or concrete blocks;
- Poor conservation and maintenance.

This situation is the same in other canals (second and third order canals in M.I.C; C.C.1.0) where this type of structures has been installed.

In order to find an explication for this situation the old operators have been contacted, who said that they know about the proper functioning of the structures at the start of the project. However, this condition changed due to several reasons:
- Complicated and tiresome operation of the structures, especially the Constant Head Orifice;

- Policy to reduce the field staff;
- Structures operation in comparison to required condition for irrigation;
- No acceptance and gradual rejection by the farmers, taking into account the relationship between the delivered water and the cost of the used water (charged volume).

The silting, seepage and leakage are important limitations for the operation of the M.I.C. Silting is caused by the physical characteristics of the Cauca River, which contains a high load of suspended sediments, which is deposited especially in the first reach of the canal (km 00+000 to km 3+700). See photo 5.5.

Photo 5.5 Silting in main irrigation canal in the RUT Irrigation District

The sediment accumulation decreases the conveyance capacity of the canal and affects the water distribution amongst the farmers and causes extra costs for conservation and maintenance. On the other hand, the sediment removal by machines has changed the bottom slope and the shape of the cross section; helping also the leakage as the canal was built on an embankment. In view of its closeness to the Cauca River, a hydraulic gradient between the canal and river develops. In this way, the installation of a sandtrap has been considered as a solution for the silting problem, for the Tierrablanca station as well as for the Candelaria one.

Leakage has been detected in the reaches from km 0 to km 2; km 8 to km 9+500, km 18 to km 19, km 26+500 to km 28+500, km 33 to km 36, and km 39 to km 42 (Blanco, 1996). In addition to the above-mentioned causes, there are two other factors that contribute to the leakage:

- From a geological point of view, sand-clay prevails in the soil along the M.I.C, especially in the natural and semi-spot dikes; helping in this way the leakage. A drilling carried out by CVC in 1971 (Muñoz, 1978) showed the following characteristics for the subsoil profile (Table 5.4).

Table 5.4 Subsoil Profile in the M.I.C Sector (Muñoz, 1978)

Texture	Thickness (m)
- clay silt with thin intercalations of fine sand	4.9
- medium consistence clay with medium to high plasticity	4.0
- soft organic clay, high plasticity	0.5
- medium to fine sand	2.0
- low plasticity silt	2.5
- sandy silt or sandy clay	3.0
- light clay sand	3.5

From Table 5.4 follows that the layout of the M.I.C was not the most appropriate one for an unlined canal from a geological point of view.

- On the other hand, after its construction the canal stayed a long time out off service, which caused cracking of the canal body; not allowing any natural sealing with silt and clay particles transported by the water flow. The leakage does no only decrease the conveyance efficiency of the canal but also contributes in a negative way to its stability.

Fish called "corroncho" has its habitat not only in the M.I.C but also in the other canals of the district. This fish builds a hole in the sidewalls of the canal where the water leaks, being a most critical situation when the canal flow is at full capacity. See photo 5.6.

The Tierrablanca Pumping Station supplies the Conveyance Canal (C.C), it is 1.4 km long, the cross section has a trapezoidal shape and a bottom width of 2.5 m; the total depth is 2.5 m (original design), and the total capacity is 6.8 m3/s. It has a small branch canal located at km 0+430, which is 1265 m long and with a bottom width and total depth of 0.60 m, the total capacity 0.21 m3/s and the canal covers an area of 54 ha. In spite of its short length, the canal is important because it supplies irrigation water to Interceptor and Conveyance 1.0 canals. It is connected with the first one by an inverted siphon (km 1+400), whereas the connection to the second is via an offtake (Constant Head Orifice) in combination with a structure with upstream water level control (km 1+000).

This canal shows that silting up is the major problem and its maintenance and conservation by machines has caused a dramatic modification of the shape of the cross sections. The offtake structure (constant head orifice) is deteriorated and the screw, flywheel, and metallic frame disappeared. See photo 5.7.

The Conveyance Canal 1.0 (C.C 1.0) irrigates the western sector of the RUT Irrigation District between the right bank of the Interceptor canal and the left bank of the Main Drainage canal. It has a total length of about 30 km. However, during the construction period the farmers rejected its construction. Their protest movement resulted in the final construction of a reach of 12 km, which influences only the southern part of the previously considered sector. C.C. 1.0 is a concrete lined canal and has a cross section in a trapezoidal shape, the bottom width is 2.5 m and the total depth is 1.9 m, the canal capacity is 3.08 m3/s and the canal covers an area of about 2000 ha. According to the original design, it has a total of 10 branch canals, some of them with branch canals at second and third order level. Nowadays only six branch canals along the canal are working; the canal has also 17 offtakes (See table 5.5).

Photo 5.6 Leakage in main irrigation canal due to fish activity

Photo 5.7 Level control and offtake structure (left) and silting up in conveyance canal 1.0 (right)

Photo 5.8 Level control and offtake structures in conveyance canal 1.0

Table 5.5 Direct Intakes and Derivation Canals in C.C. 1.0

Direct intakes (Number)	Derivation canals	Location (Abscissa)	Dimensions		Capacity (m³/s)	Length (m)
			b (m)	h (m)		
17						
	1.0 – 0.23	km 1+230	0.60	0.80	0.21	960
	1.0 – 0.54	km 1+540	1.20	0.70	0.28	674
	1.0 – 0.78	km 1+780	1.50	0.70	0.50	2,262
	1.0 – 3.20	km 3+200	1.50	0.80	0.80	3,420
	1.0 – 6.63	km 6+630	1.50	0.90	0.20	3,200
	1.0 – 11.57	km 11+570	1.50	0.85	0.70	262
Total: 17	6 (*)					

- (*) In service
- Out of service: 1.0-047, 1.0-5.20, 1.0-3.70, 1.0-5.73

The offtake structures are Constant Head Orifices that present the same problems as mentioned before. In some cases, these structures are complemented with long crested weirs for the water level control; there are five structures of this type along the Canal 1.0 (km 3+200, km 8+620, km 6+630, km 8+620, and km 11+620). See photo 5.8. In a similar way, this arrangement is also available at the branch canals of the second and third order.

The major problems present in the C.C. 1.0 can be summarized as:

-   Silting up in C.C. 1.0 and in the branch canals at lower level, especially those called 3.20 and 6.63; which are of a lower level than M.I.C and C.C;
-   Deterioration and poor maintenance of the supplying infrastructure (offtake gates and level control);
-   Leakage through dilatation joints, which are in a bad condition and through concrete lining fissures of the sidewalls. Leakage is supported by the construction of structures in the canal embankment and the location of some branch canals near the wetlands that are low-lying areas with respect to the canal bottom;
-   Deterioration of the body of the canal caused by cattle trampling;
-   Presence of vegetation that is supported by silting up, and garbage disposal by farmers;
-   Presence of solid wastes as result of the land preparation activities (soil, weeds, harvest wastes). The farmers next to a canal carry out their land preparation until the edge of the canal without respecting the land strip that has been established to protect the canal. In some cases, the surface drains (papaya tree crop) flow directly to the canal. See photo 9.

The effect of the changes in cropping pattern on the hydraulic infrastructure of the RUT Irrigation District can be clearly observed in the area under irrigation by C.C. 1.0. Since1992, sugar cane has been introduced in a gradual way; and at this moment about 30% of the area is under sugar cane. The RUT Irrigation District was projected for other crops than sugar cane, which demands a more water than vegetables, grains and fruit trees; the traditional crops in the RUT area. On the other hand, the irrigation infrastructure has not the capacity to carry the required amount of water for sugar cane; therefore, a part of the control structures are considered as an obstacle for irrigation of this crop. This is the main reason why the farmers have destructed some hydraulic works. For instance, the long crested weirs and the intake structure a farm level. In the first case, the structure prevents the fast availability of water for the sugar cane grown downstream of it. Moreover, due to poor maintenance the structure traps garbage, weeds, tree branches, etc. and the capacity of the canal decreases. This is the reason why the farmers continue with the destruction of this type of structure.

Photo 5.9 Effect of land preparation on the canal C.C. 1.0

Figure 5.10 Hydraulic structures modified by farmers in RUT Irrigation District

In relation to the intake structures at farm level, some sugar cane farmers have modified the structure in order to get more water for their crops. Normally, this structure consists of a submerged orifice where the discharge is measured with the difference in hydraulic head between the water level in the supply canal and the level downstream of the orifice. In order to increase the hydraulic head and thus to get more water, the farmers destroy the sloping apron at the downstream side of the structure. Photo 5.10 illustrates the two situations.

The Interceptor Canal (I.C) is part of the protection infrastructure of the RUT Irrigation District and it protects the area against floods from the small rivers and streams coming from the Western Mountain Range. In the past two interventions have been carried in order to supply irrigation water from this canal and by pumping from the Tierrablanca Station to the Conveyance Canal. The first intervention was a result of the unfinished construction of C.C. 1.0. After a long time after the start of operation of the RUT Irrigation District and as a consequence of a strong drought period with low crops yields, the farmers on the periphery of the command area of C.C. 1.0 requested irrigation services for their lands. Taking into account the shortage of economic resources for the enlargement of C.C. 1.0, it was decided to construct two branch canals (called 13.12 and 11.34) from the Interceptor Canal. To divert the water, two structures with adjustable water level control were built during 1972-1975 in the Interceptor Canal at km 3+655 (structure 1) and at km 13+122 (structure 2; at the intersection with the road of La Unión – La Victoria). These structures are equipped with radial gates, which can be mechanically or electrically operated. In each case, the offtake consists of a structure with constant water level control and an outlet structure with sliding gates, which allow a flow in both directions of the branch canal. In this way, the irrigation service was provided for the North-West area of the RUT Irrigation District, between right bank of the Interceptor Canal, the left bank of main drainage canal, and a few kilometres to the South of the road La Unión – La Victoria. See photo 5.11.

Irrigation water is also pumped from the Interceptor Canal in order to irrigate the area South of the canal and in the middle parts of the left bank of the Interceptor Canal[1]. To convey the water, the radial gate of structure number 2 has to be closed in order to raise sufficiently the water level. This procedure takes too much time due to the long reach that the flow has to travel from the Tierrablanca Pumping Station (about 20 km) plus the time required to fill the canal. In order to reduce the time and under the pressure of the farmers, a new water level control structure, number 3 was installed in 1995 at km 24+500. This was the second technical intervention for the Interceptor Canal.

Sedimentation, variation of the shape of the cross sections, presence of vegetation, poor maintenance of the hydraulic infrastructure, and high costs for operation and maintenance were mentioned before as the major problems of the Interceptor Canal. However, the most important restriction for the agricultural development is related to the quality of the water for irrigation. The wastewater of the Roldanillo, and Toro Municipalities are partially purified in treatment plants located near the left bank of the Interceptor Canal, and then disposed to it, from which it is used for the irrigation of the crops. See photo 5.12.

Photo 5.11 Level control structures in Interceptor Canal: left (km 24+500), right (km 13+122)

In this way, the canal water is contaminated with high content of organic matter and mineral nutrients creating favourable conditions for the growth of vegetation, which affects the operation of the canal from a hydraulic point of view, increases the maintenance costs, and reduces the conveyance capacity and efficiency of the canal given that this vegetation consumes an essential amount of water.

---

[1] In the design phase of the RUT project was considered to irrigate this area pumping water from C.C. 1.0 canal through the construction of a small pumping station. Finally, due to shortage of economic resources this alternative was discarded.

On the other hand, it is important to recognize the health risk for the consumers of agricultural products as vegetables and fruits are irrigated with this water. The irrigators have also suffered from the effects caused by the manipulation of this irrigation water, which became manifest through skin infections. The export of fruits has also been restricted due to the fact that the produce does not fulfil the quality standards from a health point of view.

Photo 5.12 Disposal of wastewater to the Interceptor Canal

5.3.3   Drainage Infrastructure

The drainage infrastructure consists of a main system and a system at farm level. The first one consists of a main and secondary drainage canals. The main drain, namely the Main Drainage Canal (M.D.C), and the secondary drains are under the responsibility of the ASORUT for their operation and maintenance, whereas the drains at farm level are under the responsibility of the farmers for their maintenance and conservation. In some cases, the drainage network at farm level does not exist. The Candelaria and Cayetana Pumping Stations are also part of the drainage infrastructure; the same is valid for two small pumping stations located near the Interceptor Canal (San Luis and Portachuelo stations).

The San Luis pumping station is located in the village San Luis and it drains excess water from the area to the Interceptor Canal. It has two units with a capacity of 350 l/s, a dynamic head of 15.7 m, a velocity of 880 rpm, a motor of 107 hp, amperage 70A and voltage 440V. The Portachuelo pumping station, which drains excess water from the Portachuelo area, has two units with tri-phase motors of 40 hp, voltage 220/440V, and velocity 1765 rpm.

The Main Drainage Canal (M.D.C.) is 26 km long, has a cross sections with a trapezoidal shape, the bottom slope varies between 0.0002 and 0.004 m/m, and the average water depth is about 3.0 m. The drain divides the RUT area almost in half, it runs through the lowest elevations and discharges by gravity into the Cauca River during low stages; otherwise drainage water is pumped by the Candelaria and Cayetana pumping stations to the river. A structure called "Ceros" is located at the downstream end of the canal, at km 00+000. "Ceros" consists of a water level control structure (Neyrpic AMIL 450D) and an outlet with a flap valve. The first structure controls the water level in the canal in order to pump water for irrigation. The second one allows the water to flow from the canal to the Cauca River and at the same time prevents that the water flows in the opposite direction during high stages of the Cauca River. The tail canal of the Main Irrigation Canal discharges into the Main Drainage Canal (right bank), about 15 m upstream of the "Ceros" structure. See photo 5.13.

Photo 5.13 Structure "Ceros". Water level control (left), outlet structure (center), and connecting canal to the Cauca River

The M.D.C plays an important role in the drainage of the RUT area. The groundwater table is influenced by groundwater flows from the Western Mountain Range and the Cauca River, by seepage and leakage from the irrigation canals, and also by deep percolation from rainfall and irrigation water. The fluctuation of the groundwater table has contributed to the process of increased salinization in certain areas, inside as well as outside the RUT area, depending on the geological and morphological nature and the chemical characteristics of the subsoil. The absence of drainage works at farm level, inefficient irrigation practices, poor maintenance of the drainage network, and the use of drainage water for irrigation have supported this salinization process. As a result, the increased salinization is a limiting factor for the agricultural development of the RUT area.

Among the major problems of the M.D.C can be mentioned:
- Erosion and enlargement of the cross section caused by flows to the drains without outlet structures, installation of pumping equipments for irrigation, and mechanically control of aquatic weeds;
- Severe landslide of the sidewalls, especially in those areas where sugar cane prevails. This is caused by the traffic of heavy vehicles transporting sugar cane from the fields to the factory;
- Bottom level above the design level due to sedimentation, which limits the free discharge of the secondary drains and contributes to the raise of the groundwater table. In some secondary drains, the outlets are plugged with sediment of the M.D.C;
- Water level control structures, which are not calibrated and allow higher water levels in the M.D.C with a similar effect as mentioned in the previous item.

A network of piezometric wells forms also part of the drainage infrastructure. It was installed to monitor the fluctuations of the groundwater table and to make decisions about the operation of the whole drainage system (canals, structures, and pumping stations). Initially the wells were installed in the Portachuelo area in 1971 and next in San Antonio and San Luis in 1975, where salinization problems had been detected. Later, a total of 189 piezometric wells were installed by HIMAT (later INAT) with the purpose to widen the study to the whole RUT area. However, inappropriate location of some them caused their destruction by the farmers, since they were considered as an obstacle for their agricultural practices. The poor conservation and maintenance, and the reduction in field staff contributed also to the gradual deterioration of these wells and their out-of-service condition.

The study mentioned above (CVC, 1986) resulted in the following results and recommendations:
-    36% (3584 ha) of RUT area has salt problems in the soil profile from 0.0 to 1.0 m depth;
-    37% of the area has a groundwater table depth between 0 – 1.5 m;
-    61% of the area has salinity and drainage problems;
-    Poor maintenance of secondary drains: presence of sediments, weeds, submerged outlets, and in some cases they have been dammed up for irrigation;
-    High salt content in the groundwater and in the M.D.C, so that the quality of this water is not suitable for irrigation;
-    Inappropriate operation of the Cayetana and Candelaria Pumping Stations, especially in view of minimum water levels in the supply canals required for pumping. The design values caused a bad functioning of the M.D.C and submerged flows in the secondary drains.

Among the recommendations can be mentioned:
-    Improvement of the operation for the Candelaria and Cayetana Pumping Stations;
-    Bottom slope correction of the M.D.C and some secondary drains;
-    Improvement of the maintenance practices;
-    Construction of three small pumping stations for drainage of three secondary drains;
-    Lining of irrigation canals (specially M.I.C) in order to minimize the contribution of seepage and leakage to the groundwater;
-    Put out-of-service of the water level control structure (AMIL gate).

After 15 years of operation it can be concluded that very few of the previously mentioned recommendations have been implemented. As a result the salinization and drainage problems increased in the RUT area and that formed a limiting factor for the agricultural productivity. The results of another study (Blanco, 1996) mention again the same recommendations and include some others, like the improvement of the network of piezometric wells, a permanent monitoring of the water quality of the M.D.C and its improvement by the flow of water by gravity from the Cauca River via the Candelaria Pumping Station, and to restrict the installation of pumping stations for irrigation in the M.D.C.

5.3.4 Management Aspects

The total area of the RUT Irrigation District is 10,23 ha of which 97% has an irrigation and drainage and flood control infrastructure. In view of the management of the district, the RUT area is divided in 5 sectors, which are defined by their geographical conditions:
-    *Sector1* is located in the southern part of the RUT Irrigation District and corresponds to the area between the right bank of the Interceptor Canal, the left bank of the main drainage canal, and the road La Union – La Victoria. In turn, the sector is divided in 4 sub-sectors (1A, 1B, 1C, and 1D) and represents 28.4% of the total area;
-    *Sector 2* is also located in the southern part of the district. The right bank of the main drainage canal, the main irrigation canal, and the road La Union – La Victoria

encloses the area of sector 2. Sector 2 is divided in 4 sub-sectors (2A, 2B, 2C, and 2D) and represents 26.9% of the area;

- *Sector 3* is located in the northern part and the area is delimited by the road La Union – La Victoria, the right bank of the Interceptor Canal, and the left bank of the main drainage canal. Sector 3 is divided in 2 sub-sectors (3A, 3B) and covers about 16% of the whole area;
- *Sector 4* is also located in the northern part and the area is enclosed by the road La Union – La Victoria, the right bank of the main drainage canal, and the main irrigation canal. It has 2 sub-sectors (4A, 4B) and covers about 13.9% of the area;
- *Sector 5* is situated in the south-western part of the RUT Irrigation District and the left bank of the Interceptor Canal and the hillside of the West Mountain range border sector 5. The area is about 14.8% of the whole area.

At the moment, four irrigation inspectors carry out the operation activities for the whole area; each one is responsible for the management of one or more sectors, and they are periodically rotated from one sector to the other.

In general, there are two cropping seasons per year; the first one (season A) starts in February-March, and the second one (season B) in August-September. Two types of crops are grown, namely seasonal and semi-permanent or permanent crops. In the first season the main crops are cotton and sorghum rotated with maize and soybean. In the latter season, the main crops are grapes, fruit trees, passion fruit and pastures. In 1992 sugar cane crop has been introduced and the area for sugar cane has increased in a gradual way.

Irrigation and Drainage Management

The procedure to establish the irrigation schedule changed after the transfer process. Before the transfer, the farmers had to go to the Administrative Center and together with the Government Agency (HIMAT) they determined the crop plan and plant dates with as deadline March 15 of each year. Then HIMAT calculated the total volume of irrigation water, established the irrigation schedule, carried out the water distribution and also advised each water user on the date of his first pre-planting irrigation.

The user in turn had to pay his fixed water fee in advance and had to maintain his part of the irrigation network in top condition. After the first irrigation the users could directly request the ditch tender for subsequent irrigation turns instead of asking the central office. It is important to note that a minimum number of the water requests were fixed by HIMAT in order to establish the smallest flow in a canal.

Nowadays, the farmer decides himself about the crop to be planted, the plant date and the moment that he requests for irrigation water. Under this situation the irrigation schedule does not exist[2], and ASORUT only asks the farmers to submit their request

---

[2] According to the Irrigation inspector the existing situation presents two problems. "The first one is that after the transfer the farmers considered themselves as owners of the district with all the rights to impose new rules and with all the freedom not to follow those imposed by the Governmental Agency during a long time. With the transfer the users lost their organizational capacity; the administration staff lost their authority for the users, who had respected the rules and regulations. The second problem is related to the change in cropping pattern. The in previous times grown grain crops allowed for the establishment of an irrigation schedule with a long irrigation interval (1 or 2 weeks), but the introduction of highly sensitive cash crops (citric and fruit trees) is not possible with the same irrigation interval, these crops require an almost daily interval".

for water at least 48 hours in advance[3]. Next, the irrigation inspector fills out a form with the information on farm codes, farmer names, amount of water to be delivered, possible days for irrigation, and sector and sub-sector. After this action, the flow in the irrigation network is determined to satisfy the various water demands of the farmers.

In this situation, some extreme cases might occur. Sometimes irrigation water has been pumped in order to fulfil the water demand of only one farmer, who is at a far distance (several kilometres) from the pumping station. These cases occur, because ASORUT tries to avoid as much as possible conflicts with and lawsuits from the farmers.

As a consequence of this operation approach, a lot of operation hours are spent to deliver the requested water to the users. As a result the inspectors use a lot of time on the operation activities and do not have sufficient time for maintenance. Before the transfer, the canals and pumping stations were stopped during 20 days for maintenance; at the moment this stopping is not possible due to the pressure of the people (threats, lawsuits, etc.) and the maintenance works are done during the wet season.

Monica's lakes or fishponds present another interesting case. The company has installed pumping equipment on the left bank of the Cauca River, in front of their sport facilities on less than 100 m from the river. However, the water pumped from this site contains a high load of sediment, which affects in a negative way the ecological conditions of the ponds. To minimize this effect they have asked ASORUT to supply water from the Candelaria pumping station that is about 7 km upstream of the fishponds. In this way, a better quality of water is supplied as most of the sediments are deposited in the upstream canal reach[4].

Before the transfer of the RUT Irrigation District the water distribution in the irrigation network was on a continuous basis (24 hours) during the irrigation season. Today, a period called "peak hours" exists during which the energy cost is very high and the irrigation water supply is stopped. This period covers 6 hours per day; namely from 10 am to 13 pm and from 18 to 21 pm.

The operation of the RUT irrigation district was designed for gravity irrigation. However, most of the farm lands are not upgraded for this condition; for this reason the system operates on double pumping, namely from the river to the canals and from the canals to the fields. This circumstance significantly affects the water delivery in the istribution system.

Each of the existing pumping stations covers a specific area for the supply of irrigation water. The Tierrablanca pumping station supplies water to the sectors 1, 3, and 5; and the sub-sectors 2A and part of 2B; Candelaria supplies water to the sub-sectors 2B (part), 2C and 2D; and finally Cayetana supplies irrigation water to sector 4.

There are some differences in the way of water distribution amongst the main irrigation canal, Interceptor Canal, and main drainage canal. In the case of the main irrigation canal, the water distribution is influenced by the structures for water level control, which divide the canal into 5 reaches. As an illustration of the present water distribution, the water management for the first two reaches of the main canal will be described.

The first reach is located between the Tierrablanca pumping station and the structure for water level control at km 6+021. In this reach there are direct offtakes

---

[3] Observations during the field survey showed that a farmer presented his request for water by cell phone directly to the field operation engineer.

[4] The water allocation is 30 l/s during 3 hours. The charged cost is Col$ 20.1 per m^3, while the actual cost according to ASORUT is Col$ 80 per m^3.

with sliding gates and branch canals provided with a simple gate (capacity of the tertiary canals is 20 to 60 l/s) or with a CHO structure (canal capacity is about 400 to 500 l/s). The irrigation water is supplied for vegetables and grain crops and the field irrigation is predominantly by gravity (furrows). In the case of vegetables, sprinkler irrigation is not applied in view of the fact that this method may create a favourable environment for diseases. Also the costs for sprinkler are high and therefore some farmers have already introduced drip irrigation.

Once the irrigation inspector knows the total discharge to be delivered to the offtakes and/or branch canals, he will decide with the operator of the pumping station on the number of pumps to be activated. The water level in the main irrigation canal is observed with a water level gauge at the beginning of the canal[5]. The main objective is to maintain an almost constant water level along the canal reach. The irrigation inspector calls this level "a good level", but sometimes it is very difficult to get this level, especially when irrigation water is supplied to the second reach, where the water level controls are not operated in a hydraulic, but in a manual way.

The amount of water to be delivered to each branch via a sliding gate or a CHO structure is estimated by the irrigation inspector according to his experience and "good judgement"; because the measuring wells in these structures do not work properly or are out off service. In the case of a CHO structure, the irrigation inspector opens the upstream gate (if it exists) completely and then he operates the downstream gate by turning the wheel in such a way that the gate screw is displaced by 4 cm for each 60 l/s.

When sprinkler irrigation is used, which occurs in a large part of the RUT area, an additional amount of water has to be conveyed to the branch canals; otherwise the pumps would not have the required depth for submergence.

In general, it can be said that the operation of the system in the RUT area is more based on the criterion to establish the proper water levels in the network than the supply of the allocated discharges. In some cases, not all the farmers along a canal request for irrigation water; however the water to be supplied to the canal must result in the before mentioned "good level".

As a consequence of this water management practice a large amount of excess water[6] flows to the drainage system and results in an inefficient use of the irrigation water at considerable energy costs. This also explains why sometimes the pumping periods for irrigation as well as drainage coincide. On the other hand, it can be said that the measuring wells installed at the offtakes do not play an important role in the operation and delivery of the water and gradually they were considered as not needed.

The irrigation management for the water supply in the second reach is very similar to the one described before. The main objective is to create water storage in the reach between the water level controls at km 6+021 and km 11+220 and to maintain an almost constant water level ("good level").

In accordance with the requests for irrigation water and the experience of the irrigation inspector, he opens manually the gate of the level control structure located at the upstream end until the required water level has been reached. The inspector controls this level by extending or shortening the length of the chain, which holds the radial gate. Then he observes the water level in the reach three times per day and carries out the required adjustments.

The difference in the water management of the two reaches can mainly be explained by the impact of the timely availability of irrigation water. As mentioned

---

[5] Normally, this water level corresponds to an elevation of 918.20 m+MSL
[6] The irrigation inspector estimates this amount of water at 40 l/s

before, the irrigation method in the first reach is predominantly by gravity; the farm sizes are small, the main crop is vegetable and family labour is common. Moreover, a delay in the irrigation service will not cause major conflicts in this reach of the canal.

The situation in the second reach is different. In this area sprinkler irrigation is predominant and the main characteristics are that the farmers use pumping equipment with a high capacity, the labour is contracted and numerous (about 7 assistants/operator per equipment), the farm size is large, and mainly cash crops are grown. The water management is crucial for these conditions, because a deficient water delivery will cause inappropriate water levels in the reach or not the right water availability that will cause extra and high costs for the farmer, as he will have to pay the lost time to the assistants/operator of the sprinkler equipment.

In some areas of the RUT Irrigation District the agency can be confident concerning the support of the farmers for a number of water distribution activities. For example, when irrigation water is delivered and rainfall occurs or the sprinkler equipment faces damage, then some farmers close the delivery structure. Some direct offtakes are operated to fill reservoirs and in some cases the farmers show their responsibility by closing of the offtake when the reservoir has been filled. In other places the irrigation inspector cannot expect any collaboration from the farmers and he has to secure the flywheels with chains.

The management for the Interceptor Canal is simpler than the one for the main irrigation canal. When sector 5 needs irrigation water, the Tierrablanca pumping station will pump water from the river and the control structure number 3 at km 24+500 will be closed. In this way the water level will be raised and water will be stored in the upstream canal reach. Then, the farmers pump irrigation water with their high capacity pumps that are installed on the left bank of the Interceptor Canal.

When sector 3 requires irrigation water the control structures 2 and 1 will be closed, the water level will raise and a flow will be established towards the branch canals. Upstream of the control structures 1 and 2, a level control structure with a float has been installed to maintain a constant water level at the beginning of the branch canal. At the moment, chains are used to operate these structures and the local people obstruct them with stones and sticks in order to obtain irrigation water in a fraudulent way.

The irrigation management of Canal 1.0 is similar to that in the main irrigation canal. Here, the structure for the water level control consists of a long crested weir and a lateral sliding gate or CHO structure as offtake. In some branch canals, the offtakes are too high and for that reason a larger discharge than the one required is needed to get the appropriate water level in the canal. This practice causes excess water, which is lost at the canal end towards the drainage system.

The main drainage canal is also used for the supply of irrigation water to the farms under its command. As mentioned before excess irrigation water is needed to supply water to the irrigation network to get the appropriate water level for irrigation. In this way, excess water flows to the drainage system and results in water storage in the main drain when the outlet structure "Ceros" is closed.

Especially during the dry season and when irrigation water is available in the drain, the farmers take advantage of this situation to irrigate their crops by using pumps on the banks of the canal. The Cayetana pumping station also pumps water from the main drainage canal to the main irrigation canal to supply water to sector 4. When the water level for irrigation is too low in the main drainage canal, irrigation water is supplied by gravity from the Candelaria pumping station.

As mentioned before, the water quality in the main drainage canal is not favourable for irrigation in view of its high salt content. However, the farmers stated that one of the reasons to use this water was related to the fact that the water is immediately available for irrigation, which forms an enormous contrast with the slow flow that comes from the irrigation network. However, the continuous use of this water has caused soil salinization problems in the northern part of the RUT Irrigation District, which are reflected in the crop production and yields.[7] Moreover, the water storage in the canal does not help to maintain the groundwater table at an acceptable depth and in this way it will increase the salinization process by capillary rise.

From a drainage point of view, the main drainage canal drains water by gravity through the "Ceros" structure when the water levels in the Cauca River are low. Otherwise, the "Ceros" structure is closed, the water level in the canal increases and at a certain critical water level the Cayetana pumping station starts its operation.

When high rainfall occurs in the southern part of the RUT area (Roldanillo Municipality) the water flowing in the main drainage canal is drained by the Candelaria pumping station to the Cauca River and in this way it will avoid the long way up to the Cayetana pumping station and it will support its drainage function.

Management at Farm Level

Before the construction of the RUT Irrigation District and during the first years of functioning, the farmers used sprinkler irrigation by using small equipment financed by INCORA and 3 or 4 farmers used one equipment set. After the levelling of the land gravity irrigation was used on a small-scale.

Nowadays, sprinkler irrigation is predominant in 65% of the area and in 58% of the farms, gravity irrigation is practised on 21% of the area and in 34% of the farms, drip irrigation is found in 2% of the area and in 1% of the farms, and the hand method (called "mateo") in 1% of the area and 4% of the farms (Burbano and Forero, 1999).

Sprinkler irrigation is very common for grain crops (corn, soybean, and sorghum) in small and large farms; drip irrigation is used for citric and fruit trees, gravity method for sugar cane, although sprinkler irrigation is also used during the first stages; and the hand method ("mateo") is applied for fruit trees (passion fruit) in small farms.

When sprinkler irrigation is used the farmer installs his pumping equipment in the irrigation canal outside his farm, because he cannot use his farm canal that does not have the appropriate depth for pumping.

The average gross volume for irrigation is about 423 m³/ha; however during the field survey values from 576 to 864 m³/ha were found, varying according to farm area, capacity and number of sprinklers and irrigation time[8]. Although these values do not give sufficient evidence about the irrigation efficiency at farm level; some management aspects affect the irrigation efficiency. Among these can be mentioned:

-     Tendency of the assistants/irrigator to operate the equipment for more hours than are needed with their aim to work more hours. This means an inefficient water use, additional costs for the farmer, and conflicts with the irrigation inspector when he denounces this type of action;

---

[7] The irrigation inspector mentioned a crop yield reduction of 50 and 58% in sorghum and sugar cane respectively.

[8] For grain crops, a farm with an area of 6 ha and with 4 sprinklers with a total capacity of 60 l/s 5184 m³ were applied during 24 hours; while for 1 ha, and 20 l/s during 8 hours 576 m³ were applied.

- The assistants/irrigator deceives the farmer when they only irrigate the periphery of the farm plot. This fringe is then in a good condition, but the inside shows a very poor irrigated area;
- The assistants/irrigator demand payment for a minimum of 2 working hours, independently of the actual irrigation time.

In some canals where sprinkler equipment with high capacity is installed conflicts concerning the water use might raise between the farmers located at the head and tail end of the canal, because the first ones have a better opportunity to use the available water than the second ones. In the specific case of the sugar cane, the conflicts have resulted in the fact that the users themselves imposed distribution procedures by force.

When gravity irrigation is used the farmer allows the water to flow trough the farm outlet, and then the water is distributed over the plots. Some farmers pump irrigation water from a branch canal to a canal at farm level, from which water is supplied to the plots by gravity.

In some cases, the irrigation efficiency is expected to be low in view of the lack of land levelling and the large water depth applied during the irrigation[9].

The sugar cane crop in the RUT area has been developed by the so-called "Colonos"[10] or directly by the sugar cane factory (2 factories exist in the area). Siphons of 1.5 l/s or with a diameter of 2 or 3 inches, and sometimes even double siphons are used to convey irrigation water to the furrows.

Nowadays, the introduction of sugar cane has increased the water demand in some areas of the RUT Irrigation District in comparison to the earlier times when grain crops were predominantly grown. Large numbers of sprinklers (3 or 4) with high capacity (120 l/s) are installed along the irrigation canals in contrast with those used for the irrigation of grain crops (2 or 3 l/s).

Drip irrigation is mainly used for citric and fruit trees crops, normally on large farms that require high investments and modern irrigation technology. The agricultural production gets a high added value through agro-industrial processing (fruit cans, juice, fruit pulp) in view of the local, regional and national markets and international prospects. This irrigation method is common on the areas of the La Union Municipality and part of sector 5, where the Interceptor Canal supplies irrigation water to the fields. Here, the water quality is a limiting factor due to the contamination of the water by the wastewater disposal from the municipalities and settlements along the Interceptor Canal. On an average, the volume of irrigation water is about 121 m^3/ha (which is equivalent of 12.1 mm) and the irrigation time is 4 hours for an irrigation interval of 1 to 3 days.

Hand irrigation method called "mateo" has been the traditional method of small farmers with small plots for fruit trees especially passion fruit. Irrigation water is supplied plant-by-plant and put into small pits excavated around the tree. On an average, the volume of irrigation water is 367 m^3/ha and the irrigation time is longer than 10 hours.

---

[9] During the field survey 40 l/s were applied during 8 hours and during 2 days on 2 ha of tomato. The irrigation supply resulted in a volume of 2304 m^3 or in 115.2 mm, being 52.6 mm per day. This 52.6 mm is very high in view of the crop and the light soil with a high infiltration rate.

[10] A farmer growing sugar cane and the product is sold to the sugar cane factory. The farmer is landowner or tenant.

Organizational Aspects

The highest authority in the RUT Irrigation District is the General Assembly in which a total of 25 water users represent all the users, namely with 5 users for each sector. The General Assembly elects a Board of Directors consisting of a president and vice president, a general manager, a treasurer, fiscal, members, secretary, representation of the users, and of the Government Agency for Land Improvement.

There are four committees for technical, educational, training and financial aspects, who advise the Board of Directors. The organization of the RUT Irrigation District after the transfer consisted of an administration, a maintenance and conservation, and an operation department. The two last mentioned departments had changed so much that they were regrouped in view of a staff reduction. After the construction of the RUT Irrigation District, the operation department consisted of 12 irrigation inspectors, who used a bike for their transportation, then this number was reduced to 6 in 1978 and the inspectors used a motorcycle, and finally the number decreased to 4 in 2001. Figure 5.14 illustrates the organizational structure of the RUT Irrigation District.

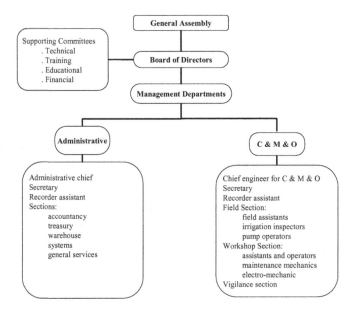

Figure 5.14 Organizational structure of the RUT Irrigation District

5.3.5 Socio-economic Aspects

In 1998, 8,892 inhabitants lived in the RUT Irrigation District. The population was made up of 62% adults; 46% men and 54% women respectively. About 58% of the population had followed a basic education, 32% a middle education and only 1.5% followed a higher education; about 10% of the population is illiterate.

With respect to public services can be mentioned that only 33% of the population has drinking water services, 57% is supplied by cistern and 10% by the district irrigation canals. Only 88% of the inhabitants have electricity service. 51% and 40% of the families use septic tanks and latrines, respectively as a system for their wastewater disposal; about 10% does not use any system.

Property Distribution

Since 1998 the distribution of the property size in the RUT Irrigation District has been stabilized. At the beginning of the RUT Irrigation District the area was conceived for small farmers according to the legal framework given by the Agrarian Reform Law, which stated that each family could receive 5 ha and that 50 ha was the limit for the maximum property size. Table 5.6 shows the present distribution of the property size in the RUT area. As can be seen 75% of the properties is smaller than 5 ha and covers 19.2% of the total area. About 22.7% of the properties have an area varying between 5 and 50 ha, which corresponds to 63.9 %, being the largest part of the area. Finally, 1.4% of the properties have a size between 50 and 100 ha, which corresponds to 15.9% of the area.

Table 5.7 shows the distribution of the property size in the different sectors of the RUT area. The largest concentration of properties (45.4%) is found in sector 2, where the average farm size is about 3.2 ha, the smallest concentration (9.7%) is found in the sector 3 and 5, where the average farm size is 8.4 ha.

Finally, it can be said that in the RUT Irrigation District the small (75% minifundio), middle (22.7%), and large farmers (1.5%) coexist next to each other. However, most of the land (63.9%) is owned by the middle farmers with a farm size from 5 to 50 ha; followed by the small farmers (19.2%), and the large farmers (16.9%).

There is serious evidence that due to economic and social factors[11] the property distribution and tenure were affected during the period 1986 – 1994. This phenomenon did not only affect the RUT Irrigation District, but the entire agricultural sector in Colombia.

From economic point of view, the so-called Economic Liberalization did not stimulate the agricultural activities due to the import of products with a lower price and the cost increase of the agricultural inputs. As a consequence, the agricultural sector did not grow during the decade of the 90's, but the agricultural area was reduced to 800,000 ha and 228,500 jobs were lost.[12] As a result, some farmers in the country stopped their agricultural activity and have chosen to sell their land, as in the case of the RUT Irrigation District they sold or rented their lands to the sugar cane cultivators.

From a social point of view, it can be mentioned that also the RUT Irrigation District has not been excluded of the impact of the drug trafficking. According to the National General Control Office[13], the groups involved in drug-traffic own about 4 million ha, which is equivalent to 48% of the productive lands of the country. This figure might be larger due to the fact that the official property registers do not show who the real owners are. It is very common that a person is used to play the role of a fake owner. The above-mentioned area belongs to 403 municipalities, which are

---

[11] It is the case of the Alejandria Farm as a result of the purchase and grouping of 18 properties. Before the 18 properties were cultivated with grains, now the Alejandria Farm is cultivated with sugar cane.

[12] Agricultural National Survey, 2001

[13] El Tiempo Newspaper (02-09-2003

located in the most fertile zones of several departments of Colombia; one of them is the Valle del Cauca Department. These illegal groups took advantage of the crisis within the agricultural sector and they bought the lands from the farmers for large amounts of money. In some areas, like in the coffee zone, the Antioquia, Tolima and Valle del Cauca Departments they bought the lands by using treats, pressure and violent.

Table 5.6 Distribution of Property Size in the RUT Irrigation District (ASORUT, 2002)

Area range (ha)	Properties		Users		Area	
	Number	%	Number	%	(ha)	%
0 - 5	1424	75.7	1147	75.2	1963.5	19.2
5 – 20	312	16.6	255	16.7	2993.5	29.3
20 – 50	114	6.1	99	6.5	3533.4	34.6
50 – 100	26	1.4	24	1.5	1619.8	15.9
> 100	1	0.1	1	0.1	104.9	1.0
Total	1882	100	1526	100	10215.1	100

Table 5.7 Distribution of Properties and Area per Sector (based on ASORUT, 2003)

Sector	Irrigable Area		Properties		Ratio (ha/property)
	(ha)	%	Number	%	
1	2900.6	28.4	459	24.4	6.3
2	2750.3	26.9	854	45.4	3.2
3	1637.1	16.0	183	9.7	8.9
4	1415.4	13.9	201	10.7	7.0
5	1511.7	14.8	185	9.8	8.2
Total	10215.1	100.0	1882	100	

With this strategy these illegal groups had as objectives to legalize their money, protect it against devaluation, and to get politic and economic power in their region. As a consequence the agricultural area was reduced, the land property concentrated and it increased the costs, and finally a forced displacement of farmers took place.

Some evidences of the above-mentioned impacts are also found in the RUT Irrigation District; among which can be mentioned:

-   Fictive owners and confusing property documents;
-   New owners of large areas who live in the main cities and are represented for ASORUT by their administrators;
-   Farm lands given up for recreational activities;
-   Conflict areas, where the water users feel themselves supported by an "important personage".

Agricultural Development

At the beginning, the potential area of the RUT project (16,712 ha) represented 4.4% of the flat area of Valle del Cauca Department with a land use of 60% grass, 29% annual crops, 6% permanent ones, and 5% other uses.[14] The construction of the RUT Irrigation District caused a large impact on the agricultural development of the region. According to the statistical data for the period 1985 –2003 the grain crops, cotton, fruit trees and lastly the sugar cane crop have been the dominant crops in the RUT area; some of them with added value through agro-industrial processing.

---

[14] National Agricultural Survey, 1958

Traditionally the grain crops have been grown in the area with different intensities during the study period. On an average, 34% of the area was cultivated with grains during the period 1985-1989, 56% in the period 1990 – 1996, and 35% in the 1997 – 2003; the national and international market policies explain the fluctuating behaviour of the grain crops. Soybean, corn and sorghum have been the grains planted traditionally in the area and the most important crops were soybean during the period 1985-1990, sorghum in 1991-1996, and corn in 1998-2003; the latter being helped by the so-called "productive chain"; which included harvest, industrial process, and marketing (agriculture by contract). Other characteristics of the grain crops are that the cropped area is increased during the second season of each year.

Cotton is only planted during the first season of the year according to the recommendations established for the growth of cotton in Colombia. The crop was important in the RUT area up to 1996 and covered an average crop area of 38%; after that year the area dramatically decreased to an average value of 3.8%, and it is an insignificant crop in view of the area for the year 2003. Cotton is rotated with grains for the second season of the year.

Grape has been a traditional crop in the RUT area, especially in the La Union Municipality where a factory exists to produce wine for the national market. However, the crop area is considered to be small; namely 4.3% during the period 1985-1992; 7% in 1993-1998, and 4.8% for 1994-2003. Fruit trees present a similar behaviour as the grape crop; 270 ha covering 2.5% of the area were planted during the period 1993-1996; while an average area of 500 ha that corresponds to 8% of total area was planted during the period 1997-2003. Grape and fruit trees offer a great potential for the agro-industrial development of the RUT area; however, the high investment and production costs, production technology, and (small-scale farmers) minifundio characteristics of the RUT are limiting factors for the development of these crops.

Grass appears in the RUT Irrigation District in 1993 covering 4.8% of the area, and it shows an important increase in 1998, the covered area is then 8%. Other crops like vegetables, melon, passion fruit, and papaya trees are grown in the RUT Irrigation District on a minor scale. Sugar cane crop presents a special case in the RUT area. On an average 63 ha, equivalent to 0.6% of the total area, were planted during the period 1993 – 1996, but in the period 1997 - 2003 the cropped area was abruptly increased to an average of 2,600 ha that corresponds with 30% of the total area. Economic and social factors, which caused the crisis for the agricultural sector are the expansions of the sugar cane in the RUT Irrigation District.

In general terms can be said that grains, sugar cane, fruit trees and grass are the main and dominant crops in the RUT area; of which the passion fruit, grape, papaya, guava, melon and citric are the most important fruit trees. From the year 2000, the average composition of the crops in the entire RUT area mainly consists of 39% grains, 28% sugar cane, 16.8% fruit trees, and 8.1% grass. Figures 5.15 and 5.16 show the crop distribution for the RUT area during the period 1985 – 2003 per yearly season and the growth of the cropped area for the sugar cane. On the other hand, the main crops mentioned above show a spatial distribution over the different sectors of the RUT area. Table 5.8 shows the spatial distribution of the cropping pattern for the year 2000 and the seasons A and B. From this table can be seen that the grain crops are dominant in the sectors 1 and 2 with 36% and 59% of the area respectively, sugar cane dominates in the sectors 4 and 3 with 37% and 40% of the area, and the fruit trees in sector 5 with 41.8% of the area. Grass crop is only important in sector 1 with 16.6% of the area.

Some farmers tried to explain the tendency for the grain crop and they said that it corresponds with a "easy culture", because the grain crop demands less efforts and investments and it allows the farmer to get money for his subsistence but not sufficient to cover the acquired costs such as the payment of the water fees. This situation is more common for small farmers than for the large ones.

Table 5.8 Spatial Distribution of the Cropping Pattern in RUT for the year 2000 (Based on ASORUT, 2003)

Crops	Season A						Season B				
	Sectors (% Area)						Sectors (% Area)				
	1	2	3	4	5		1	2	3	4	5
Cotton	7.8	1.6	35.1	2.4	7.6	Sugar Cane	23.2	10.0	37.4	40.3	18.2
Sugar Cane	23.1	10.9	14.8	40.2	19.9	Grass	16.6	6.1	1.2	5.1	8.2
Grass	14.6	6.0	9.0	5.5	8.6	Grains (*)	36.0	59.0	34.7	33.7	10.7
Grains (*)	35.5	54.4	24.0	36.0	14.6	Fruit Trees	6.7	12	14.4	7.3	41.8
Fruit Trees	7.8	13.0	7.5	8.2	44.6						

Season: A
  - Grains: corn, sorghum, soybean
  - Fruit Trees: Sector 1: passion fruit, grape; Sector 2: passion fruit, grape, melon, guava; Sector 3: passion fruit, grape, melon, papaya tree; Sector 4: passion fruit, grape, guava; Sector 5: passion fruit, grape, melon, guava, papaya tree, soursop

Season: B
  - Grains: corn, sorghum, soybean
  - Fruit Trees: Sector 1: passion fruit, papaya tree, grape; Sector 2: passion fruit, grape, citrus, guava; Sector 3: passion fruit, papaya tree, grape, guava, melon; Sector 4: passion fruit, grape, guava; Sector 5: passion fruit, papaya tree, grape, melon, guava, soursop

Table 5.9.1 shows the net income per ha for the main crops in the RUT Irrigation District. As can be seen a large difference exists between the grain crops and the fruit trees. However, the fruit trees demand an investment greater than the grains and this presents a limiting factor for the small farmers given their small economic resources together with a low technology and small land area (minifundio).

Table 5.9.1 Profitability Indicators of the Main Crops in the RUT Irrigation District, year 2002 (Based on ASORUT, 2003)

Crop	Yield	Value (Col$/ha)		
	(kg/ha)	Production cost	Production value	Net income
Sugar cane	105,000	1,545,389	3,150,000	1,604,611
Grain crops:				
- Soybean	2,500	1,584,812	1,700,000	115,188
- Sorghum	6,500	1,601,309	2,275,000	673,691
- Corn	6,500	2,499,309	2,990,000	490,691
Fruit trees:				
- Grape	16,000	7,820,993	12,800,000	4,979,007
- Passion fruit	25,000	7,835,392	12,500,000	4,664,608
- Melon	40,000	10,257,948	20,000,000	9,742,052
- Papaya tree	120,000	32,101,000	72,000,000	39,899,000
Cotton	3,000	2,807,797	4,200,000	1,392,203

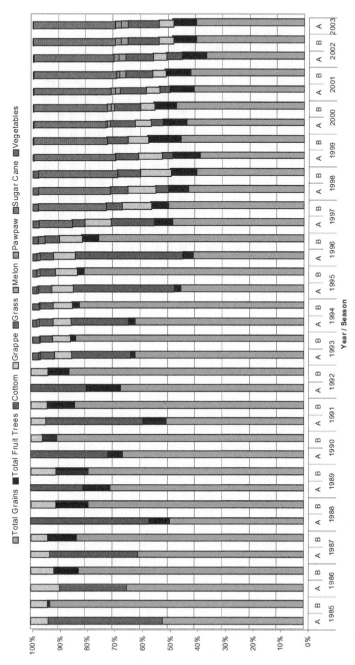

Figure 5.16 Crop distribution in the RUT Irrigation District for the period: 1985 – 2003

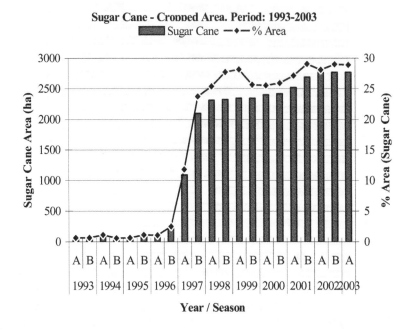

Figure 5.17 Cropped area evolution with sugar cane in the RUT Irrigation District.

Table 5.9.2 (cont) Profitability Indicators of the Main Crops in the RUT Irrigation District, year 2002 (Based on ASORUT, 2003)

Crop	Area (ha) – Season		Net Income (Col$)	
	A	B	$/ha	Total
Sugar cane	2767		1,604,611	4,439,958,637
Soybean	487	390	115,188	101,019,876
Sorghum	743	810	673,691	1,046,242,123
Corn	2161	2456	490,691	2,265,520,347
Grape	523		4,979,007	2,604,020,661
Passion fruit	446		4,664,608	2,080,415,168
Melon	229		9,742,052	2,230,929,908
Papaya tree	221		39,899,000	8,817,679,000
Total	7577	3656		23,586,000,000

Table 5.9.2 clearly shows that the RUT area produces on an average Col$ 2,099,687 per ha per year. The table also clearly presents the major impact of sugar cane and fruit trees on the total production of the RUT Irrigation District.

The lack of mechanisms for marketing regulation is a serious constraint threatening the present high profitability of the fruit trees. Overproduction and the lack of storage centres cause an abrupt price fluctuation of some products. The case of the papaya tree can be given as an example; the price has varied up to 60% due to an excessive offer.

On the other hand, the middlemen affect the profitability of the crops and play an important role in the production process. In spite of the presence of credit entities, the farmers do not request this service in view of the many requirements set by the entities. For this reason, they go to the middleman, which supplies them agricultural inputs and establishes the selling price of the products. At the end of the season, he discounts the value of all given inputs from the production value, and sometimes the profit margin for the farmer is very low.

Community Participation

Lack of "sense of belonging" is the most common criticism of the users of the RUT Irrigation District. Some historic reasons explain this situation:
- The farmers considered the RUT Irrigation District as a project imposed by the government; in view of this idea the community even opposed the construction works which resulted in some unfinished works (for example the Canal 1.0);
- Discomfort of the some users due to fact that their farms were divided by the canal alignment during the construction stage;
- Insufficient or low payment or non-payment at all of the farmers for the land acquired by the government for the construction of the main works (canals, road, etc);
- Distrust and lack of credibility for the investment project of the government;
- A paternalistic culture by the government, which developed for a long time the implementation of the land improvement projects and assumed all the financial and management responsibilities.

The users expressed their low "sense of belonging" through the lack of participation in the decision-making processes, by breaking obtained compromises (payment of bills, maintenance of canals), and rejection of improvement programs and projects proposed by the administration staff.

The users consider the participation in meetings as a loss of time. When they attend a meeting they do not participate in an active way due to their feelings of discomfiture or fear and they prefer to express their opinions in small groups outside the general assembly. In view of these observed difficulties, the representation of the water users to the general assembly is carried out by 5 users per sector, with a total of 25 users for the whole area.

The RUT Irrigation District is located in an area that is covered by the influence and activities of educational (universities, agricultural technical school) and their supporting institutes from 5 departments surrounding the RUT Irrigation District. These institutes have developed several supporting, development, and research activities, but in an intermittent and non-coordinated way that most of the time resulted in poor outcomes that have a very small impact on the RUT Irrigation District. For this reason, the users are reluctant to follow any training and improvement program and they say to be tired of them. However, it is important to recognize that some farmers in the area have developed successful productive agricultural systems based on their own experience.

ASORUT is seen as a bureaucratic entity by the user, which inherited the administrative style from the old government agency (Himat, later INAT). Even for the user it is not clear how the administration of the district works and which activities belong to the responsibility of the ASORUT after the transfer process. This and the ignorance about the functioning of the district complicate the relationship between the water users and the administrative staff that results in permanent conflicts.

In relation to the gender aspects there is not evidence of a wide participation of women in the decision-making processes within ASORUT. However during the development of the Participative Diagnosis Workshop a women potential group was identified, which included different educational grades (from professional to technical) and they showed a high sensibility for social and environmental aspects.

A population sample of 91 water users showed an age distribution as follows: 11% between 20-35 years, 29% between 36-50, 44% between 51-60, and 17% between 66-80. A great worry exists in the RUT area in view of the fact that there is not a change in generation or a new generation of young farmers to develop the agriculture in the future. Taking into account the low profitability of the agricultural activities during the past years, most of the farmers are sending their youngster to the main cities to study a profession that is not related to the agricultural sector.

Financial Aspects

After the transfer process the subsidies from the government side were dismantled and the costs for operation and maintenance had to be covered by the collection of water fees. The revenues of the RUT Irrigation District are based on the water tariff and the user's fund. The first one is composed by the fixed and volumetric tariff. Fixed tariff corresponds on the payment that each user must reimburse per improved area ($/ha), while the volumetric tariff is related to the water volume used for irrigation.

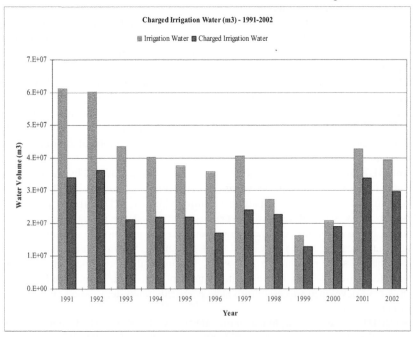

Figure 5.17-Charged irrigation water in the RUT Irrigation District (Based on ASORUT, 2003)

The volumetric tariff represents a crucial point for the financing of the RUT Irrigation District. The irrigation water volume delivered to the users is difficult to measure with a certain degree of accuracy given the lack of measurement structures at farm level. For users with sprinkler irrigation the water volume is determined according to the capacity (l/s) of the equipment and the working hours (delivery time). Conflicts with the irrigation inspectors are related to the estimation of the working hours because some users damage the hour's meter of the equipment. Some farmers declare to the irrigation inspector a certain number of working hours but in practice that number is much greater.

For gravity irrigation the water volume is determined according to the opening of the delivery gate, the water level in the supply canal, the delivery time and the experience of the irrigation inspector to estimate the discharge. In the case of sugar cane sector the volume is based on the used siphons according to their number and capacity (l/s). Figure 5.17 shows the charged water volume for the period 1991 – 2002.

In general, the determination of the delivered volume of water depends on the experience of irrigation inspector, and in case of conflicts with the user an underestimated value of the discharge might be expected. In these circumstances and in view of the characteristics of the water distribution based on pre-established water levels in the supply canals, it is expected that the charged water volume is smaller than the pumped one; causing a negative impact on the revenues of the RUT Irrigation District.

During the last two years the debt of the water users by tariff payment has been increased abruptly and has become about 8% of the total budget; this might be due to the crisis in the agricultural sector, the low profitability of the crops, and the low financial capacity of users. Some of these reasons explain the present situation, which will threat the financial sustainability of the RUT Irrigation District.

The level of the tariffs is considered as too low, and season after season it is intended to establish a more realistic structure for the tariff founding and to decrease the constraints from users side due to their low financial capacity. In this way, low tariffs and low collections have caused a shortage of the economic resources for operation and maintenance of the hydraulic infrastructure of the RUT Irrigation District, causing its deterioration, impacting the water distribution, and conflicts among the different involved actors.[15] Table 5.10 shows the evolution of tariff value for the period 2000 – 2003. Table 5.11 shows the farm size used as criterion to establish the value of fixed tariff for the year 2003.

Some users do not pay the bills arguing that they do not use water for agricultural activities. In some places the delivery of a bill causes auger for the user. The Candelaria sector is characterized as a conflict area; an area that is constituted by small farmers (minifundio) that cultivate at small-scale grains and fruit trees. There, the bills are not paid, the farmer avoid the contact with the irrigation inspector, and irrigation water is taken for a long time but they declare less time, and in conflict period they bring up the support from an "important personage" of the area.

Another way of non-counted water is that taken by users when they have not asked for water request. These users wait for water supply to a user located at the downstream end of the canal, which has requested for a water supply and is punctual with his

---

[15] Some farmers say that the progressive deterioration of the infrastructure is a way to pressure economic resources for rehabilitation from the government. Others say that if the RUT Irrigation District is completely deteriorated or collapsed, the government will take it over again

payment. When irrigation water is running in the canal the water is taken in a fraudulent way.

Table 5.10 Evolution of the Tariff Value in the RUT Irrigation District (Based on ASORUT, 2003)

Year	Tariff Value (Col$)	
	Fixed ($ / ha - semester)	Volumetric ($ / m³)
2000	62,172	16.60
2001	65,281	17.43
2002	72,189	19.17
2003		21.00

Table 5.11 Fixed Tariff based on the farm size; year 2003 (Based on ASORUT, 2003)

Area range in ha	Fixed ($ / ha - semester)
0 – 2	35,905
0 – 2 (sector 5)	25,133
2 – 5	71,809
2 – 5 (sector 5)	50,266
5 – 15	73,622
5 – 15 (sector 5)	51,536
15 – 40	73,719
15 – 40 (sector 5)	51,603
> 40	75,888
> 40 (sector 5)	53,120

5.4   Transfer Process and Impacts

The second phase of the transfer process in Colombia started with the transfer of the RUT Irrigation District in 1990, more than 15 years after the transfer of the Coello and Sandaña Irrigation Districts in 1976 (phase I). The transfer process took only 3 months, during which the stages concerning promotion, evaluation of district conditions, preparation of supporting plan, negations between agency and water users association, and signing of the administration contract were carried out. The transfer process took a short time due to the fact that some of the essential activities were not implemented; amongst them can be mentioned:
- Creation of the water users association, because it existed in some form from the beginning of the RUT project;
- Training of farmers' representatives and management staff;
- Review of the operation and maintenance procedures and plans;
- Repair or improvement of the secondary network;
- Participation of farmers in the decision-making concerning the prioritization of the improvement works;
- Any investment from the farmers for the improvements was not required;
- Decision about who will be responsible for the future rehabilitation works.

The transfer process was mainly focused on activities concerning the organization of the existing water users association, consideration of the water charges, and repair or improvement of the intake and/or main canal. During the transfer process, farmers agreed to take over the management and to give up the government subsidies, which were between 60 to 80%. Most of the subsidies were related to the relatively high costs of pumping. Four years before the transfer took place the main pumping stations were completely rehabilitated, and the negotiators of the government agency convinced the water users that the system had been improved. Therefore it was not necessary to include this item in the negotiation agenda. However, it would appear that the farmers underestimated the future energy costs and decided to give up the governmental subsidies.

Another element of the transfer negotiation dealt with the future of the agency's staff. Farmers agreed to keep most of them, although a reduction of about 10 to 12% took place. Finally the negotiations dealt with personnel training prior to or immediately after transfer. However, since most of the district personnel continued to work in their jobs, the training issue did not carry much weight and very little training was given as part of the transfer process.

Amongst the powers delegated and the functions transferred to the water users association, ASORUT can be mentioned:
- Legally recognized entity with the authority to manage the water diverted to a defined service area;
- Non-profit, quasi-municipal entity with legal rights of way for the irrigation canals, drains and structures;
- It has the right to select its own leaders, to make rules, and impose sanctions up to the maximum penalty of fines for damages to property, cessation of the water service, and taking the violator to court;
- Authority to make operation and maintenance plans and budgets, and to set water charges. Initially, this function was shared with the government agency, but after the Land Improvement Law 41 of 1993 was enacted this function falls under the total responsibility of ASORUT;
- Authority to hire or release management staff;
- Right to make contracts and to raise additional revenues, but not for profit;
- No right to make profit.

The transfer process was a main component of the government privatization policy, which included amongst others the elimination of all agricultural subsidies. This resulted in a sharp reduction of the profitability of crop growing. The water users claimed that they cannot be any longer self-sufficient in the management of the system; and the RUT's administration requested the government for the restoration of the subsidies. Although the government did not comply with this request, it allocated some funds to ease the system's burden due to the energy cost.

Nowadays, outstanding bills have been increased; many users who were paying their debts on time keep falling behind because of the unpaid interest fee of the past years. This situation has caused the water users to request the government to take over again the management of the irrigation district. The administration of the irrigation district has tried to cut down costs. However, old energy subsidies have been eliminated and the irrigation system no longer gets special rates. There are, however, some different energy rates depending on the time of the day. At present, the peak-hour periods exist during 6 hours per day (10-13 and 18-21) during which the operation of the pumping stations is stopped in view of the higher energy cost. However, the low electricity rates very seldom coincide with the pumping needs of the system.

After the transfer several changes in management practices were introduced in order to increase the efficiency, especially by reducing the expenditures. This concern resulted in various actions such as the simplification of the organization structure, merging or abolishing positions, reducing staff, increasing work hours and service areas, streamlining procedures for fee payment, and introducing computerized management information systems. In relation to the staff reduction, the RUT Irrigation District decreased 3 years after the transfer the number of staff by 18%, causing also an increase of the service area by 18% per staff member. At the moment, merging sections has caused another staff reduction. Especially the operation, maintenance and preservation sections were merged and the number of irrigation inspectors has been reduced from 6 to 4.

The International Water Management Institute, IWMI, carried out a research to determine the impacts of the irrigation management transfer in 5 irrigation districts in Colombia including the RUT Irrigation District. The study period (1987-1995) included 4 years before and 5 years after of the transfer year (1990). The results of this study (Vermillion and Garces-Restrepo, 1998) in relation to the RUT Irrigation District are summarized as follows:

Farmer Perceptions about District Performance Before and After Turnover

For most of the farmers of the 5 irrigation districts, the most significant changes were not related to the district performance. On an average, 82% of them did not detect major changes in aspects related to adequacy of the water delivery, fairness of water distribution, adequacy of maintenance, financial management, overall administration, and effectiveness of communications. It could be said that 30% of the farmers in the RUT Irrigation District felt that the overall administration has been improved after the transfer, and they were equally split on the topic whether the communication between the farmers and the district staff was effective or not;

Cost of Irrigation to the Government and the Farmers

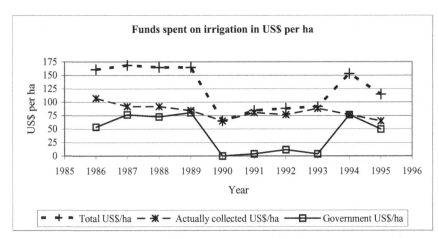

Figure 5.18 Funds spent on irrigation in the RUT Irrigation District (Adapted from Vermillion and Garcés, 1999)

Before the transfer the average irrigation budget was about 56% of the total budget contributed by the farmers (US$ 93 per ha), while the contribution of the government was about 44% (US$ 74 per ha). The total budget was US$ 167 per ha. After the transfer and during the following 5 years, the average contribution of the farmers was increased to 87% of the total budget (US$ 77 per ha) and for the government decreased to about 13% (US$ 14 per ha). However, during the period 1991-1994 the average contribution of the government was 6% and increased to 40% in the year 1995 when the RUT Irrigation District fell in a serious crisis and the government allocated some additional funds. From these data clearly follows the impact of the transfer on the reduction of government expenditures for the recurring irrigation budget.

On the other hand, the average total irrigation budget decreased about 46%, namely from US$ 167 per ha before to US$ 91 per ha after the transfer. During the 5 years following the transfer the contribution of the farmers varied from US$ 83 to US$ 65 per ha, a reduction of 22%; this as a consequence of the change in the tariffs as determined by the administration in response to the elimination of the government subsidies.

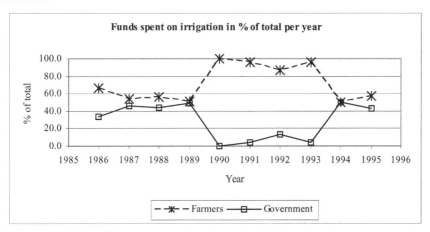

Figure 5.19 Division of funds spent on irrigation in the RUT Irrigation District (Adapted from Vermillion and Garcés, 1999)

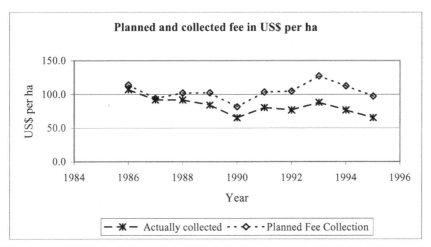

Figure 5.20 Planned and collected fees from the farmers in the RUT Irrigation District (Adapted from Vermillion and Garcés, 1999)

Fee Collection Rate

The fee collection rate is the ratio between the collected water tariffs. The charged tariffs and the average fee collection rate was 91% before the transfer and 70% after it.

It showed also a decreasing tendency during the following 5 years, namely it was about 67% in 1995. It seems that this tendency has been maintained because now the RUT Irrigation District has a very high debt from the farmer's side that requires a great effort from the administrative staff to reach payment agreements with the farmers.

Operation and Maintenance Budgets and Financial Self-sufficiency

The total variation in the costs for operation and maintenance before and after the transfer was about 42%. Before the transfer the operation and maintenance cost was stable at US$ 163 per ha and dropped by 62% to US$ 63 per ha at the moment of the transfer, and then it was increased by 51% to US$ 95 per ha for the year 1995. On the other hand, the total variation in the financial self-sufficiency[16] was about 35% between 1985 and 1995. During the period before the transfer the self-sufficiency was maintained at a stable value of about 51%, next it has been increased to 110% at the moment of the transfer, and then it decreased to 69% in the year 1995. Here, the operation and maintenance cost was higher than the fees collected and the government had to provide financial assistance to the district.

Quality of Operations

The irrigation water supply supplements the effective rainfall during all the months of the year. The Relative Water Supply (RWS)[17] and the Relative Irrigation Supply (RIS)[18] are indicators, which give information concerning the availability of irrigation water to satisfy the actual crop demand. In the RUT Irrigation District values larger than 2 are observed, which suggest a poor water management meaning that more water is pumped to the district than what is required for the irrigation of the crops. After the transfer the RIS values showed a decreasing tendency due to the high pumping costs and the poor financial situation caused by the elimination of governmental subsidies and the decline in revenues from the water charges.

The Capacity Utilization Ratio (CUR)[19] gives an indication of the extent to which the canal delivery capacity is being attained. The RUT Irrigation District showed a gradual improvement of this ratio until 1993; at that time the CUR reached a value of 1.25. The rise was mainly related to the shift to the cultivation of crops with higher water requirements. The decline after 1993 was apparently related to the decline in RIS in recent years that was caused by the financial crisis mentioned before.

---

[16] Financial self-sufficiency is the ratio between collection of the fees and the operation and maintenance budget.
[17] RWS is the sum of irrigation supply (IS) and total rainfall (TR) divided by crop water requirement (CWR).
[18] RIS is irrigation supply (IS) divided by the difference between crop water requirement (CWR) and effective rainfall (ER).
[19] CUR is the ratio of the annual peak demand for irrigation to the maximum canal carrying capacity.

From this point of view, the transfer of the management itself did not bring any clear and significant change in the quality of the irrigations operation for the RUT Irrigation District.

Impacts on the Sustainability of the Infrastructure

A functional structure is defined as a structure that can presently perform its basic design functions and that shows no sign of losing this capacity within about one year. About 50% of all the structures in the RUT Irrigation District has been inspected and 82% has been found in a functional condition. For the canal reaches a dysfunctional condition means that the reach is unable to convey at least 70% of the desired capacity, while nearly dysfunctional means that the canal reach is most likely to become dysfunctional within one year. In the RUT Irrigation District about 10% of the canal length has been inspected and about 83% and 17% has been found in a functional and nearly dysfunctional condition respectively. The study determined that an increase of 2% of the current annual maintenance expenditures was needed to cover the total accumulated or deferred maintenance requirement. This increase did not include the condition of the pumping stations in the area, which will present a significant additional cost requirement. In view of these observations it seems that the maintenance appears to be sustainable in the RUT Irrigation District from the point of resource mobilization.

Impacts on the Productivity

In the RUT Irrigation District, the gross value of the output (GVO) per unit of land (US\$/ha) did not change significantly and it kept an almost stable value after the transfer. The value of the output per unit of water (US\$/m^3) has been improved after the transfer; this in turn can mainly be attributed to the decline in the RIS, which has been influenced by the financial pressures caused by the transfer (smaller water orders due to the higher pumping cost). The increase in the output value was also influenced by a shift from traditional to higher value crops such as sugar cane.

5.5   Conclusions

The RUT Irrigation District is one of the most important irrigation systems in Colombia in view of the very suitable characteristics of its available natural resources (climate, soils, water) for the development of the agriculture; the strategic geographical location of the district, and the potential for further agro-industrial development in view of national and international markets.

Nowadays, the RUT Irrigation District shows a growing crisis, which is a serious threat for the sustainability of the whole system, when the necessary measures are not taken on short notice. External and internal factors; with their social, economic, environmental, technical, and institutional characteristics have contributed in different grade to this situation.**External Factors**

The control of the external factors is not only a matter of the RUT's community, but also the RUT's administration can undertake the necessary actions together with the local, regional and national institutes to formulate and implement plans to control most of these external factors and to minimize their impact. Amongst the external factors can be mentioned.

Management Transfer

The government's policy to transfer the management of the irrigation system to the water users association has resulted in large benefits for the government, because its participation in the coverage of the irrigation costs was highly reduced after the transfer. In the RUT case, the government eliminated all the subsidies, and as a consequence the RUT's administration has to increase the tariffs for the provision of the irrigation water service.

In response to this fact, the water users did not pay their water bills and demanded less water, which caused an increasing reduction in the fee collection and put in danger the financial self-sufficiency to cover the management costs. About 13 years after the transfer, the water users are not paying the water tariff and they have a high debt with the RUT Irrigation District that causes an increase in postponed maintenance of the hydraulic infrastructure and has a negative impact on the supply of irrigation water. A large part of the users did not understand the meaning of the transfer and they still believe that the Government is in charge of the management of the district through the Board of Directors of the Association.

Economic Liberalization of Colombia

This event did not stimulate the agricultural activities due to the import of products with a lower price and due to the increase of costs for the agricultural inputs. In response, some farmers stopped their agricultural activity, rented their lands to sugarcane cultivators, sold their farms and in some cases it resulted even in the presence of illegal groups.

Environmental Aspects

The Cauca River is the main water source for the RUT Irrigation District. However, the water quality, from a physical and chemical point of view, has been deteriorated due to the disposal of wastewater from the cities along the river course in the Cauca and Valle del Cauca departments. Moreover, the basin of the Cauca River and its tributaries are seriously deteriorated by deforestation, which has caused a high sediment load in the Cauca River.

A similar situation occurs along the Interceptor Canal where wastewater from the municipalities along the canal is discharged into the canal; moreover, a high sediment load transported by the small streams coming from the Western Mountain Range is deposited in the canal. As a consequence, the costs for operation and maintenance are increased and the water quality is a serious constraint for the marketing of the agricultural products.

Internal Factors

The internal factors are related to the quality of the operation to provide the irrigation water service to the users and to the relationship between the different actors involved, the organizational and financial aspects and community participation, and the environmental impacts as a consequence of the agricultural practices.

Operation of the System

Farmers request at any irrigation water due to the fact that planting plans and irrigation schedules do not exist in the RUT Irrigation District. This situation has caused an inadequate operation of the system; under these circumstances the pumping stations become very inefficient and require a high-energy consumption, which has a negative impact on the total operation cost.

Hydraulic Infrastructure

The RUT Irrigation District is more than 30 years old. A rehabilitation program is required for the irrigation and drainage network in order to reduce the water losses from seepage and leakage. The hydraulic control and regulating structures have to be improved in order to optimize the water distribution to the users.

Cropping Pattern

During the last decade the cropping pattern in the RUT Irrigation District has changed. The introduction of cash crops like fruit trees and sugar cane has caused several abrupt changes in the water management and present structure of the water management needs to be adapted. Fruit trees require irrigation with shorter irrigation intervals, while the sugar cane demands larger amounts of water than the amounts required by the traditional crops.

Financial Aspects

The low profit from the agricultural activities is one of the reasons why the farmers cannot pay their water bills and why the fee collection rate during the past years has decreased dramatically. Most farmers cultivate grain crops, which demand less effort and investment expenses than cash crops; however the profitability of the grain crops is much less than that of cash crops like grape, passion fruit, melon and papaya. Especially the farm size, minifundio condition, low technology, and low payment capacity for credits are some of the major constraints for the small farmers, which prevent them to cultivate highly profitable crops.

Social and Organizational Aspects

It appears that one of the crucial elements that contribute to the RUT's crisis is the lack of any sense of belonging of the water users in the district. This situation has some historical explanations, amongst others the fact that the farmers still consider the RUT Project as an imposed project, even so much that they opposed in the past the construction of some works.

This lack of sense of belonging was even more evident after the transfer of the management when the water users association received the responsibility for the district management. At many occasions the water users have shown by their lack of participation in decision-making, a low interest to take part in any of the development programs, no follow-up of compromises obtained; several conflicts with the district's staff, an unreasonable use of the hydraulic infrastructure, and low credibility to the RUT's administration. These are all examples of the lack of interest, attachment and cooperation, the lack of any sense of belonging.

Environmental Aspects

Given the natural geomorphologic and environmental characteristics of the RUT area, the area is very vulnerable to the effects of salinity and a shallow groundwater table. After the construction of the RUT Irrigation District this vulnerability has been put at risk by agricultural practices like an inefficient use of irrigation water at farm level, a poor performance of the drainage system, lack of land improvement at farm level, and use of drainage water with high salt content for irrigation. These facts have contributed to the deterioration of the water and soil resources and are threatening the environmental sustainability of the RUT area and have caused a low profitability of the agricultural activities.

In addition to the external and internal factors above mentioned the following considerations play also an important role and should be taken into account at the moment to consider an improvement proposal for the RUT Irrigation District.

The association ASORUT has mainly focused its activities on the operation of the system for the distribution of irrigation water to the users. For the present conditions within the district, ASORUT has to play a more inspiring role, for example by taking the leadership for the development of integral programs to achieve improvement of the socio-economic conditions of the farmers based on the profitability of their agricultural activity. In this way, programs involving the strengthening of the organization, rehabilitation and modernization of the hydraulic infrastructure, development of productive projects including all phases of the production chain, environmental protection, and the representation before local, regional, national, and international institutions should be considered as part of its new role.

Two types of farmers co-exist within the RUT area, namely the small and middle farmers and the larger ones. Each one of them has its well defined, but different socio-economic and cultural characteristics. A future process to improve the economy, based on the development of a profitable agriculture must take into account this difference in order to reach a harmonious and fair development for both the two sectors. Moreover, a new and wide participation approach is needed to involve all the different actors, to achieve a well-supported formulation of the proposed interventions and to make decisions in such a way that they will be profitable and have a large chance for a successful management and implementation.

Nowadays an irrigation system is considered as the result of a historic and social process (Sabatier and Ruf, 1991 cited in Apollin and Eberhart, 1998), and not only as the result of technical interventions. Therefore, the physical design of an irrigation system must reflect all the conditions of the social environment. It is also the result of physical interventions, crisis, conflicts, agreements and consensus; and socio-economic and agricultural changes that might cause a malfunction of the system

From a hydraulic point of view, the RUT Irrigation District has been designed to be operated as a gravity system under the assumption that the farmers would adapt their irrigation practices to these gravity conditions. Finally, the farmers chose their irrigation practices in a different way and some of them use sprinkler methods. For sprinkler irrigation the network has to work with other and higher water levels than those established in the design phase. As a consequence the irrigation system must convey much more irrigation water than that is needed by the crops and it result in extra water losses and inefficient water use. This example clearly shows the impact of social requirements on the functioning of the system, especially when the technology selection does not take into account these social aspects.

6

Implementation of the Framework in the RUT Irrigation District

In order to formulate a proposal for an integrated and sustainable management plan for ASORUT an analytical framework for integrated water resources management (van Hofwegen and Jaspers, 1999) was adapted. For the case of ASORUT the analytical process was mainly concentrated on the assessment of the present situation, the formulation of the desired management situation and the required interventions. These activities were done by well-prepared interviews with farmers, administrative, technical, and field personnel from ASORUT, representatives of the municipalities and representatives of public and private institutions related to the land improvement sector.

In view of an integrated and sustainable management plan for ASORUT two workshops have been held and were carried out with the broad participation of all the actors involved. The first workshop has been held with the purpose to diagnose the current situation of ASORUT, to identify and analyze the different factors influencing its present critical situation and to classify future trends for an integrated and sustainable management. The aim of the second workshop has been the presentation of and discussions concerning the proposal for the most needed integrated and sustainable management and the formulation of the required interventions. The interviews and workshops were the main tools to obtain an integrated, participatory, bottom-up approach. Finally, during this process in which all the actors participated the necessary steps have been taken to come from the identified "present management situation" to a "desired integrated and sustainable management situation". As a result of this stepwise process, the major elements and concepts were obtained for the formulation of a few scenarios for an integrated and sustainable management and for the identification of the required interventions.

6.1   Evaluation of the Present Situation

At the moment, the RUT Irrigation District is in a very critical condition due to the impact of several, sometimes combined factors, which effects were evident at various moments during the lifetime of the irrigation district. In order to understand the current situation of the RUT Irrigation District it might be constructive to present a short historic account of the development process of the irrigation district, all the way from the start until the present time. This historic explanation will take into account the fact that the present state of the irrigation system is not only the result of technical designs and interventions, but also the result of historic and social processes. Moreover, this state is the final result of crises, conflicts, agreements and consensuses and socio-economic and agricultural changes that have resulted in the malfunctioning of the system (Apollin and Eberhart, 1998).

From the start, the development of the RUT Irrigation District has met a strong opposition from the local community that has considered the implementation of the irrigation infrastructure as a project imposed by the National Government. However, the intervention of the Government corresponded with the, at that time typical and internationally accepted policy for land improvement projects. During the 1960s and 1970s this type of projects was developed with minimal participation of the local community and under almost total responsibility of the Government. This meant that the implementation and development of land improvement projects were considered completely as mere physical interventions limited attention to the social, cultural, and economic requirements within an acceptable social context.

In this way, the RUT Irrigation District was developed with a very small sense of belonging from the farmers, which considered the project mainly of interest for the Government. Therefore it was according to them the responsibility of the Government to carry out all the administration, operation and maintenance activities. The impact of this attitude was not serious during the administration of the district by the Government (the period 1971-1989), because the government agency assumed that it was in charge of the whole management of the district that included the operation and maintenance with financial resources provided by the government and based on a paternalistic policy for the land improvement sector. The low sense of belonging of the farmers was increased by their discomfort due to the negative impact that was caused by the canal alignment on their farms, insufficient payment for the land acquired by the Government for the construction of the main works, doubts about and lack of credibility of the investment project, and the paternalistic approach of the Government during the development of the land improvement projects.

The period after the transfer in 1989 marked a new stage for the RUT Irrigation District. After the transfer, the Water Users Association ASORUT became fully responsible for the administration, operation and maintenance activities covering all the associated costs without the participation of the Government, which eliminated all the subsidies. The transfer took a very short time (3 months) and it was mainly focused on activities related to the organization of the water users association, on considerations concerning the water charges, and repair or improvement of the intake and/or main canal.

More than 14 years after the transfer the farmers still mention that this transfer process has not been carried out in a systematic and gradual way allowing them an appropriate adaptation to and training for the new responsibilities of the organization. Amongst the farmers a feeling exists that the transfer process was developed too rapid without sufficient assistance and support from the Government. The farmers believe that the Government was more interested in an as-fast-as possible decentralization process of the management of the irrigation district with the purpose to satisfy the compromises obtained from the international financial entities for the land improvement sector. They believe that only training was given in some, merely routine aspects for administration, operation and maintenance of the district, for which it was assumed that the farmers had already a wide experience and for this reason the Government also considered that any additional training was not necessary.

The critical situation that ASORUT faces today can be understood and interpreted as a logical result of the historic development of events, especially those linked with the delegation of the administrative activities to ASORUT through the policy of management transfer. As a result, the administration changed from an authentic governmental administration to a private management, in which the users via their organization are fully responsible.

From an analysis of the role and the performance of the water users organization after the transfer process, it can be concluded that the management of the district by the different administrative entities has been based upon a management style that is similar to the one carried out previously by the governmental agency. The main focus of the present and previous management has been the service to provide irrigation water, from which they could obtain the revenues (water tariffs) for the sustenance and financing of the organization. The present management organization has acquired and inherited this specific management aspect from the government agency. By taking this aspect as a reference point for the management, the users in the district believe that the irrigation transfer process brought only a change of actors within the management. However, the main difference is that before the transfer the public entity could count for all the necessary technical and financial resources on a paternalistic and subsidiary Government, while the private management organization after the transfer can only count for its financial resources to cover the costs for administration, operation and maintenance on the water charges to be collected from the water users. After a few years, the financial resources were insufficient and even depleted as they were affected by a low paying capacity of the farmers and agricultural activities with a low profitability.

In these financial circumstances, one of the main difficulties met by the organization was the impact of the gradual decrease and after some time the total termination of the Governmental subsidies, which were given to cover part of the costs for operation and maintenance. As a solution, the organization increased the water charges, which caused a negative impact on the financial status of the farmers and it resulted in the start of outstanding bills, which were increasing in time and caused serious and negative impacts on the technical and socio-economic conditions of the organization. Thus, the traditional circle of an organization with this type of conditions started to get shape and it consolidated after some time with the following main characteristics: low collection of water charges, temporal and spatial suspension of the maintenance activities for the irrigation infrastructure, a low quality of the provided irrigation services, an inefficient use of the irrigation infrastructure and irrigation water, conflicts amongst water users and between them and the administrative and technical staff of the organization, and finally as a closing of the circle: the failure of the acquired agreement with the farmers, namely the agreement concerning the payment of the water tariffs. This situation caused a continued process to postpone maintenance activities of the hydraulic infrastructure and that resulted in its deterioration and an inefficient operation of the supply of irrigation water.

In these circumstances, ASORUT is suffering a process of destruction at different levels, which has resulted in a state of virtually collapse of the organization. This fact may lead to the following conclusion, namely that the initial objective of the delegation of management from the government to ASORUT, which included the administration, operation, and maintenance with the aim to supply irrigation water, resulted in a limited scope and ASORUT was not able to achieve and develop sufficient managerial capacity to enlarge its range of activities to fields beyond the mere operational aspects.

In this way, the various boards of directors focused their actions only on the development of day-to-day activities, in particular they looked for solutions for the most pressing conflicts related to the operation and maintenance, directing its attention to find the most appropriate executive rules, some of them of a coercive (using force) type in order to recuperate the every time growing outstanding bills. Cost reduction was another mechanism used by the management to lessen its financial burden. Some changes in the organization were carried out, such as the simplification of the organizational structure, merging of or abolishing positions, reduction of the staff, increasing the working hours and service areas, streamlining the procedures for fee payment, and the introduction of computerized management information systems. At the moment, the merging of sections has caused a new reduction in staff members. Especially, the operation, maintenance and conservation sections were merged and the number of irrigation inspectors has been reduced from 6 to 4; this decision has a large, but negative influence on the quality of the maintenance and operation activities.

So far, this management approach of ASORUT clearly shows that the management is only a simple provider of irrigation water without considering what is happening with the water at farm level and whether it contributes to the success of the production system of the farmers. The lack of this last aspect has resulted in not well-developed relationships or not very positive cooperation between ASORUT and the farmers; even more under the current circumstances most of the agricultural undertakings has to be considered as a highly risky activity with a low profitability. In other words, ASORUT is far behind the considerations, analyses and formulation of proposals, which could contribute to the solution of the socio-economic problems. The gap is even growing with time and has a very harmful impact on the well being of the farmers, especially due to an unprofitable agriculture.

In the framework of this management concept, the various boards of directors of ASORUT have not been able to formulate development-coherent plans for the short, medium, and long term. These plans will have to include all the important social-economic components and to consider an integral, participative, and collective approach, based upon the involvement of all actors directly engaged in the management (farmers, technical and administrative staff) and the ones indirectly engaged (stakeholders, supporting private and public entities, and the society at large). The plans will have to focus on the most productive projects based on an irrigated agriculture and taking into account all the stages of the production chain (production, post-harvest management, added value, agricultural industry, and marketing) and need to look for farming activities under the criteria of profitability, efficiency, competitiveness, and sustainability.

The low sense of belonging was increased after the transfer. The users did not see ASORUT as a supporting entity that helped to develop the agricultural activities in an efficient and profitable way; on the contrary they saw the organization as a bureaucratic entity that was an extension of the Government Agency and also the role of ASORUT was not clear for them. They expressed their low sense of belonging in several ways; through their lack of participation in the decision-making processes, by breaking the achieved agreements (payment of bills, maintenance of canals), by conflicts with the district's staff, by an excessive use of the hydraulic infrastructure, by rejection of suggested improvement programs and projects proposed by the administrative staff, and by giving a low credibility to the RUT's administration.

A weak leadership of the direction also follows from the fact that they were not able to propose and implement any development project based on irrigated agriculture, which could improve the socio-economic conditions of the farmers. The leadership in the past has mainly come from the prosperous farmers and private organizations, which had a large power to call people, especially medium and large farmers, together for a formal meeting to implement specific projects. This situation is a clear example of the discrepancy of interests between the two parties: the ASORUT organization focused on the provision of irrigation water that is paid through a tariff and the farmers who expect actions from the same organization to support them in the development of projects that will improve their standard of living.

ASORUT clearly shows the characteristics of an organization isolated from the local, regional and national context and with a very weak linkage with public and private institutions. The influence area of ASORUT is within the direct jurisdiction of the three important municipalities, namely Roldanillo, La Union, and Toro, and within the indirect jurisdiction of La Victoria, Zarzal, and Bolivar municipalities. However, the development policies of these municipalities do not show any formal development project or cooperation agreement with the RUT Irrigation District; regardless of the fact that the municipalities benefit from the economic activities within the irrigation district. Examples of these benefits are amongst others the generation of employment for the rural people, of agricultural businesses in the urban areas, the creation of revenues for the municipalities through taxes, an increase in the value of properties, and development of infrastructure for public services.

This lack of cooperation is a clear indicator of the low capability of the board of directors for inter-institutional relationships. This management attitude has resulted in some problems that are caused by external factors, such as the specific case of the canal interceptor, the water quality, maintenance of mutual public works, and the lack of development projects with international cooperation for the modernization of the irrigation district.

The interceptor canal shows a high sediment load and contamination of the irrigation water due to discharge of domestic wastewater from the municipalities of Roldanillo, La Union, and Toro. The high sediment load seriously influences the water distribution efficiency of the canal and has increased the maintenance cost due to the continuous extraction and disposal of sediment and the control of the aquatic vegetation; while the contamination by the wastewater has restricted the opportunities to sell agricultural products (fruit and vegetables) to national and international markets.

ASORUT is bearing the total costs for the maintenance of the canal in spite of the fact that the sedimentation is caused by an external factor being the deforestation of the adjacent catchments. In this case, the board of ASORUT has not been able to establish agreements and/or contacts with the local and regional environmental authorities to get compensation, agreements to share the maintenance costs or to implement jointly plans for reforestation and reduction of the erosion process on the middle-long term. The regional environmental authority has offered some possibilities to implement reforestation plans, but these should be requested, formulated and developed by the parties interested; so far ASORUT has not shown any leading role in this perspective.

ASORUT shows a similar attitude in view of the quality of the irrigation water in the interceptor canal that is affected by the wastewater; domestic and industrial wastewater and sediment load in the water pumped from the Cauca River. So far, ASORUT has poorly acted before the responsible environmental authorities in order to convince and to press the municipalities and institutions to establish and improve the treatment of the wastewater flowing to the ASORUT area. The Roldanillo, La Union, and Toro municipalities need to improve the efficiency of their treatment plants in order to minimize the negative impact of their wastewater on the water quality in the interceptor canal. In the case of the Cauca River, ASORUT would have to request financial compensation in view of the negative impact of the high sediment load in the Cauca River that is pumped together with the water and is deposited in the irrigation network causing high costs for the canal maintenance.

The protection dike is one of the components of the large land reclamation project developed for the plain area of the Cauca Valle Department under the direction of the Corporation of Valle del Cauca (CVC). Hence, this project is of regional interest and not only for ASORUT and for that reason the costs for improvement and maintenance of the dike should not be the responsibility of ASORUT alone. Identical situations exist when the internal road network is considered. This network is not only used by the farmers but also by other users from outside the area, therefore the maintenance costs would have to be shared with the local administration of the benefiting municipalities.

The project for the modernization of the irrigation infrastructure will be mentioned as a last example of the poor management capability of the organization. This project was been considered in the framework of an international cooperation agreement with the aim to perform a feasibility study for the modernization of the irrigation district. However, the management board of the organization did not carry out its expected role in view of the importance of the project and its functioning was even that poor, that other parties had to take the responsibility of the project. This was one of the main reasons why the project did not achieve the expected results.

The present crisis of the RUT Irrigation District is a serious growing threat to the sustainability of the whole district and system when the necessary measures are not taken quickly. The impacts of the present crisis are manifested in the poor conditions of ASORUT; namely in the technical, economic, social, environmental, and institutional parts of the organization. Figure 6.1 shows the inter-relation of factors influencing the current situation in the RUT Irrigation District.

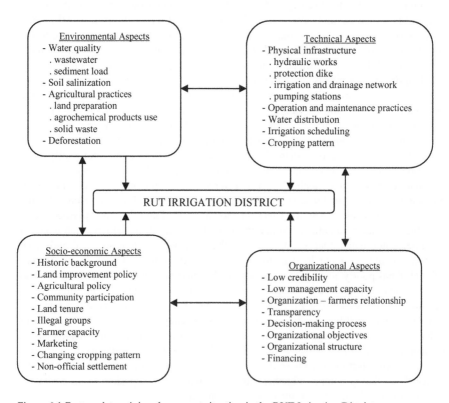

Figure 6.1 Factors determining the current situation in the RUT Irrigation District

From a technical point of view, the main problems concern the current state of the hydraulic infrastructure and the operation and maintenance practices. The pumping stations play an important and crucial role in the supply of the irrigation water. However, their lifetime has risen above the normally accepted lifetimes for pumps. The pumping stations do not have any reliable device to measure the discharge; they are operated in an inefficient way causing high-energy costs, and operation of the pumps at full capacity is restricted by low water stages of the Cauca River in the dry season.

The present irrigation network shows a low conveyance and distribution efficiency due to water losses by leakage and seepage that are helped by unlined canal reaches, cracking of canals lining, holes built by fishes in the sides, poor condition of hydraulic structures, and leakage of water at the tail end of the irrigation canal network. Silting, sedimentation and aquatic vegetation affect the canal operation and decrease the conveyance capacity of the canals, the efficient use of the available water, and increase the cost for maintenance. The shape of the cross sections together with the bottom slope has changed due to inappropriate maintenance practices.

The drainage network shows at several places deterioration and decline of its function. At some locations erosion occurs or the cross sections have become larger and larger due to the flow of water to the drains without any protection or outlet structure. At some places the farmers have installed pumping equipment for irrigation which has weakened the drains and the dikes. The mechanically removal of aquatic weeds is also a cause of the decline of the drains; sometimes severe landslides at the sides of the drains occur that are caused by traffic of heavy vehicles. Due to sedimentation the bottom levels of the drains become higher than the design level, which hampers the free discharge of secondary drains and contributes to a raise of the groundwater table. Some water level control structures are not calibrated resulting in higher than required water levels that have a similar effect on the groundwater tables as the sedimentation. The deterioration and destruction of the network of piezometric wells installed to monitor the fluctuation of the groundwater table have a negative impact on the functioning of the drainage network.

Many hydraulic structures function poorly, water level control structures are not operated in a correct way or according to their design, no attention is paid to the hydraulic principles of special structures, no applicable knowledge is available concerning the function and existence of the small wells in old times used for the measurement of discharges and poor conservation and maintenance. Some original structures have been modified in a physical and operational way, especially in view of the introduction of a new crop in the area, namely the introduction of sugar cane.

The marginal dike has some sections, which are highly vulnerable due to erosive actions of the Cauca River. The height of the dike forms a serious risk as it is possible that water flows over the crest during high water levels in the Cauca River. The unofficial settlements and building activities on the strip of land between the left bank of the river dike and the right bank of the main irrigation canal present another risk factor.

From an operational point of view, the distribution of the irrigation water is affected by changes in the cropping pattern, by a lack of irrigation schedules and cropping plans. At the moment there exists an incompatibility between the design criteria for the water distribution and the present operation of the system. During the last decade a dramatical change has occurred within the cropping pattern, namely the introduction of fruits and the sugar cane. Sugar cane covers almost 30% of the area today and this has serious impacts on the water management and the hydraulic infrastructure. Fruit trees require irrigation supply with short irrigation intervals, while sugar cane demands larger amounts of water than the amounts required by the traditional crops. Irrigation scheduling and cropping plans do not exist and the farmers decide themselves about the crops to be planted, the planting date and the moment that they request for irrigation water. This situation has become more common after the transfer of the irrigation management. Originally, the operation of the RUT Irrigation District was designed for gravity irrigation. However, most of the lands have not been levelled or upgraded for this condition and that has affected the water delivery in the distribution system. In general, the operation of the system is based more on the criterion that the proper water levels in the network should be maintained than that the allocated discharges should be supplied. As a consequence of this water management practice a large amount of excess water flows to the drainage system and that causes an inefficient use of the available irrigation water at considerable energy costs.

From an environmental point of view, the major problems are related to the physical-chemical and bacteriological conditions of the irrigation water. The domestic and industrial wastewater and the deforestation process of the adjacent catchment basins are affecting the water quality of some irrigation canals and the Cauca River. At several places in the district the salinization process is intensified by the unofficial reuse of drainage water for irrigation, inefficient irrigation practices, lack of a drainage system at farm level, and the natural process of salinization caused by the groundwater flow coming from the mountain range west of the irrigation district. At the moment 35% of the area is affected by salinization.

The use of agro-chemical products for agriculture and the inefficient management of solid waste in the area constitute a serious risk from an environmental point of view. Also the excessive use of agricultural machinery for the land preparation is affecting the chemical and physical properties of the soils.

Several socio-economic components seriously contribute to the critical situation in ASORUT. Two types of farmers co-exist within the RUT area, namely the small and middle farmers and the large ones. Each one of them has its well defined, but different socio-economic and cultural characteristics. The introduction of crops with a high profitability like fruit crops is not possible for the small farmers in view of their small economic resources together with a low technology and small land area (minifundio). The lack of instruments for marketing regulation and prevention of overproduction and the lack of storage centres form a serious constraint threatening the high profitability of the fruit trees. The ambiguous role of the middlemen affects also the profitability of the crops and plays an important role in the production process. Land tenure and the presence of illegal groups are factors that have influenced the social-economic environment and caused a reduction of the agricultural area, resulted in a concentration of land property, in the creation of conflict zones, and under some conditions the forced displacement of the farmers.

As an organization ASORUT has mainly focused on operational aspects of the system without major interest in the contribution to improved socio-economic conditions of the farmers. They did not assist the farmers through the formulation, implementation and development of integral programs based on a profitable agriculture and the active participation of all the actors involved. In the eyes of the farmers the credibility of ASORUT is low and they see the organization as an extension of the previous governmental agency. Also the role of the organization after the management transfer is not clear for them. This condition has been supported by the lack of transparency in the decision-making process and has made corruption easier and at some moments the influence of power became large. The low capability of the management in view of the inter-institutional and other problems has been recognized by several actors at local, regional and national level. In general terms, it can be said that ASORUT has not achieved a sufficient level of organization that is in line with their substantial responsibilities given by the government during and after the management transfer and that should allow for an autonomous and self-sufficient organization, being capable to monitor, interpret, realize and satisfy the interests of its direct and indirect beneficiaries.

6.2   Expectations for an Integrated and Sustainable Management

The society at large has great expectations for an integrated and sustainable management of the RUT Irrigation District in spite of the present critical status of ASORUT. These expectations are mainly based on the current potential for the Colombian agriculture, the role of the Government for the land improvement sector, and the socio-economic importance of the RUT Irrigation District for its directly and indirectly influenced area.

Current Expectations for the Colombian Agriculture

From a historic point of view, the development of the agricultural sector had a large impact on Colombian economy, but during the last decade its contribution has decreased and is becoming less important than other economic sectors like the services sector and the secondary ones. In view of this development, the contribution of the agricultural sector to the gross national product has varied with (in percent): –0.05, 3.85, 0.68, 0.52 and 2.37 for the years 1994, 2000, 2001, 2002 and 2003 respectively; while the variation for the total nation product has been (in percent): –4.20, 2.92, 1.39, 1.62, and 3.74 for the same years (National Planning Department, 2004).

The governmental policies for the agricultural sector and the actions of illegal groups have influenced in a negative way the performance of the sector. However, the agricultural sector is still and remains to be a very important component of the national economy to that extent that the recovery and strengthening of this sector are part of the current policy of the Colombian Government. As a result of this policy the cropped area during 2003 was increased by 4%; the yield of cotton, yellow corn, cassava for agro-industry, rice, and long term crops (African palm, rubber, and fruits) increased by 30, 34, 40, 11, and 4%, respectively; while the import of milk, meat, cotton, and cocoa was reduced; the increase of exported African palm, sugar, fruits and coffee was 84, 63, 15, and 13% respectively; and during that year 327,000 new jobs were generated. Finally, from a financial point of view the Agricultural Financial Fund (FINAGRO) increased its allocation by 53% and the Agrarian Bank by 55%, with an increase of 62% for the small producers.

On the other hand, the Government considers globalisation of the economy as a strategy for the reactivation of the economy through international trade agreements in which the agricultural sector will play an important role. Amongst the most important trade agreements can be mentioned the Free Trade Agreement (TLC), Nations Andean Trade (CAN), Free Trade Area within the Americas (ALCA) and the Andean Trade Promotion and Drug Eradication Act (ATPDEA).

As can be seen, the agricultural sector has a favourable future, which is supported by the Governmental policy that focuses on the reactivation of the agricultural sector as development motor for the strengthening of the whole economy of the country. On the other hand, the future international trade agreements offer a great opportunity to the agricultural sector with the incentive to develop an efficient, profitable and competitive agriculture. In this context, the economic interests of the RUT Irrigation District can be helped because the RUT Irrigation District offers very good conditions for the production of fruits and vegetables, which have a preference for the international markets.

Role of the Government

The policy for the transfer of the irrigation management started and was endorsed by the enactment of the Law 41 of 1993 for the land improvement sector and of the recent decree 003 of 2004 by which rules were set for the organization of the sector and with organizational guidelines for the users organizations including the basic regulations for the functioning of the small, middle and large irrigation districts. In other words, the communities through these organizational forms are in charge of the management, the formulation, implementation, starting off, administration, operation, and maintenance of the land improvement projects under a legal framework established by the government. The latter mainly executes supporting and accompanying tasks in order to assist the organizations in the performance of their functions.

The Departmental Agriculture Secretaries and the Municipality Units for Agricultural Technical Assistance (UMATA), are in charge of the governmental policy at departmental and municipal level. These organizations play nowadays an important role as they are the link between INCODER and the community organizations that will address the available resources for investment in an efficient way.

In this context, the organizations at community level need a strong organizational structure, empowerment by the communities, a high and autonomous management capability, and a strengthening of the inter-institutional relationships in order to access and use the available resources. These characteristics are necessary in view of the development and management of land improvement projects in an efficient and sustainable manner.

Importance of the RUT Irrigation District

As mentioned before the RUT Irrigation District is the most important land improvement project in the northern part of the Valle del Cauca Department. The economy of the northern part of the department is based on the agricultural sector, which contributes 45% to the gross departmental product. The sixteen municipalities belonging to the northern region supply about 50% of the food requirements of the department and 12% of the fruit production in Colombia. In this context, it can be said that the RUT Irrigation District has been a development motor for the economy, not only for the northern region of the Valle del Cauca Department but also for the adjacent regions belonging to the Risaralda and Quindio Departments.

The infrastructure for irrigation, drainage and flood control allowed a shift from subsistence agriculture to more intensive agriculture. It allowed the introduction of cash crops within the existing cropping patterns that were changing with time and according to social and economic factors. Sorghum, corn and soybean are the traditional crops, followed by cotton and fruits on a minor scale; from the middle of the 1990s an important expansion of the sugar cane crop occurred.

The RUT Irrigation District is one of the most important irrigation systems in Colombia in view of the for agriculture very suitable and ample available natural resources (climate, soils, and water), the strategic and geographical location of the district, and the potential for further agro-industrial developments in view of national and international markets. The RUT Irrigation District offers very promising expectations for the regional socio-economic development based on an efficient, profitable, and competitive agriculture that the region and country will allow to make use of the future challenges that are established for the agricultural sector by the framework of the global economy.

In the previous sections it has been mentioned that there are sufficient reasons to undertake a modernization program for the RUT Irrigation District, which will help the community involved in a recovery of the socio-economic development through the practice of irrigated agriculture under criteria of efficiency, equity, profitability and sustainability. The development program needs to be guided by the community itself as the leader of the organization. Essential requirements for a community-controlled organization are a strong management capacity, leadership, inter-institutional relationships, and transparency in order to make the decision process clear and supported by all users. The program needs to be focused on the development of productive projects that will have irrigated agriculture as their pivot and that involves all the phases of the production chain: namely the production, post-harvest improvement processes, agro-industry, and marketing.

The development program needs to be formulated and executed on basis of an integral and sustainable management approach. An integral management approach means the full participation of all the farmers involved as the first and most important beneficiaries; of the technical and administrative staff; the local, regional and national authorities for the land improvement sector; and representatives of the public and private supporting entities. An integral approach needs to be based on the knowledge of all the physical aspects of the natural resources and their conservation, on a reliable socio-economic development, emphasis on a demand-driven and demand-oriented approach; and the decision-making process will have to take place at lowest appropriate level. On the other hand, a sustainable management approach refers to the physical and organizational infrastructures needed for the management, administration, operation, and maintenance of the system during its lifetime according to established specifications and rules, and giving to the users the expected benefits without causing irreversible damage to the natural environment.

6.3   Scenarios for an Integrated and Sustainable Management

Proposals of the scenarios for an integrated and sustainable management of the RUT Irrigation District will have to take into account the potential internal and external factors and their influence on the management process. Internal factors can be directly managed by ASORUT as an organization, they can be controlled by the organization and they will determine its weaknesses and strengths. The external factors are outside the range of action, but their effects will have an impact on the organization and they constitute the threats and opportunities for its further development; the control of these factors is outside the reach of the organization. Figure 6.2 describes the characteristics of these factors.

Strengths
- **Environmental**
  . sufficiency of water resources
  . land resources
  . good climate conditions
- **Technical**
  . existing of hydraulic infrastructure
  . practical experience in water management, irrigation methods, agricultural practices, and cash and traditional crops management
- **Socio-economical**
  . high impact on the local and regional economy
  . community with agricultural tradition
  . generation of direct and indirect jobs
  . organizational experience
  . agro-industry processes experience
  . good road network and proximity to urban centres

*Weaknesses
- **Environmental**
  . low water quality
  . increasing process of soil salinization
  . inappropriate agricultural practices
  . deforestation process
  . solid waste management
- **Technical**
  . obsolete hydraulic infrastructure
  . inappropriate operation and maintenance
  . inappropriate water management
  . non-efficient irrigation water use
  . poor irrigation practices at farm level
  . lack of training for technical aspects
- **Socio-economical**
  . land tenure and changing of the land use
  . presence of illegal groups
  . non-official settlements
  . low women participation
  . low sense of belonging by farmers
  . absence of a new generation of young farmers
  . low financial capacity of the farmers
  . non-profitable agriculture practice
  . non-regulation of marketing and prices
  . accessibility to credit entities
- **Organizational**
  . lack of organizational objectives
  . non-autonomous organization
  . non-financial self-sufficiency
  . low management capacity
  . low capacity for inter-institutional relationship
  . low credibility before farmers and external entities
  . low transparency for making decision process
  . low capacity for conflict resolution

Opportunities
- Future expectations for the agricultural sector
- Flexible legal framework for the sector
- Decentralization of the government policy
- Possibilities for international trades
- Financial resources availability
- Possibilities for international, national, regional, and local cooperation
- Existing supporting private and public entities
- Possibilities for forward markets
- Development of parallel agro-business

Threats
- Environmental degradation due to external factors
- Non-flexible government policy for the water resources management implemented by autonomous Regional Corporation (water rights)
- Pressure by other users sectors
- Action of external actors (illegal groups)
- Changing governmental policy for the sector
- Agreement conditions for international markets
- Management agency different to the water users association (ASORUT)

Figure 6.2 Factors influencing the potential development of the RUT Irrigation District

The weaknesses correspond with the specific characteristics of the organization, which reflect its current precarious condition and constitute the elements that have to be changed in view of a more appropriate future condition. These characteristics can be of a technical, environmental, or socio-economical type and their removal depends on the staying time within the organization; the longer the staying time, the harder its removal. The strengths are also specific characteristics of the organization and contribute in a suitable way to a required condition. The opportunities and threats exist in the adjacent context of the organization and their characteristics can determine in a high degree the dynamics of the development of the organization. The opportunities give power to the organization, increase its possibilities for development, while the threats produce a contrary effect.

Two possibilities or scenarios can be considered for the short, middle and long term in view of an integrated and sustainable management for the RUT Irrigation District. Both could lead to conditions for an integral and sustainable management, but at different level and with different impact due to the fact that the main actors are different in both cases. Each one of these scenarios will present different characteristics in view of their nature, actors, objectives and way for their implementation.

6.3.1 Direct Governmental Intervention Scenario: DGIS

The first scenario will be called the "Direct Governmental Intervention Scenario" (DGIS) in which the government would perform the central role. This scenario has as starting point the current critical condition of the RUT Irrigation District and its continued deterioration supported by the low management capability of the present organization to produce the necessary changes that will lead to a gradual physical and organizational recovery. An external factor could also accelerate and could be a vehicle for this scenario; an example of this is the action of illegal groups in the area. The development of this scenario can be considered in three stages: one stage corresponds with an accelerated deterioration process, followed by another stage that corresponds with the government direct intervention, and finally the stage with an implementation of the necessary reforms inside the organization.

The accelerated deterioration process would be characterized by the aggravation of the current conflicts faced by the organization with special emphasis on those concerning the technical, socio-economic, and organizational aspects. Within this context and from a technical point of view it is expected that several aspects will show an increase, for example an increase of the deterioration of the physical infrastructure caused by untimely, deficient and non-appropriate practices for maintenance; an increase of inappropriate operation practices for the irrigation infrastructure causing high operation costs and dissatisfaction with the service for the irrigation water supply; and increase of the conflicts between the organization and farmers reflecting the fact that the farmers do not fulfil the acquired agreements, for instance on the maintenance of minor hydraulic infrastructure and the fee payment.

From a social-economic point of view it is expected that some factors will show an increase, for instance an increase in non-profitable agricultural practices that will cause a low willingness and capability of the farmers to pay the required water fees; an increase of outstanding bills by the fact that the water users are not paying for the service of the water supply and that will result in a decreasing financial capacity of the organization. Other factors will result in changes in land use with a key expectation for the expansion of extensive crops like sugar cane or others non-agricultural uses; major and critical changes in the land tenure; lack of incentives for capital investments especially for cash crops, agro-industry process, and marketing; increase of favourable conditions for the presence of illegal groups that cause unemployment and poverty; and increase in the low sense of belonging of the farmers.

From an organizational point of view, ASORUT will show a very low management capability and its credibility will be at its lowest level for the farmers, government and supporting public and private entities and that will result in a greater than before isolation of the organization within the national, regional and local context. As a consequence, ASORUT will become every time a less autonomous organization, which will be highly dependable on external parties and which will have none financial self-sufficiency, a poor capability for the establishment of inter-institutional relationships, and lack of authority for conflict resolution. On the other hand, ASORUT will continue to increase its lack of transparency for the decision-making process combined with corruption behaviour and some board members of the direction will carry out their political activities in order to support their aspiration to obtain administrative positions in the municipalities within ASORUT's area of influence. In this way, ASORUT will focus its main actions only on those activities that are required to operate the district at a very low performance level and in this way these actions may directly and in an extreme way contribute to the total collapse of the organization.

This situation will be the starting condition for the second phase of this scenario in which the government will take over the total control of the organization and will undertake the necessary actions to find a new organizational model for the management of the system. In this case and taking into account the past experiences with the delegation of the administrative responsibilities to the water users association the government may decide for a concession model, which has recently been considered during the changes of the legal framework for the land improvement sector (Resolution 1399 from August 11, 2005).

According to the Colombian Law a Concession Contract contains the provision, operation, exploitation, organization and partial or total management of a public service; or the construction, exploitation and partial or total conservation of a work or good for the public use or service. It also includes all the necessary activities to be developed by the concessionaire for the appropriate provision or functioning of the work or public service and the remuneration for the control and care of the work or service by the concessionary entity is arranged by rights, tariffs, tax, and valuation, amongst others.

Experiences in concession contracts for the land improvement sector do not yet exist in Colombia; but there exist successful experiences in other sectors, such as the vital infrastructure sector. However, in the year 2004 this management model has already been mentioned for the land improvement sector as an alternative for some irrigation districts that are under non-sustainable conditions, as in the case of the Maria La Baja Irrigation District.

The scenario here considered seems to present the appropriate conditions for the establishment of a concession contract that is very much helped by the existing conditions of the area: natural resources, climate conditions, infrastructure, facilities, productive activities, etc. Furthermore, it is supposed that this concession model will have a favourable reception by the large farmers in ASORUT, which have the available economic resources, technology, high credit capacity, and willingness to undertake new and to enlarge their own production activities by paying for the investments needed, and they will show the willingness to develop the appropriate organizational forms as well. An example of this attitude has been their acceptance and agreement for the modernization project of ASORUT that has recently been considered through international cooperation; these large farmers were at the head of the project preparations by playing a much more active role than ASORUT itself.

From its side, the government will offer the appropriate conditions for the establishment of the concession model, especially in view of credit facilities and technical support, being a partner during the implementation process and a mouthpiece to express the views before national and international organizations.

In this context, this management model by concession could reach technical and financial sustainability. At middle term the initial investments will be reflected by an improved condition of a part of the hydraulic infrastructure, followed by a rehabilitation and modernization plan for the long-term; improvements of the operation and maintenance practices, better supply of the irrigation water to the users, and a reduction of the conflicts amongst each other. All this will cause a healthier condition from a financial point of view for the organization and an improvement of the socio-economic conditions for a part of the benefited users, especially the large farmers and producers.

The social-economic composition of ASORUT shows that in the area and at the moment there co-exits two types of farmers (producers) namely: small-middle farmers and large ones. As shown before 92.4% of the land properties are smaller than 20 ha and that corresponds with 92% of the farmers, which are considered as small-middle producers and they cover 48.5% of the total area. For the small producers the figures show that 76% of properties are smaller than 5 ha and that corresponds with 75% of the farmers and they cover about 19.2% of the total area. At this point a new question arises: is this scenario also valid for the small and middle farmers?

In fact during the last years the management of the RUT Irrigation District has been influenced to a great extent by the large farmers, who proposed and implemented several management policies in view of the fact that they were looking for their main benefits. This has caused a situation in which the group of small and middle farmers has been moved to the edge of the decision-making process and this group has increased over time. Under these circumstances a management by concession could deepen the differences between the large and small-middle farmers and that might seriously endanger the social sustainability of the concession model.

It is expected that a close relationship between the large farmers and the concessionaire will develop and be maintained as long as it will satisfy as much as possible their mutual interests. For example, this relationship might result on the one hand in the fact that the large farmers receive their irrigation water on a timely and reliable way in order to develop their highly productive and profitable activities; while on the other hand, the concessionary will be well paid as provider of the irrigation water service. It is expected that this type of relationship will not develop in the case of the small and middle farmers, who have other socio-economic conditions than the large farmers; conditions that are less attractive for the concessionaire and the concession system. Along these lines it may be expected that the concessionaire will have special interest to optimise his function as irrigation water provider and will mainly address his activity to those users who are able to pay for a more efficient service.

As a result, this scenario will result in a management organization that will concentrate its service as irrigation water supplier on the large farmers, which present a strong socio-economic entity as a result of their most profitable agriculture practices. On the other hand, the presence of the small farmers in the area will create a critical socio-economic condition. As a consequence of their subsistence agriculture practices their financial and economic capacity will be very low. Under these conditions, the differences between the existing two farmers types will be deepened and major social conflicts might be expected. An increase of the poverty level, changes in the land use, concentration of the land property by large farmers, presence of illegal groups, un-official settlements, and migration of small farmers to urban areas will be some of the expected and not wanted social impacts.

As a conclusion, this scenario of management concession is not expected to contribute to an integrated and sustainable management on the long-term. This management model will present economic, financial and technical sustainability in the short and middle term, but its social sustainability will be seriously endangered due to the deepening of the differences between the small-middle farmers and large ones under this concession model. The model does not contribute to an integrated management because the expected benefits will create social inequalities by strengthening the large farmers and excluding the small-middle ones.

6.3.2   Scenario for a Participatory and Communal Approach: A New Role for ASORUT

In the previous scenario the government plays a very essential, maybe the most important role being the central actor; in this new scenario the community consisting of the water users in the RUT Irrigation District will play the main role, while the water users association as representative of the users will head the development process. The fear of the community, which is fed by a most probable, but unwanted intervention of the government, which could give the system management to another private organization, forms the main incentive to reflect on the present critical situation and management condition of the ASORUT organization. This reflection is supported by the poor socio-economic conditions of the small and middle farmers and by the pressure of public and private entities, local as well as regional, to undertake the necessary actions for the in their eyes more than needed re-organization of the association.

In order to obtain a sustainable management this development process needs to follow an integral and participatory approach involving all the internal and external actors: namely the farmers as first beneficiaries; the technical and administrative staff; representatives of each municipality located in the influence area of the irrigation district; the land improvement sector; other supporting private and public entities. As a final result of this process, the factors that might limit or endanger the sustainable development of the organization will have to be identified and classified according to their importance; moreover, guidelines will have to be given for the future physical and non-physical interventions required obtaining a sustainable management.

Within this participatory scenario, ASORUT is an organization with a defined and clear statement about its future mission, which is aimed at the improvement of the quality of life of the farmers through a management of the irrigation district under principles of equity, competitiveness, sustainability and multi-functionality in line with the legal framework established by the government. This means that the activities of ASORUT will have to be broad and will be focused on enhanced socio-economic conditions for its beneficiaries allowing them to improve their living standard in an integrated and sustainable way, the role of ASORUT should not be limited to a job of a simple administrator who is only responsible for the hydraulic infrastructure needed for the irrigation water supply.

In view of these goals, the future management of ASORUT needs to have a well-developed ability to formulate, plan and implement development-coherent programs for the short, medium, and long term. These programs will have to include social-economic, technical, and environmental components and have to be based on an integral, participative, and communal approach, which will have to include the participation of all the directly and indirectly involved actors. These integral development programs will have as main pivot the promotion and practice of irrigated agriculture under criteria of profitability, efficiency, competitiveness, and sustainability, and that also take care of the protection and conservation of the natural resources, and will include all the stages of the agricultural production chain, namely production, harvest and post-harvest improvement, added value, agricultural industry, agro-business and marketing.

When ASORUT intends to look for better socio-economic conditions for its users, it has to show more than sufficient interest in the strengthening of its organizational infrastructure in order to be able to offer an efficient irrigation water service, to develop training programs for its users and administrative and technical staff, to give up-to-date support to the users to improve the agricultural practices, to promote programs that will guarantee the social security of its users and families, to promote the improvement of the physical infrastructure through rehabilitation and modernization programs, to encourage actions for the protection and conservation of the environment, and to maintain and strengthen inter-institutional relationships.

The integrated, participative and communal approach will be the key in this scenario to solve the discrepancies between the small-middle and large farmers and to get the necessary cooperation of the supporting public and private entities. As was mentioned before the most recent agricultural projects have been established and carried out under the responsibility of the large farmers and they are the sole and direct beneficiaries of the economic returns, especially as they have created conditions that exclude the small and middle farmers from the benefits. These conditions have even more accentuated the socio-economic differences between the two farmer groups.

Small and Middle Farmers

Within the scenario for a participatory and communal approach, the organization of ASORUT will have to pay special attention to the small and middle farmers and to strengthen the capability of them to be more competitive and economical. As a strategy, ASORUT will have to identify, formulate and implement profitable and fruitful agricultural projects in a cooperative and combined way, grouping the small and middle farmers together according to their preferences and in this way overcoming the restriction imposed by their small farm size (minifundio) and taking advantage of the scale economy. The production of each project will be brought to the facilities located at the administrative centre of ASORUT where extra value can be added to the products through some crucial physical transformations, classifications, packing, etc. Next, ASORUT will be responsible for the marketing of the products, which will be transported to consumer centres at local, regional, and national level. For these activities, forward or private futures contracts have to be prepared by ASORUT from the start of the project activities. In this way ASORUT can obtain favorable selling prices and assure the volume and quality of the products. These contracts can be negotiated with large department stores, large farmers who need a certain production volume for their marketing, with agro-processing industry, or the National Agricultural

Stock Exchange. At the end of these activities, ASORUT will look after the distribution of the revenues amongst the participants of the project.

It is obvious that in this scenario ASORUT plays a very crucial role during the planning and development of the project cycle. For this role ASORUT needs the capability to identify, select, implement, and control money-generating projects, to give support to the farmers in all kind of technical aspects at farm level (land preparation, crop selection, irrigation and agronomic practices, post-harvest management for agricultural products), financing and marketing.

As a legal organization, ASORUT will act on behalf of the farmers before public and private institutions in order to get support to develop programs in a mutual way. Moreover, ASORUT will be able to comply with the economic resources supplied by the government through the Agricultural Financial Fund (FINAGRO) and its different credit lines for the financing of commercial projects. Then, ASORUT will transfer the credit directly to the users through supervised or controlled credits, for example as agricultural inputs, technical assistance, machinery rent, etc; services which can be provided by the organization or others recognized entities at a reasonable cost. It means that technical and professional personnel are available from the organization or can be contracted from other organizations to carry out the required activities or services.

The system of controlled credit supply will reduce particularly the possibilities of: money misuse by the farmers as they will receive the credit directly from the credit entity; price speculation for agricultural inputs by suppliers; negative actions and impacts of the middlemen.

The commercial projects will be mainly developed for the traditional crops like grains, vegetables and fruits for which the farmers have already the necessary knowledge and sufficient experience. The limitation of the small farmers to grow fruit trees is due to the small land sizes and high investment costs and might be overcome through a natural grouping arrangement. Other commercial projects can be established in a parallel way on the short and middle term, which can be a good alternative especially for women and youth. As example can be mentioned the growth of ornamental plants like the heliconias, which possess a high potential for the national and international markets within the bio-trade program promoted by the environmental authority; the bamboo growth in view of creating small handicrafts enterprises, and the agro-tourism at major scale and long term. The establishment of the small plots for vegetable growth and the breeding of animal species will contribute to the family food security and the generation of additional income. For the establishment of alternative productive projects as those above-mentioned financing resources exist coming from government through credit entities, and supporting entities like National Apprenticeship Service (SENA). This organization offers training courses for the formulation, implementation and assistance for these projects. Figure 6.3 illustrates the flow chart of activities for ASORUT corresponding to its new role.

Large farmers

As was mentioned before, 7.6% of the land properties are larger than 20 ha and cover 51.5% of the total area. The lands belong to 8.1% of the farmers, which are considered as large farmers or producers. Some of them cultivate grains, especially yellow corn being the provider of raw material for the national agro-industry, other large farmers are cultivators of sugarcane whether as direct growers or leasing their lands to the sugarcane factories; and the remainder of the large farmers grows fruit trees, some of them in connection with the agro-industry for the production of wine, juices, and canned fruits.

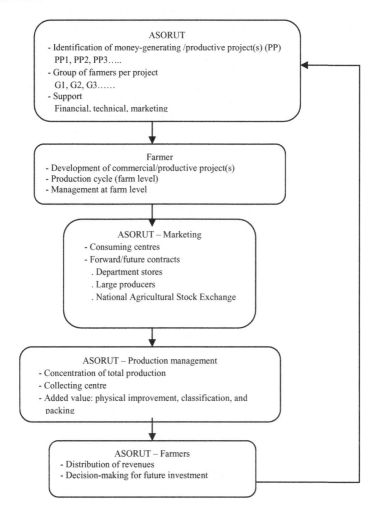

Figure 6.3 Flow chart of activities for a participatory and communal management

The relationship between ASORUT and the large farmers is totally different from the one mentioned before, namely the relationship between ASORUT and the small and middle farmers. In view of the high financial, economic, and technological capabilities of the large farmers they will be independent in many aspects. Of course they will accept advices from ASORUT as a special advisor when they discuss measures to improve and modernize the existing production systems. Moreover, ASORUT can be their intermediary before private and public, national and especially internationally institutions for the formulation, assessment, implementation and development of projects for the agro-industry and in the search of new marketing channels. In this specific management framework it is expected that the large farmers who grow sugarcane can find better conditions and incentives, and change the present land use to a more intensive one by using cash crops like fruit trees.

The large farmers expect from the irrigation system of ASORUT a timely, reliable and sufficient supply of irrigation water as a very crucial requirement in view of their high investments for the growth of cash crops. Rehabilitation and modernization of the hydraulic infrastructure and optimization of the operation of the irrigation water supply is a key issue for ASORUT in order to meet and match the expectations of the large farmers. If the organization is able to assure a reliable irrigation water supply under quality and quantity requirements then it might be expected that the large farmers are willing to endorse and pay for rehabilitation and modernization projects in the area.

Another important issue is the relationship between the large farmers and the small- and middle ones. Now, ASORUT is a liaison for the two parties and some of the economic feasible projects are developed amongst them in a mutual and equitable way. Some small-middle farmers producing grains (yellow corn) and fruits are providing their products to the large farmers in order to increase the production volume of the latter and to achieve the before agreed quota for the selling of these products.

Expectations

Some major and new opportunities might be expected as a consequence of the implementation and development of this participatory and communal approach for an integrated and sustainable management. Some of them are as follows:

- The sense of belonging from the farmer's side will increase on the short and middle term because the farmers will have improved their standard of living as a result of a major change in their income coming from the successful development of some new commercial projects. Their active participation will be recognized and encouraged in the future by the ASORUT organization that will give the farmer a growing feeling of the importance of their role;
- The farmers will have an enhanced opinion concerning the organization and they are willing to participate in an active way in the development of programs and projects proposed by the ASORUT organization. The farmers will see that the organization as an entity pursues the well-being of all of them; as a consequence conflicts will reduce and an improved environment will be created for conflict resolution;
- The practice of commercial agriculture and corresponding business will reduce the poverty and unemployment, increase the standard of living of the people, create a favourable environment for the conservation and protection of the natural resources especially the efficient use of the water resources and suitable land use. In this way the growing area of the sugarcane can be stopped and hopefully reduced; then the area for cash crops can be increased in a gradual way;
- The management capability of the organization has to be strengthened and ASORUT will not only have to provide an efficient irrigation water supply to all the farmers, but it should also be able to formulate, implement and develop profitable agricultural projects while the organization looks for improved socio-economic conditions of the farmers and takes advantage of the secondary development opportunities within a new economic perspective and within the existing juridical-legal environment;

- The community within the irrigation district will obtain more and more authority in a gradual way and they will be able to develop a mutual decision-making process in which the socio-economic and cultural diversity of all its members are recognized. In this way, the major differences between small and middle farmers and the large ones can be brought down resulting in a harmonic development with and between the two parties;
- ASORUT will become an organization that is recognized at local, regional and national level by all the public and private entities working within the land improvement sector due to its sound management of the state resources allocated for its administration. The organization might even become an example and model organization for other organizations working in the same field. It will be an organization with a high ability for inter-institutional relationships that will allow the interaction with public and private institutions in order to get the required approvals for and to obtain the resources to develop the necessary, commercial programs for its beneficiaries;
- The physical infrastructure of ASORUT will be used in a more efficient way; more attention will be given to the Administrative Centre that can be used for many other purposes, for example for training courses for the farmers and their families, for agricultural and related exhibitions and fairs, for agricultural business, as a collecting centre for agricultural products where they can be improved and obtain added value;
- The marketing of agricultural products will be supported, because this new management scheme will allow for the sowing and planting schedules, establishment of guidelines for volume and quality of production, for price setting and regulation, and a reduction of the influence of the middlemen.

As a general and final conclusion it can be said that this scenario for a new role of ASORUT is the most suitable one for the irrigation district, the water users and farmers in view an integrated and sustainable management. Its implementation and successful development will create new conditions that will empower the community and allow them to participate in an active way in the decision-making process, while they take their responsibilities as they are the central actors for the necessary change processes.

6.4   Required Physical and Non-Physical Interventions

One of the results of the second participative workshop was the identification of the required physical and non-physical interventions in the irrigation district in order to reach a model for an integrated and sustainable management. Physical interventions are directly related to the improvement of the physical infrastructure (hydraulic works), operation and maintenance of the hydraulic infrastructure, and irrigation and drainage methods and practices. The non-physical interventions are related to the socio-economic, environmental, organizational and legal aspects. The proposed interventions are considered with respect to technical, environmental, socio-economic, organizational and legal aspects

6.4.1 Interventions for Technical Aspects

The most important interventions are related to the pumping stations, the protection dike, the water distribution network, the hydraulic structures, and operation of the irrigation and drainage system. For each item several issues will be considered and the required interventions are proposed.

Pumping Stations

The main topics related to the pumping stations include instrumentation of the stations, alternative energy sources, sediment control, and operation of the pumps under critical conditions in the Cauca River, and rehabilitation, modernization and replacement of the pump units.

-   The instrumentation of the pumping stations needed concerns especially the installation of reliable discharge measurement devices in order to be able to quantify the volume of pumped irrigation water in an accurate way and to assess the performance (efficiency) of the pumping units;
-   The existing measuring system is based on a calibration curve that relates the discharge with the water level in the Cauca River. This system is already obsolete for a very long time; it was established 30 years ago and has never been verified, especially in view of the changed natural conditions of the Cauca River. During the fieldwork in 2005 a verification test was carried out by using a calibrated current meter. The results showed a difference of 40% for the discharge; the measured discharges were much larger than the values given by the calibration curve for the same water level;
-   The Cayetana Pumping Station has a broad crested weir to measure the discharge when this pumping station delivers water for irrigation purposes. However, a field survey indicated that the calibration curve of the weir is used in a wrong way; when the operator records the discharge he follows a incorrect procedure introducing in the calibration curve of the weir the opening value of the blades of the pump (Kaplan type) in operation, which results in a wrong reading of the discharge on the calibration curve;
-   The use of another energy source might be a feasible alternative to be considered in order to reduce the electric energy costs for the operation of the pumping stations, especially during the "peak hours" (6 hours per day) during which the present energy costs are very high and the irrigation water supply is stopped. Natural gas, wind, and solar energy are energy sources that can be considered in the future. Another alternative to decrease the operation costs is to replace the water supply by pumping by a supply by gravity. This, however, requires a long conveyance canal that will divert water from the Cauca River far upstream of the ASORUT area;
-   The construction of sand traps downstream of the pumping stations will be a substantial measure to improve the physical quality of the irrigation water that is pumped from the Cauca River. The sand traps will reduce the maintenance costs of the irrigation canals; improve the conveyance capacity and the water distribution;
-   A lowering of the bottom of the suction chambers of the pumps will result in an improved operation of the pumping stations, also during low water levels in the Cauca River and especially during the dry season. This aspect is very important to guarantee a timely and reliable water supply of irrigation water;

- The pump units are already much older than their lifetime and replacement is needed.

Protection Dike

The protection dike protects the area against floods that might be caused by high water levels in the Cauca River.
- *Slide stability.* Along the total length (44 km) of the protection dike some highly vulnerable sectors have been observed that are a result of the erosive action of the Cauca River. Los Millanes, La Bodega, and Remolinos are sectors in which protection works like sheet piles, rock lining, and groins have been built on the left riverbank in order to improve the stability of the dike;
- *Heightening.* Several points along the river dike have been identified that are too low and that need to be heightened in order to prevent overflowing during high water levels in the Cauca River. This risk must be controlled because it would seriously affect the economy of the region in a negative way. During past critical situations the water users in the area put sand bags on the top of the dike in these critical reaches with the purpose to increase the height of the dike as a temporal protection measure;
- *Settlements.* Relocation of the unofficial settlements that are located on the strip of land between the left river bank and the right bank of the main canal (marginal canal) is required. There, a number of small farmers have established their houses and crops causing serious risks to the stability of the river dike as well as to the main irrigation canal.

Water Distribution System

These interventions are addressed to the irrigation canal network for the water distribution namely marginal irrigation canal, conveyance canal, canal 1.0 and interceptor canal.
- *Water losses.* The main aim of the proposed interventions is the reduction of water losses by seepage and leakage. Canal lining at some places, improvement of the existing dilatation joints, control of the concrete lining fissures in the canals, control of unauthorized fishing and cattle herds (trampling) and the correct maintenance of hydraulic structures are some of the interventions needed;
- *Canal maintenance.* Canal maintenance will reduce the silting up of the interceptor canal by a course correction of the small streams that come from the hilly areas. This correction will be possible by the construction of some groins in the streams. The implementation of reforestation programs in the watersheds of the small streams is another and very essential intervention at middle and long term;
- *Weed control.* A good and appropriate maintenance will support the activities that concentrate on weed control in the dilatation joints and the removal of sediments in the canals that will limit most of the aquatic vegetation.

Hydraulic Structures

Rehabilitation and modernization are required for the major hydraulic structures installed in the main distribution system. Most of them are deteriorated and cannot be operated anymore in the way as they have been designed. They have been modified in view of the mechanical operation by the "irrigation inspectors" who nowadays use chains and flywheels and operate the structures according to their experience and good judgement. Examples are the four structures in the main irrigation and the conveyance canal, which are designed for downstream flow control in these canals.

Similar situations can be observed at the regulation and control structures, type Constant Head Orifice (CHO), and the corresponding water level control structures along the conveyance canal 1.0. The three water level control structures of the interceptor canal need also improvement and modernization in order to assure the correct operation of the canal, which should be based on its double function for irrigation and water interception. In the long term, a decision has to be made concerning the future separation of these two functions in view of a more proper operation of the interceptor canal.

On the other hand, all the existing offtake structures were designed with a measuring device to determine the discharge under the sliding gate. Each offtake has two small wells which are connected by a piezometer to the upstream and downstream water level, respectively. From the static heads, upstream and downstream of the gate, follows the discharge and the gate can be regulated according to the water needs. This provision has been designed for all the offtakes built during the initial works. At the moment most of the small wells do not work, because they are filled with soil, covered with vegetation or concrete blocks. A rehabilitation of these measuring devices is needed in order to be able to measure the amount of water supplied, to implement the volumetric tariffs in a reliable way and to improve the water supply in a more efficient way.

Water Management

The water management is a crucial issue in view of a sustainable management of the RUT Irrigation District. After the transfer of the management from the government to the water users association the water management practices changed in many ways from those followed by the governmental agency. The transfer caused modifications to the framework of the organization as well as to the operation of hydraulic infrastructure for the supply of the irrigation water. Although in a slow way, these modifications have resulted in a very inefficient performance of the organization, in deterioration of the physical infrastructure of the irrigation and drainage system and in conflicts related to the operation, maintenance and administration of the system that caused so far an unsustainable development of the irrigation district.

The improvement of the present water management needs modernization and adaptations that have to go from a rigid water distribution system to a much more flexible one that will be adapted to the changing conditions that are inherent to the modifications in the cropping pattern. Before the transfer the cropping pattern was mainly based on grain crops for which the cropping schedule, plant dates and irrigation schedules were established by the governmental agency without any or major interferences from the farmers. The design operation schedules for the water distribution within the network were based on a continuous supply (24 hours) during the whole irrigation season. The establishment of this kind of irrigation schedule was possible due to the grain crops on the fields that allow for a relatively long irrigation interval (1 or 2 weeks). After the transfer of the management the cropping pattern drastically changed due to the introduction of highly sensitive cash crops (citric and fruit trees), which require a totally different irrigation schedule, because the irrigation interval of these crops is almost daily. Moreover, the introduction of sugar cane also caused some significant changes in the traditional water management practices.

Differences in interests and the lack of a clear policy from the side of ASORUT in view of these changing conditions caused an improper water management that resulted in inefficient water use, misuse of the hydraulic infrastructure, inappropriate operation of the pumping stations, high energy costs for the operation of the pumping stations, loss of authority by the administration towards the users, and in some cases water management in a occasionally arbitrary way and sometimes even by force.

It is expected that the cropping pattern in the future will have a tendency to change towards grain crops, sugar cane and fruit trees and that the area cropped with sugar cane will slowly decrease on the long term as it will be replaced by more intensive crops like vegetables and fruit trees according to the prospects of the new scenario for a Participatory and Communal Approach: a new role for ASORUT. In this way, the water distribution system must be modified in such a way that it is able to satisfy the irrigation requirements and that it will take into account the particular characteristics of the various crops selected for the new cropping patterns. This means that the irrigation infrastructure should be able to supply water in a flexible way and that it has the necessary provisions to distribute the irrigation water in an equitable, sufficient and timely way and at the same time that it will deliver the water by taking into account the different irrigation intervals of the crops.

During the second participative workshop, the water users developed and proposed a plan to divide the area in sectors for an improved irrigation water distribution. This would allow for the establishment of an irrigation schedule for the whole area; and cropping schedules and growth calendars, especially for the grain crops. The establishment of the boundaries of the irrigation sectors or blocks will depend on several factors, for instance on the command area of an irrigation canal of the main system, namely the main canal, interceptor canal, and canal 1.0; the practical capacity of the pumping stations; the present division in five sectors in view of the management of the irrigation district. These sectors are mainly defined by management aspects and geographical conditions. Moreover, the boundaries of the new sectors will be influenced by the future envisaged development of the land use (cropping pattern) in the various sectors.

Several possibilities to divide the area in sectors can be considered for the current condition and without major modifications of the existing hydraulic infrastructure. For instance the whole area could be divided into two irrigation blocks, namely block one will be conform the grouping of sector 2 and 4 under the command of the main canal. The other block will be formed by the grouping of the sector 1, 3 and 5 under the command of irrigation canal 1.0 and the interceptor canal. Previously, the duration of the irrigation period had to be defined by taking into account the irrigation intervals (irrigation frequency) of the crops and next the irrigation period of each block is defined according to its command area. According to the cropping pattern for each sector and the related crop water requirements, the required discharges for each irrigation canal are determined and an operation schedule for the pumping stations has been established. The required discharges for each of the irrigation canals will be compared with their design discharges and from this comparison follows whether the division in sectors is a feasible alternative.

The water distribution block by block (sectors) means significant organizational changes for the water management. ASORUT as the irrigation agency will be responsible for the water management at the main irrigation level that will have to result in the supply of irrigation water on a timely, sufficient, equitable and efficient way; while the management at farm level or the level of a group of farms will be the responsibility of the farmers (water users organization at sector level). A proper management at this level will benefit from the establishment of storage reservoirs that are especially needed for the farms with highly sensitive cash crop (citric and fruit trees). Appendix V shows some technical alternatives for the irrigation sectorization.

From a functional point of view the physical interventions for the hydraulic infrastructure have to be considered at the middle and long term in order to improve and maintain the technical sustainability. One of the interventions is related to the interceptor canal, which was mainly designed to protect the area against floods from the small rivers coming from the Western Mountain Range. However, after a long time after the start of operation of the RUT Irrigation District the interceptor canal was also used (1972-1975) to supply irrigation water for some areas of sector 1 and 3 by pumping water from the Tierrablanca Station to the Conveyance Canal. In 1995 water was also supplied to sector 5 as a result of the unfinished construction works of irrigation canal 1.0. The first intervention was possible through the installation of structures with adjustable water level control with radial gates, which can be mechanically or electrically operated. In this way, the irrigation service was provided for the Northwest area of the RUT Irrigation District. The second intervention was the construction of a new water level control structure that will allow the irrigation of the Southwest area (sector 5) of the RUT District.

In this way the interceptor canal has a double function: flood protection and supply of irrigation water for some areas of the district. The two functions obstruct each other; the control structures for irrigation are an obstacle for a quickly evacuation of the streams coming from the Western Mountain Range and in some cases the structures are even blocked by tree-trunks, stones and branches. Moreover, the high sediment load of the streams and the removal of this sediment from the interceptor canal, the aquatic vegetation, and the insufficient and improper maintenance cause an inefficient irrigation operation. Therefore, the two functions of the interceptor canal have to be separated in the future to obtain a more efficient operation. The irrigation function can be replaced by an extension of the existing irrigation canal 1.0 in the same way as was considered during the design stage. The water supply for sector 1 and 3 could be provided by irrigation canal 1.0 and the interceptor canal would maintain its original function, being the drainage of the water coming from the western Mountain Range.

Another important intervention to be considered at the mid term is the present and future function of the Cayetana Pumping Station. Initially, this pumping station was designed to pump excess water from the main drainage canal to the Cauca River. However, the Cayetana pumping station also pumps water from the main drainage canal to the main irrigation canal to supply irrigation water to sector 4. As was mentioned before, this supply of irrigation water is possible, due to the fact that extra, excess water is needed to raise the canal water level and to supply water to the irrigation network at the proper water level for irrigation. In this way, the excess water flows to the drainage system and will result in water storage in the main drain in case the outlet structure "Ceros" is closed. The water quality in the main drain is not favourable for irrigation in view of its high salt content and its continuous use has caused soil salinization problems in the northern part of the RUT Irrigation District. Moreover, the water storage in the canal will not maintain the groundwater table at an acceptable depth, being a low level, but the higher water table will increase the salinization process by capillary rise.

Therefore the function of Cayetana Pumping Station for irrigation is not very suitable and practical, as the excess water pumped by this station causes negative environmental impacts. If additional irrigation water is needed three other interventions can be considered: namely increasing the conveyance and distribution efficiency of the current distribution system, especially of the Marginal Canal; a redesign of the Cayetana Pumping Station in such a way that irrigation water can be taken from the Cauca River and/or an increase of the Candelaria Pumping Station with one or two pumping units. However, taking into account the number of pumps per pumping station, their respective capacity, and the total irrigation area of the irrigation district; gross calculations show that the total installed capacity of 9.6 m³/s of the Candelaria and Tierrablanca Pumping Stations corresponds with an irrigation gift of 0.96 l/s.ha or a water requirement of 8.2 mm/day. This means that the alternative to improve the conveyance and distribution efficiency should be seriously considered in combination with actions to improve the application efficiency at farm level, together with an optimization of the operation of the Candelaria and Tierrablanca Pumping Stations. In this way the Cayetana Pumping Station could be used for drainage as was considered at the start of the design stage.

Land levelling is needed in all the cases that the irrigation system is operated as a gravity system, in the similar way as the original hydraulic design. As was mentioned before, the present operation of the RUT system is primarily based on the criterion to establish the proper water levels in the network than to supply the allocated discharges. This means that the first criterion needs excess water to establish these water levels in the canals. This operation practice facilitates the storage of extra water in the main drainage canal, which is also used for the supply of irrigation water to the farms under command of the drainage canal or by water pumped by the Cayetana Pumping Station to the Main Irrigation Canal. In principle, this practice should be avoided as much as possible in view of the low water quality of the extra water that accelerates the salinization process. Thus, land levelling and the improvement of all the hydraulic structures for flow control contribute to a better operation of the irrigation system as a gravity system.

In the framework of this research a detailed study concerning the Optimisation of Water Management in the RUT Irrigation District has been carried out (Otero, 2002). The study states that the present irrigation supplements the rainfall without defined criteria for the operation of the pumps and it is likely that more water than required is supplied to the area, affecting not only the pumping costs for irrigation but also those for drainage. To investigate these operation aspects the actual situation was evaluated based on an overall water balance at scheme level and an analysis of the fluctuation of the groundwater table in the area. Consecutively, a water balance at field level using the WASIM model was used to determine the irrigation requirements, yield reductions and drain flows based on a 20-year simulation. Hydrodynamic simulations for the main drain using the drain flow from the water balance and one-, two- and three-days rainfall with a return period of 5 years provided information on the required pumping capacity for drainage and the extent of the areas inundated. Appendix VI contains a summary of this study.

The WASIM program (Hess et al., 2000) was used to simulate the daily water balance at field level for maize and sugar cane and for three soil types. The water balance was studied for three scenarios, namely: the present situation in the District; without any irrigation to find the yields under rainfed conditions; and with irrigation for no yield reduction.

The hydrodynamic response of the main drain to different rainfall conditions and the pumping operations was simulated with the hydrodynamic model DUFLOW (IHE,

1998). The main drain collects water from runoff and sub-surface flows, and conveys the excess water to the northern part of the District where the outlet is connected to the Cauca River. The simulations with DUFLOW covered the following scenarios:

- A one-day rainfall of 80 mm/d, which corresponds to a return period of 5 year, and the present agricultural practices without pumping operation;
- A one-day rainfall of 80 mm/d, but now with pumping operation;
- A two-day rainfall of 50 mm/d, which corresponds to a return period of 5 year and with pumping operation;
- A three-day rainfall of 37 mm/d, which corresponds to a return period of 5 year and with pumping operation.

The most important conclusions and recommendations of the study were as follows:

- The present operation of the pumping stations can be improved. According to the water balance at field level, the areas planted with sugarcane and maize require on average 16% less water than the mean annual volume of water pumped into the system for this cropping pattern. This reduction in irrigation supply will also result in 23% less sub-surface water to be drained off;
- Shallow groundwater affects a large area of the district during the growing season. The areas most affected are located in the North near the outlet, in the South and in some spots in the centre of the district. The situation could be improved by a better drainage system at field level or by changing the cropping pattern with crops that have shorter root depths. Crops like maize, sorghum and even soybean can be a rainfed crop, but other crops like sugarcane should be irrigated;
- The groundwater levels calculated for the present irrigation conditions will be harmful, mainly in the sandy clay loam soils. This soil type should be clearly identified in view of possible improvement of the drainage at field level and the necessary periodical maintenance of the drains;
- Irrigation has a high impact on the drainage flows; the supply should be limited to or be less than the potential evapotranspiration. Any excess water will flow almost directly to the main drain or raise the groundwater table;
- Irrigation and effective rainfall result in an almost uniform sub-surface flow that ranges between 0 and 2 mm per day. The pumping operation for drainage is unpredictable since it depends mainly on the runoff and the water levels in the Cauca River. Pump operation at La Candelaria station should start from water levels at 910.5 m+MSL, while La Cayetana Station should start at 910.0 m+MSL. The present pumping capacity cannot prevent inundations and high groundwater for 1- and 2-day rainfall with a return period of 5 year (exceedance of 20%).

6.4.2 Interventions for the Environmental Aspects

From an environmental point of view, the interventions concern the preservation and improvement of the quality of the water, the appropriate disposal of solid waste, the protection and conservation of the watersheds and micro-watersheds, the control of the salinization process, and the formulation of plans for the environmental management.

Water Quality

One important restriction in view of agricultural development is the quality of the irrigation water that is threatened by contamination by domestic wastewater from the municipalities of Roldanillo, La Union and Toro. The water is partially purified in treatment plants and then carried to the interceptor canal, from where it is used for irrigation. The canal water is contaminated by a high content of organic matter and mineral nutrients causing health risks for the consumers of the agricultural products and the irrigators, and limiting the export of fruits and vegetables due to the fact that the produce does not fulfil the international quality and health standards.

The efficiency of the treatment plants of the municipalities needs to be increased in order to improve the quality of the water in the interceptor canal so that it can be used in a safe and economic way for the irrigation of the crops. It has to be mentioned that ASORUT is not responsible for the improvement of the treatment plants; therefore, the main action of ASORUT in view of the required improvements will be a constant pressure on the Departmental Association of Water Supply Systems and Wastewater Disposal of the Valle del Cauca Department (ACUAVALLE) that is directly responsible for the treatment of all the wastewater of the municipalities in the department. ASORUT will also have to look for and promote feasible treatment alternatives of the wastewater coming from small villages, settlements, and farms located within the command area of the irrigation district. Septic tanks, wetlands, lagoons are, amongst others, some of the technical alternatives to be considered.

Moreover, it is advisable that ASORUT promotes the reuse of domestic wastewater after it has been treated in view of an increased demand for irrigation water. Some data from the wastewater treatment system of the municipality of La Union give a clear idea concerning the potential of the reuse of the outflow once it has been treated. The outflow of the treatment plant is about 5,200 m^3/day, which is equivalent to 60 l/s; assuming an irrigation gift of 1 l/s.ha, an area of 60 ha could be irrigated. This figure can be increased when using drip irrigation, for instance for fruit trees. Assuming that the water costs are Col\$ 20 per m^3, the total cost would be Col\$ 3,120,000 per month. The nitrogen concentration of the wastewater is about 30 mg/l and the amount of nitrogen in the total outflow is 156 kg/day, which is the equivalent of 7 bags of urea per day (23 kg of nitrogen per bag). The seven bags per day mean 210 bags per month at a monthly total cost of Col\$ 8,085,000. These figures might be an incentive for the improvement of the wastewater treatment plants and reuse of the treated water (Peña and Muñoz, 2004, Alonso, 2004).

The Cauca River is the main source of irrigation water for the RUT Irrigation District. However, the quality of the water pumped into the district is characterized by a high sediment load and contamination by wastewater disposals and chemical wastes from the municipalities and industrial areas upstream of the RUT District. Also in this case ASORUT is not responsible for the improvement of the water quality, but it needs to take all necessary actions to convince and to force the environmental authority in charge of the water quality of the Cauca River, namely the Regional Autonomous Corporation of the Valle del Cauca (CVC). This authority will have to impose the necessary regulations to treat the wastewater before it is conveyed to the river; to enforce the guidelines and laws concerning the treatment of the solid waste of the municipalities, and to promote and take all the necessary actions to keep the deforestation process under control within the Cauca River basin.

Solid Waste

Garbage, household waste and construction rubble coming from the municipalities are dumped into the canal network. This has reduced the conveyance capacity of the canals, increased the maintenance costs, and affected the optimal operation of the irrigation network. ASORUT has to look for cooperation agreements with the local administration of the municipalities so that they will enforce the rules for the most appropriate solid waste disposal in the district.

The improper disposal of solid waste at farm level, like agro-chemicals, will cause a continuing contamination of the water and soil resources. Moreover, this incorrect disposal might increase the heath risk caused by improper handling of the chemicals. It is the responsibility of ASORUT to plan and launch intervention actions for a more proper management of the solid waste at farm level by, for example, containers, bags, cans, etc. In some cases, the containers and cans of the agro-chemical products can be used again and can be brought to the farmers by stores in close coordination with ASORUT. Disposal of the waste after harvest is another issue to be considered; treating the soils by adding compost is an activity to be promoted.

Watersheds

The streams coming from the Western Mountain Range transport high loads of sediment, mainly due to deforestation and erosion processes taking place in their basins. Among the streams intercepted by the interceptor canal are the Roldanillo and Toro rivers and the Rey, San Luis, La Union, and San Antonio streams. The interceptor canal is affected in a negative way by these small rivers and streams, which carry large amounts of sediment.

It is the responsibility of ASORUT to propose and stimulate actions by the Municipality Units for Agricultural Technical Assistance (UMATA) and by the Majors Offices of the municipalities in order to formulate and implement plans for the improvement, conservation and protection of the watersheds of these small rivers and streams in a cooperative way. The likely reduction in sediment load to the interceptor canal will contribute at the midterm to a reduction in costs for preservation and maintenance, an improved operation of the network and to the financial sustainability of the RUT District. The Regional Autonomous Corporation of the Valle del Cauca (CVC) promotes reforestation plans and provides technical and financial support, the latter only on request, to be formulated and developed by the parties interested. Within the RUT area, reforestation with native species is recommended for several functional uses in agricultural practices, like wind barriers to control erosion and the establishment of bio-marketing, for example the growing of bamboo.

Soils

The process of salinization in some areas of the RUT Irrigation District is helped by the reuse of drainage water for irrigation, inefficient irrigation practices at field level, lack of an adequate drainage system at farm level, poor maintenance of the drainage network, and the natural process of salinization caused by the groundwater coming from the Western Mountain Range. Most of the salinization processes can be controlled and limited by some land improvement activities at farm level, for instance by the implementation of surface and/or subsurface drainage, restrictions on the installation of irrigation pumps in the Main Drainage Canal, and optimization of irrigation practices. Moreover, the rehabilitation, operation and maintenance of the previously installed network of piezometric wells will help to monitor the fluctuations of the groundwater table and to make decisions concerning the operation of the whole drainage system. Permanent monitoring of the water quality of the Main Drainage Canal is also needed to analyze the salinization problems along the drain.

Another point of concern is the excessive use of agricultural machinery for land preparation. This intense use will have to be limited and if possible avoided in view of its effect on the chemical and physical features of the soils that might cause degradation, hard pan, and waterlogging, amongst others. Burning of the waste after harvest is another, not-recommended practice as it will seriously affect microbiological activities in the soil.

Environmental Management

ASORUT will have the task and obligation to promote activities in view of the improvement and sustainable use of the natural resources for agriculture by the enforcement of the laws and regulations established by the responsible environmental authorities. In this way, ASORUT will have as its main activities, amongst others: the formulation and implementation of an environmental management plan, development of the monitoring activities based on environmental indicators, to implement and execute the essential activities as mentioned in the environmental management plan, to realize in a gradual way new agricultural practices in view of the environmental sustainability, to support the training of the community in environmental issues and aspects that are directly and indirectly related to the command area of the district, and to support the environmental authorities in the development of reclamation and conservation activities within the watershed basins.

Environmental education of the users is an important topic in order to raise the environmental conscientious, especially amongst children and young people by taking advantage of the presence and action of the various educational institutions in the region and district. ASORUT could propose to the local educational institutions to include in their educational curriculum courses concerning ecology, environmental preservation, and renewable natural resources, to promote the development of inter-disciplinary studies and the development of environmental field activities with the participation of the community, and to promote campaigns for the popular environmental education through rural and urban means as well. ASORUT needs to also strengthen the participation of women in the development of the proposed environmental education projects, because during the development of the Participative

Diagnosis Workshop a potential group of women was identified, which included various educational grades (from professional to technical) and the group showed a high awareness for social and environmental aspects.

In general terms, ASORUT needs to promote, implement and be in favour of clean production processes during all stages of the agricultural production chain. The cooperation agreement with the Autonomous Regional Corporation CVC for the implementation of organic agriculture must be strengthened. The attractive results obtained from the experience with small pilot projects must be spread amongst most of the farmers in order to capture their attention given its environmental and economic benefits. Organic agriculture raises the productivity of low-input agricultural systems, provides market opportunities, offers the opportunity to discover, through blending of traditional knowledge and modern science, new and innovative production technologies, promotes the national and international public debate on sustainability by creating awareness of environmental and social concerns that merit attention.

6.4.3 Intervention for Legal and Organizational Aspects

Based on the present agreements and resolutions, ASORUT will have as one of its main tasks: the update of its internal regulations in line with the legal framework in force, to maintain a continuous and permanent verification of these regulations, and to evaluate the impact of the application of the rules. Moreover, ASORUT is responsible for the transfer of the knowledge and the sharing of the regulations with the farmers, the other users and actors involved with the purpose that they all know their rights and duties and that they can and will perform their respective role in view of a sound management of the irrigation district.

From an organizational point of view, ASORUT must reflect on and update its present organizational structure. The new organization needs to match with the latest proposed role for ASORUT within the *Scenario for a Participatory and Communal Approach: a New Role for ASORUT* discussed before in view of an integrated and sustainable management. The organizational structure will have to be in line with the basic tasks, namely the legal, organizational and operational functions. The legal function will have to facilitate an environment for the establishment of policies on water and on institutional development, and of normative and executive legislation. The organizational function involves coordination, planning, decision-making in the irrigation systems, and reducing problems and conflicts between the users through a platform for a coordination of water use and water allocation. The operational function will focus on the use or control of water for irrigation in order to fulfil the specific needs and demands of the farmers.

Therefore, the highest authority in the RUT Irrigation District will be the General Assembly, which will be constituted by all users preferentially, all of them with voting right and participating actively in the process of decision-making. However, the fact that the whole district will be divided into 5 sectors is also recognized in the formation of the General Assembly, namely that it will have representatives of each of the five sectors. In this latter case, ASORUT is responsible for an equitable and participative participation of the users at sector level.

A modification of the configuration of the Board of Directors is suggested. In addition to the members that form the traditional Board of Directors, namely a president, vice-president, general manager, treasurer, fiscal expert, secretary, and representative of the users, it is recommended to include the contribution of representatives of the land improvement sector (INCODER), the local administrations and the other supporting cities. They can participate in the General Assembly and the meetings of the Board of Directors with a right to speak, but without a right to vote. The representation of INCODER can be delegated to the Departmental Agriculture Secretary and his task will be to assure the transparency and accountability for the decision-making process and to enforce the sectoral policy. The representation of the local administrations (3 municipalities) can be delegated by the Major to one of the Municipality Units for Agricultural Technical Assistance (UMATA), and its main task will be that the municipalities are more involved and that more effective compromises will be obtained between the municipalities and the irrigation district. The representation of the supporting entities (private or public) can be delegated to the Chamber of Commerce, the environmental authority, or some of the academic institutions existing in the area. The main task will be to contribute to a strengthening of the inter-institutional relationships.

In addition to the four existing committees (technical, educational, training and financial), that advise the Board of Directors a social committee is recommended. This committee could act as a department within the organization of ASORUT, once it has become fully mature. Its task will be to maintain a close contact with the farmers and their families by identifying and implementing projects for the improvement of the standard of living and by improving the farmer-ASORUT relationship. In addition, an extra department is recommended for the management of future production projects; its task will be the identification, formulation and implementation of production projects, the management of the post-harvest products, the marketing and the management of the farmers-ASORUT relationship in view of these production projects.

From an operational point of view it is recommended to separate the operation and maintenance tasks and to place them in two different departments, each one with its own facilities and personnel. Additionally, the creation of a section within the operation department is recommended in order to monitor and evaluate the quality and quantity of the natural resources (soils and water) within the district. Figure 6.4 shows a chart of the suggested organization for ASORUT.

6.4.4 Interventions for Socio-economic Aspects

The interventions for socio-economic aspects concern the financial, economic and social components which were examined and discussed by the participants during the first participative diagnostic workshop.

Financial Aspects

One of the major problems faced by ASORUT concerns the financing of the administration, operation and maintenance. Basically, the revenues of ASORUT are based on a water tariff, which consists of a fixed and volumetric tariff. The first one corresponds with the payment that each user must pay per area (Col$/ha), while the volumetric tariff is related to the volume of water used for irrigation. The volume of irrigation water delivered to the users is difficult to measure with a certain degree of accuracy due to the lack of proper measurement structures at farm level. Sometimes the determination of the volume of water is a source of conflicts and an underestimated value of the volume might be expected. Therefore, it is expected that the volume charged is smaller than the one really pumped; causing a negative impact on the revenues of ASORUT and a shortage in the economic resources for operation and maintenance of the hydraulic infrastructure.

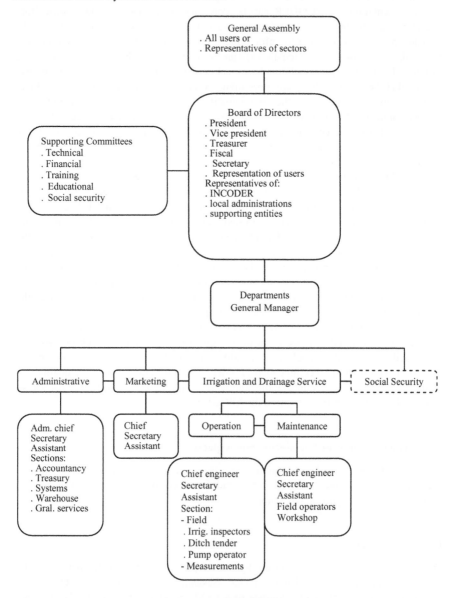

Figure 6.4 Suggested organizational structure for the RUT Irrigation District

As a first intervention, a study to determine the present tariff structure is required. An important starting point for the establishment of the tariff structure is that the farmers are willing and have shown their willingness to pay for the service that supplies irrigation water, because until now agriculture activity is for them a profitable economic activity. After the transfer of the management to the association, the water fees were increased as a way to balance the elimination of the subsidies from the government. This increase resulted in a reduction in the water demand from the farmer's side. After that the tariff values have been increased according to the consumer prices indicator. Since 2003 some trials have been carried out to establish a differential tariff for the fixed one that only takes into account the area size.

The study to determine a differential tariff structure will have to take into account factors such as: farm size and location, productivity, profitability and land use (cash crops or traditional ones), condition of the hydraulic infrastructure for the irrigation and drainage service, quality and quantity of natural resources (water and soils). In other words a rural stratification is needed as a requirement for a differential tariff. In this context, subsidiary tariffs could be considered and expected from the higher strata to the lower ones, as is the case in the urban drinking water sector.

The improvement of the hydraulic structures that will guarantee more accurate measurements of the volume of irrigation water delivered at farm level is an essential requirement for the establishment of a reliable volumetric tariff. The tariff will have to be supported by incentives for an efficient use and water reuse, and fines by an inefficient use of irrigation water.

The amounts outstanding present a serious threat to the financial sustainability of the RUT Irrigation District. The recovery of the money depends on economic and social factors. Actions to improve the economic condition of the farmers will help to increase the payment of the debts, especially in the case of the small and middle farmers. For this reason, the promotion of profitable agricultural practices is a good mechanism to raise the financial capacity of the farmers. On the other hand, the current land tenure is another factor influencing the debts and their payment. An important part of the debts corresponds to some big landowners, whose main activity is not just the legal agricultural activity, but the illegal one, this is the case of the drug traffickers. The payment of their debts depends on their willingness to pay, their continuing stay in the area, and their juridical status.

Some actions can be carried out by ASORUT with the purpose to improve the financial situation. The costs for operation and maintenance can be shared with institutions that affect these costs in a direct or indirect way. For example, some hydraulic structures will have to be considered as works for the public interest in such a way that ASORUT is not the only one responsible for its maintenance. An example is the protection dike, which is part of a larger project for reclamation and protection developed by CVC a long time ago. The maintenance costs of this dike will have to be shared with the Autonomous Regional Corporation (CVC).

Similar situations can be observed for the maintenance costs of the road network within the RUT Irrigation District. The roads are not only used by the farmers, but also by other users from the three municipalities located within the influence area of the irrigation district. In this case the administration of the municipalities will have to contribute to the maintenance costs of the road network. As was mentioned before, the irrigation water quality is affected by the high sediment load and by contamination from the disposal of domestic and industrial waste and wastewater into the main water courses. This situation seriously influences the maintenance and operation costs of the irrigation network of the RUT Irrigation District. For this reason, ASORUT will have to claim economic compensation form CVC, local administrations and ACUAVALLE; these institutions are responsible for the awareness, control, protection, and conservation of the natural resources under their jurisdiction. The water concession given by CVC to ASORUT needs to be agreed upon in terms of quantity and quality.

ASORUT will have to consider other possibilities for additional economic resources different to those coming from water fees. The new scenario proposed for ASORUT (Participatory and Communal Approach: a New Role for ASORUT) will give them new possibilities to get financial resources on the short term, namely funds coming from the participation in the identification, formulation, and implementation of agricultural production project; obtaining new credit lines, provision of technical services for the development of agricultural activities, and marketing of agricultural products. On the mid- and long-term, ASORUT can obtain more economic income from other, but parallel money-making projects, like agro-tourism, small rural enterprises, agro-industrial processing, and training programs for other irrigation districts. Clear rules, accountability and transparency are the main requirements to develop these activities.

Economic Aspects

The aspect that requires the most care is the strengthening of the productive agricultural chains; it should be developed according to the existing natural environmental conditions within the ASORUT area, namely grains, fruits, vegetables, cotton and grass. Some of them offer good perspectives in view of international markets (fruits and vegetables) and will involve some added value through agro-industrial processes within the area. The grains, especially corn, offer good possibilities to meet the local and national demand and they can provide raw material for the national industry (corn flour, concentrated food for animals). Sugar cane forms the basis for an important production chain in the Valle del Cauca Department that consists of an important agro-industrial cluster. However, sugar cane is not a very suitable crop to be grown with the prevailing environmental conditions of ASORUT and the introduction on a large and increasing scale has had a very negative impact on the physical infrastructure. It is expected that under the new management conditions of ASORUT, the area with sugar cane will slowly decrease and will be replaced by other intensive crops like vegetables and fruit trees. ASORUT is also responsible of the support for the introduction of new and promising agricultural chains, such as in the case of chilli, cacao, tobacco, stumps, and forest for functional use.

Husbandry and animal breeding (minor species) and the establishment of market gardens need to be promoted in order to recover and maintain the traditional domestic practices that have been tools that contributed to the food security and the improvement of the domestic economy. This kind of activities will facilitate the participation of the women and young people and will assure a new farmer generation.

Marketing is a key factor in view of the socio-economic sustainability of the community. ASORUT will have to promote and support the privately negotiated contracts for future delivery of a commodity or product. These contracts will assure favourable selling prices, volume and quality of the agricultural products. These private futures contracts can be negotiated with large department stores, with large farmers who need a certain production volume for their marketing, and with the agro-processing industry or the National Agricultural Stock Exchange. This is an appropriate marketing mechanism especially for small and middle farmers, who will be represented then by ASORUT during contract negotiations with the large parties.

Social Aspects

The major effort of ASORUT needs to be focused on the recovery and strengthening of the sense of belonging of the farmers. This might be possible when the main objectives of the organization are in line with the interests and expectations of the farmers. They expect from the organization for instance, development proposals based on profitable agriculture practices and with the aim to improve their standard of living in an efficient and equitable way. If the organization is capable to satisfy the interests of the farmers, it might be expected that they will be willing to participate and support the development process within ASORUT. Transparency, accountability, clear actions, administrative efficiency, and equity are requirements demanded by farmers to create confidence in the organization. In this sense the organization needs to develop sufficient capacity in order to receive and interpret the wishes and demands of the beneficiaries by using approaches for local initiatives and incorporating participation and subsidiary principles. At this point it is important to mention the definition for capacity development as given by the United Nations Development Programme: "Capacity development is the sum of all efforts needed to nurture, enhance and utilize the skills and capabilities of people and institutions at all levels – locally, nationally, regional and internationally – so that they can better progress towards sustainable development. Developing capacity involves empowering people and organizations to solve their problems, rather than attempting to solve the problems directly. When capacity development is successful, the result is more effective: people and institutions are better able to provide products and services on a sustainable basis".

Appendix VII shows a summary of the required interventions for the RUT Irrigation District.

6.5   Monitoring and Analysis of Short Term Impacts

The main physical and non-physical interventions have been identified for the RUT Irrigation District with the intention to select the most appropriate model for an integrated and sustainable management; that model follows the "Scenario for a Participatory and Communal Approach: a New Role for ASORUT". The interventions identified involved various technical, environmental, legal and organizational, and socio-economic aspects, which have already been fully explained in the previous sections.

Some of the interventions will take place on the short term and others will take place on the middle and long term; their implementation of the interventions in the time will depend on factors such as complexity, need for implementation, availability of financial funds and human resources, and especially on the willingness of the members of the organization and the supporting entities to undertake the required interventions.

The implementation of the interventions to reach the selected management model for the RUT Irrigation District will have as starting point the current status of the organization, which corresponds with a condition that is near to a fully collapse of the organization. In order to reverse this situation a transition stage will have to be considered before the implementation of the initial phases of the selected management model. The main focus of the activities to be executed within the transition stage will be a detailed consideration and contemplation of the selected management model by the organisation itself.

The transition stage will take place on the short term and the main attention will go to the development of the strategic planning process that is considered as the process of thinking and decision-making which does not only involve the internal relationships but also those between the organization as a whole and its 'transactional' environment (Ackoff, 1981 cited in Malano and Hofwegen, 1999). According to this vision, the general process involved in the formulation of a strategic plan consists of:

- Review of the purpose of the organisation;
- Inventory of the 'transactional' environment in which the organisation must perform its current tasks, maintain its position and functions, react on the risks, challenges and constraints;
- Identification of realistic, achievable but challenging objectives and goals with time based targets;
- Formulation of strategies, programs and activities directed towards the achievement of the agreed objectives and goals;
- Inventory of the resources of the organization that are at its disposal including their strengths and weaknesses;
- Formulation of the action plan that results from the above discussed analysis;
- Monitoring and evaluation of the progress towards the achievement of the objectives and goals.

The implementation of the activities within this transition stage can not be carried out by ASORUT in an isolated way. In the framework of an integral and participatory approach, this strategic planning process needs the involvement of all the direct and indirect actors; that include the farmers; the administrative and technical staff from ASORUT; and the representatives of the supporting public and private entities; local and regional administration; government agency for the Land Improvement Sector, and the academic sector. Given the low credibility of the existing organization in the eyes of the farmers and other supporting institutions, the role of a process facilitator is considered as a very important matter. Taking into account the recent experiences in the district, it is evident that this role could be performed by the three main institutions within the region, namely the Agriculture Secretary of the Valle del Cauca Department (public sector), the Chamber of Commerce (private sector) and the Valle University (academic sector). These institutions showed leadership and credibility at the eyes of the stakeholders in recent critical moments suffered by the organization.

During the implementation of this transition stage the participants will have to formulate some key questions, such as: where are we now; where do we want to be; and how and when do we get there? The answers will result in a broader appreciation of the current and future issues that will confront the irrigation and drainage organization (Malano and Hofwegen, 1999).

The strategic planning process will give the following outputs:

- *Mission Statement*: This statement will result in a comprehensive vision about the future of the organization. It is a concise statement that sets out the final goal and the essential tasks of the organisation. According to Goodstein et al., 1993 (cited in Malano and Hofwegen, 1999) by formulating its mission, the organisation must answer four main questions: what function(s) does the organisation perform; for whom does the organisation perform these functions; how does the organisation carry out these functions; why does the organisation exist?
- *Strategic Objectives*: The goals of the strategic plan should reflect the priorities that are identified as the critical success factors. They must be stated as explicitly as possible, as they will become the strategic guide fir the whole management

process. Specific objectives and targets must be attached to each goal to provide a clear quantification of what has to be achieved;

-   *Operation Plans*: The main product of the strategic planning process are the operation plans, the management plans that provide the blueprint towards the achievement of the planned main goals and that form a rational basis for the formulation of forward and annual budget estimates;

-   *Performance Monitoring*: Monitoring of the progress of the plans and the adjustment of the strategies should be an on-going management activity. The program structure gives a first and clear outline for the monitoring activities.

In order to reach the objectives of the transition stage a monitoring and evaluation system is needed to watch and supervise whether the general process as presented in the formulation of the strategic plan is being carried out in a proper way, and whether the components of the process really contribute to the final achievement of the envisaged results or goals. Monitoring and evaluation are considered as resourceful management tools. They can be seen as a lessons-learning and learning by doing, and they can be perceived as an evolution process that has to be developed and refined gradually.

Monitoring is gathering and transmission of information in order to asses performance in general. Monitoring follows the outputs and enables the management to make timely adjustments in inputs in order to stay "on track". These activities require the monitoring of the most critical indicators that measure the achievement of the strategic goals. Evaluation is the processing of the collected information in order to examine the goals and strategies, to compare the results of monitoring and observations with the present standards or targets and asks whether the strategic plan is "on track". Evaluation should be viewed as a continuous process of internal analysis and adjustments based on an effective feedback.

6.6   Conclusions

A review of the development of the RUT Irrigation District during the past decades will help to understand the critical situation that the organization has to handle at the moment. Its current condition reflects the dynamics of the different stages of the development process that the organization has passed through from its start. The present RUT Irrigation District is the result of a process of social and historic progress. The realization and development of the irrigation system, the management by a government agency, the management transfer process, and the management by a water users association are the most relevant stages that have occurred during the lifetime of the RUT Irrigation District. In each of these stages interventions, crises, conflicts, agreements and compromises took place within a very specific socio-economic, cultural, environmental, political, and legal context, which contributed to and shaped the present and future functioning of the organization.

The process of the irrigation management transfer has been a step that clearly marked and greatly changed the future course of the organization. The transfer process marked the handing over of the irrigation management from the government to the water users association, by which the organization changed from a subsidiary administration to one where the farmers will be fully responsible for the administration, operation and maintenance and the related costs. ASORUT as organization had not faced a gradual institutional development in line with the new responsibilities as they have been assumed before the transfer. After the transfer the organization focused its activities mainly and only on the process of distributing irrigation water for agriculture without taking into account the conception and implementation of a strategy in view of the social-economic development of the farmers and the improvement of the their standard of living. As a consequence, ASORUT became a non-autonomous and non-self-sufficient organization that lacked credibility, authority and confidence, that showed a low management capability that resulted in an unsustainable management and brought the organization to a condition near to collapse.

The lack of sense of belonging at the side of the farmers has been identified as a key factor for the present critical situation of ASORUT. This social malfunctioning from the side of the users can be closely related to the source of troubles of the organization during its early stages. The self-respect and confidence of the small farmers were also affected by the historic development of the production systems (the green revolution) and the initial concepts of the irrigation district had a negative impact on the traditional agricultural practices. The strengthening of this sense of belonging of the farmers is a main challenge for the administrators of ASORUT and the supporting entities, especially a change in attitude of the small farmers is very important as they together form the largest group within the organization.

The present expectations for the Colombian agriculture in view of international marketing and agreements, the new role of the Government in relation to the land improvement sector, and the socio-economic importance of the RUT Irrigation District are some driving forces that will deal with actions for a new management model of the organization. Therefore the modernization of the physical infrastructure, organizational strengthening and institutional development are requirements for the full development of the potential of the available land, water and human resources.

A bottom-up approach based on decision-making at the lowest level (subsidiary principle) is recommended as it will incorporate the active and wide participation of all, directly and indirectly involved actors, allowing for the formulation of actions and interventions in a mutual way. This approach will form a sound base for the implementation, creation and sustainable management with a high degree of success. Moreover, it will clearly show the concept that an irrigation and drainage system is the final result of a process of social-historic creation.

By using the bottom-up and participative approach, two possible scenarios can be identified in view of a sustainable management of ASORUT. Each scenario will have different characteristics for actors with the leading role, special interests, and conceptual approaches. The first one is called the "Direct Governmental Intervention Scenario", in which the government performs the central role by introducing a management model by concession, which has recently been considered and approved by the new, legal framework in Colombia. This scenario will not contribute to an integrated and sustainable management on the long-term, because the social-economic sustainability will be seriously endangered due to a serious deepening of the differences between the small-middle farmers and the larger ones. The model does not contribute to an integrated management, because the benefits expected with this new scenario will create unacceptable social inequalities as the large farmers will receive most of the profits and the small-middle ones will be excluded from the benefits.

In the second scenario that is called the "Scenario for a Participatory and Communal Approach: a New Role for ASORUT", the community consisting of the water users in the RUT Irrigation Disctrict will play the main role. This scenario is based on an integral and participatory approach involving all the internal and external actors and based on new developed guidelines for the future interventions needed for a sustainable management. This scenario will give an organization with a platform based on enterprise criteria, improving the quality of life of the farmers through a capable and highly respected management of the irrigation district that functions under the principles of equity, competitiveness, sustainability and multi-functionality.

This second scenario is the most suitable one for the irrigation district, the water users and farmers in view of an integrated and sustainable management. It will create new conditions that will empower the community and allow them to participate in an active way in the decision-making processes, while they take their responsibilities as they are the central actors for the necessary change processes.

7

Evaluation

7.1   General

Irrigation Development and Expectations

Within the Latin-American context the development of irrigated agriculture in Colombia is not so significant in spite of its high potential of lands (6.6 million ha) that are suitable for mechanized agriculture and modern irrigation. At the moment only 12% of the potential area has an irrigation and drainage infrastructure, of which 62% has been developed by the private sector. In Colombia the major irrigation development took place in the 1960s and 1970s with an average annual growth of 6.6% and 3.2% respectively, being almost zero during the period 1990 – 2000 for which the public sector concentrated its resources on the rehabilitation, completion and enlargement of the existing irrigation districts.

Before the 1980s the Colombian Government was fully responsible for the irrigation development following the at that time typical and internationally accepted policy for land improvement projects that was characterized by a minimal participation of the local community and under almost total responsibility of the Government. This meant that the implementation and development of land improvement projects was mainly considered as mere physical interventions and with a limited attention to the social, cultural, and economic requirements within an acceptable social context. As a result, the irrigation development policy caused a very weak sense of belonging from the side of the farmers, a no-payment attitude for the irrigation water services, a paternalistic culture of the Government for the irrigation development and the responsibility of the Government to carry out all the administration, operation and maintenance activities. These characteristics of the land improvement policy determined and contributed in a high degree to a very difficult implementation of the new land improvement policy after the 1990s.

From past experiences several constraints can be identified that hampered efficient investments in irrigation and drainage projects. From an institutional point of view, the main actions of the central and regional agencies have been implementing activities; they did not efficiently use the investment resources, because of the scarce contact with the community that would have to benefit from these developments. In addition these agencies did not carry out an efficient control and supervision of the private irrigation expansions. Colombia has two possibilities to develop irrigated agriculture. One is through enlargement of the existing irrigation districts as most of their potential irrigation area is not yet fully developed. This is the case in the Rio Recio, Coello, Saldaña, and Zulia Irrigation Districts. The experience of the Coello Irrigation District is an example of this possibility where the water users organization USOCOELLO was able to enlarge the district through the execution of the Cucuana Project in 1998. The other possibility is through the construction of new projects which have already been considered after the 1990s. Examples are the cases of Ranchería, Triangulo del Tolima, and Ariari Projects.

In the future it is expected that the Government policy promotes the irrigated agriculture development because the agricultural sector is still and remains to be a very important component of the national economy, and besides the globalization of the economy is a strategy for the re-activation of the economy through international trade agreements in which the agricultural sector plays an important role.

Institutional Reforms

The same development that occurred in other countries, occurred in Colombia at the beginning of the 1990s, namely the country faced a strong economic crisis, which caused the need for institutional reforms of all economic sectors. Thus, the reforms for the land improvement sector led to a change in the role of the Government and the implementation of the program for the transfer of the irrigation management from the Government to water users organizations. These changes are clearly noticeable in the decentralization of managerial functions of the land improvement sector and the transfer of the responsibilities of the irrigation management to private organizations with special emphasis on water users associations. After the transfer, the users organizations are fully responsible for the administration, operation and maintenance of the irrigation districts, including the associated costs and it would also result in a very large or total reduction of the participation of the government by subsidies. In this way, the government changed its role from a paternalistic executor and administrator of land improvement projects to an entity that supports and accompanies the water users organizations in their management activities with the aim to develop land improvement projects and to participate in the whole project cycle.

The policy for the institutional reform was endorsed by the enactment of the Law 41 of 1993 for the land improvement sector. This law set up the rules by which the land improvement sub-sector is organized and it defined land improvement as: "The infrastructure works devoted to endow an area with facilities for irrigation, drainage and flood control to increase the productivity of the farm sector. It is considered as the public sector and defines users as every juridical or natural person".

The institutional arrangement considered the land improvement sector conform the regulations formulated by the Ministry of Agriculture; the Superior Council for Land Improvement (CONSUAT) became in charge of the application of the policies for the land improvement sub-sector; the Government Agency (National Institute for Land Improvement, INAT) became the executive entity to develop the land improvement policy with other public and private entities and the National Fund for Land Improvement (FONAT) became the administrative entity for the financing of irrigation, drainage and flood control projects.

However, after seven years of land improvement programs within the new institutional framework, strong criticism rose in view of the role of the governmental agency (HIMAT later INAT) concerning the non-efficient management of the economic resources and an almost zero-development of the land improvement programs launched from 1990 by different governmental programs. These criticisms were among others related to the management of external resources for the land improvement program, the costs for the functioning of the agency, and the quality of the agency's performance.

INAT was considered as a demanding institution in view of the high costs required for the execution of its assigned functions, its performance was measured and evaluated as low in view of the area improved for irrigation in comparison with the assigned national budget and also its performance was low as a mediator between the water users associations and the Superior Council for Land Improvement for the development of land improvement projects.

In general it can be said that the land improvement programs showed a slow development. For instance, the large irrigation projects Ranchería, Triangulo del Tolima and Ariari are since 1994 the most important projects within the different governmental programs. Their potential area covers about 64,242 ha or 60% of the total target area for the projects under design and construction for the period 1994 – 1998. However, 10 years later, these projects have still the same status, namely design for Triangulo del Tolima and Ariari, and design and partial construction for the Ranchería project.

As conclusion, it can be said that the institutional framework, of which INAT was a component, was a significant factor in the high costs of the land improvement works in the public sector. This fact caused a very low institutional credibility, no interest for private investments, a serious obstacle for rural economic growth; corruption in the processes of hiring and executing the works, and finally the results of the construction works did not match the users needs (Torres, 1999).

This situation required and caused a change in the institutional framework for the land improvement. As a consequence, changes were introduced at the beginning of 2003. Thus, INAT, the Colombian Institute for Agrarian Reform (INCORA), the National Institute for Fishing and Aquaculture (INPA), and the Co-Financial Fund for the Rural Investment (DRI) were suppressed and integrated in a new organization called Colombian Institute for Rural Development, INCODER. As a first result, INCODER allowed a staff reduction of 56% for the agricultural public sector.

This merging process was completed with the delegation of most of its former functions to regional and local institutions, as in the case of the Departmental Agriculture Secretaries and the Municipality Units for Agricultural Technical Assistance (UMATA). These organizations at regional and local level play nowadays an important role as they are the link between INCODER and the community organizations that will address the available resources for investment in an efficient way. INCODER is now charged with the execution of the policy for rural development by facilitating the access to production factors, strengthening the regional entities and their communities and promoting the voicing of the institutional actions in order to contribute to the socio-economic development of the country. The creation of INCODER also allowed the implementation of the decentralization and delegation policy that delegated functions to regional and local institutions and contributed even more to the simplification of the organizational central structure of the land improvement sector. In this context, the organizations at community level need a strong organizational structure, they need to receive the necessary power by the communities, they need to have a high and autonomous management capability, and need to be strengthened in view of the inter-institutional relationships in order to access and use the available resources. These characteristics are necessary in view of the development and management of land improvement projects in an efficient and sustainable manner.

The Agreement 003 of February 2004 was considered as a key element within the new institutional framework by which organizational guidelines were established for the users' organizations. Now, a water users association is defined as a legal person with private rights, with a corporative character and special objectives and it is a non-profitable association; its main objective is the administration, operation and maintenance of an irrigation district under competitiveness, equity, and sustainability principles. The law authorizes also the water users associations to promote activities in order to reach a major social and economic development, and to carry out actions in relation to the defence and conservation of the available water resources.

However, this Agreement included a controversial aspect concerning the so-called "weighted vote" by which the influence of the user on the decision-making in the general assembly depends on the size of the area that he/she owns. This direction was not followed by the large majority of the water users associations during the process to elect a new board of directors in 2004. This new rule caused great worry amongst the small farmers, because they believe and argue that the new rule will give extra power to the large farmers, especially in the decision-making process at administrative level and in the water resources management. As a consequence, the article related to "weighted vote" was suppressed and the Infrastructure Office of INCODER evaluated the contents of the Agreement 003, and as a result proposed to the Director Council a total elimination of this agreement. Thus, a new regulation has been proposed for the administration, operation and maintenance of the irrigation districts including the transfer forms (delegation, administration, and concession) and considering that INCODER must not have direct interference in the water users organizations, but only in the management of the infrastructure (Resolution 1399 of 2005).

Irrigation Management Transfer

The transfer process is based upon a legal rule in the country's constitution called the "delegation of administration", by which a public good could be turned over to a private corporate entity (water users association) for administration on behalf of the Government. Law 41 determined that after a transfer the control over irrigation districts, finances, operation and maintenance procedures and personnel would be placed in the hands of the water users associations.

The implementation of the transfer program showed a clear difference between the first phase of the transfer program and the following ones. The first one was characterized by the fact that the farmers themselves requested the Government to transfer the management of the Coello and Saldaña Irrigation Districts, while the government agency was against the transfer because they believed to loose the political management of the land improvement sector. A reverse attitude could be observed for the following phase of the transfer program that was promoted by the Government, while the users offered a conditional acceptation to the transfer proposal. In this case, the transfer program was considered as a strategy of the government to deal with the fiscal crisis of the 1990s and to obtain the acquired compromises with the international banks, which insisted on the implementation of the policy of decentralization and delegation of functions. The farmers said that the realization of the transfer process has been very fast and that for that reason the support and training by the Governmental Agency were insufficient.

The current status of the transfer program shows that 67% of the systems (covering 64% of the total area) have been transferred. However, it seems that there is no incentive for the transfer of the systems that have only a drainage infrastructure, because 90% of the total area with irrigation and drainage infrastructure has already been transferred, while just 13% with only a drainage infrastructure have been transferred so far.

The transfer program resulted in a reduction of the governmental financial burden, because the total costs for administration, operation, and maintenance were assigned to the users association of the irrigation systems. The partial and sometimes total elimination of the subsidies for the transferred systems also reduced the irrigation costs for the Government.

The transfer program helped also the further development of a new institutional structure for the land improvement sector and contributed even more to a major cost reduction for the Government and facilitated the implementation of the policy of decentralization and delegation for the land improvement sector. The removal from power of several institutions and the fusion with other institutions in the current INCODER is an example of that policy.

The impact of the delegation of functions towards regional and local institutions can not yet be evaluated as these institutions have only recently started to assume their delegated tasks. However, scepticism exists amongst the leaders of the users organizations concerning this type of delegation, because they believe that the regional and local institutions did not show in the past any effective presence in the irrigation districts; moreover, they are influenced by the political sector and maintain a very low credibility due to a high corruption level and a very unclear transparency in their activities.

In the framework of this PhD research a representative sample of five irrigation districts that were transferred, have been visited and their current status was evaluated from a qualitative point of view. The results show some varied and different characteristics concerning the organizational situation. Three of them present a relevant and stable organization with different grades of development, the Coello and Saldaña Irrigations Districts more than the Rio Recio Irrigation District. Their stable organization is based on financial self-sufficiency as a result of the provision of adequate irrigation services and on a concentration of their main activities on the operational aspects within the practice of a profitable agriculture around a single crop, namely rice. For that reason it can be said that the management transfer contributed in a high degree to the consolidation of the production system, where the different phases of the production chain were developed by special parties and that allowed them to satisfy their wishes and interests and to devote their efforts to specific activities.

Maria La Baja Irrigation District showed a situation with a very poor organization at the extent that it has been shut down by the Government. Today the district is in a process of reorganization. For this district the management transfer accelerated the process of deterioration that already existed before the transfer and it brought the organization very quickly into a serious crisis. The existence of a traditional culture in view of the subsistence agriculture supported by a paternalistic attitude of the government and the lack of a clear vision concerning agricultural entrepreneurships contributed to the negative impacts of the management transfer here.

Finally, the Zulia Irrigation District shows since five years a great dynamic development, following a very critical condition immediately after the management transfer. The district has a special organizational model based on an integral and participatory approach around a profitable agricultural activity that is focused on rice production and on a clear vision for the future development of the organization and the district.

A more detailed analysis of the previously presented cases results in the description of three crucial stages for the organizational development of the post-transfer period, namely adaptation, maturity or full development, and consolidation. Adaptation is a critical phase where the organizations with best capabilities will adapt themselves in a good condition to the responsibilities demanded by the management transfer. The maturity or full development phase is the period in which the organization has developed its full capabilities and looks after the overall development of its beneficiaries based on a profitable irrigated agriculture practice under equity, efficiency, competitiveness and sustainability criteria. In this phase leadership and an integral and participatory management approach are the most important requirements. The consolidation phase is the situation in which the organization as a whole has achieved its maximum development condition to the extent that some specific activities of the production chain can be carried out by other parties while the organization maintains the integral management approach.

The case of the Zulia Irrigation District shows that the sustainability of an organization requires a management based upon an integral and participatory approach that has a clearly defined statement about its future mission. The vision needs to be based on the improvement of the quality of life of the farmers through a management of the irrigation district under principles of profitability, equity, competitiveness, sustainability and multi-functionality in line with the legal framework established by the government. In the framework of this concept, the organization has to be able to formulate development-coherent plans, which include all the important social-economic components and consider an integral, participatory and collective approach, based upon the involvement of all actors engaged in the management (farmers, technical and administrative staff) and the ones indirectly engaged (stakeholders, supporting private and public entities, and the society at large). The plans will have to focus on the most productive projects based on an irrigated agriculture and taking into account all the stages of the production chain (production, post-harvest management, added value, agricultural industry, and marketing).

Lessons Learned

From the previous experiences several lessons can be learned. One of them is that the management transfer must not be understood as a simple transfer of responsibilities to the users associations with the purpose to free the government of its financial burden and to contribute to a lighter fiscal deficit. The transfer of the management must be given in the cases that favorable conditions exist which allow a further strengthening of the management capability of an organization to exploit the potential of the existing socio-economic, natural and human resources in view of the improvement of the standard of living of the users. If favorable conditions do not exist the organizations that manage the system will show after the transfer a low performance and will have the whole financial burden and fiscal deficit that before were suffered by the government.

On the other hand, the experience of the management transfer in Colombia opened the way to consider a wider concept of the role of an irrigation system. This should not only be considered as a simple hydraulic infrastructure that provides irrigation water to the farmers, but also as an important component of a production system whose final objective is to contribute to the improvement of the living conditions of the farmers through irrigated agriculture under criteria of profitability, equity, efficiency, and an integral and participative management approach.

7.2   Evaluation Related to the Management of the RUT Irrigation District

The scenario "Scenario for a Participatory and Communal Approach: a New Role for ASORUT" is considered as the most suitable one in view of an integrated and sustainable management of the RUT Irrigation District. This scenario will result in the fact that the authority is really given to the community and this is expected to lead to an autonomous and self-sufficient organization that is able to make the essential decisions concerning the future organizational set-up of ASORUT and its further development. This will be strongly supported by the proposed bottom-up, integrated and participatory management approach.

However, it is too soon to evaluate the impact of the implementation of the proposed scenario on the management development, because ASORUT is an organization that is now in a critical situation as has been extensively explained in the previous chapters. When the current status continues the organization is expected to collapse in the very near future. Therefore, the stakeholders of ASORUT and the different actors involved will have to make very vital and wide-reaching decisions on the short term to safeguard the organization and the physical infrastructure. In the presented scenario the current situation is considered as a "transition stage", which bridges the present degenerating organization and the new one that will be characterized by a dynamic behaviour and a capability to interpret and satisfy the wishes of the stakeholders in view of the sought improvement of their social-economic living conditions. Moreover, this dynamic attitude of the new organization will be able to maintain and develop stable and lasting contacts with supporting public and private organizations in the land improvement sector. It will also take advantage of the available opportunities for further development as they are offered within the new management framework and the recent governmental policy for land improvement.

This transition stage meets the necessary conditions that are needed to mobilize the opinion related forces that are in search of an enhanced condition for a sustainable management of the RUT Irrigation District. The transition stage will start from the present low organizational level of ASORUT that is a result of the different impacts due to various socio-economic and political events during the lifetime of the organization and for which ASORUT had not the adequate answers and adaptation capability to tackle the new circumstances and lately acquired responsibilities after the transfer of the irrigation system management.

The meagre state of the organization is reflected for instance in the poor socio-economic condition of the farmers, especially the small and middle ones, in the poor management capability of the organization and in a low credibility in the eyes of the farmers and the society at large. Other weak points are an inadequate leadership, a low capability for inter-institutional relationships, an ongoing destruction of many parts of the physical infrastructure, the increase of various negative environmental impacts, and increased social instability. Some external factors that have contributed to this state of almost collapse are the agricultural policy of the 1990s (Economic Liberalization) and the activities and deeds of some illegal groups, especially the paramilitary factions and drug traffickers.

However, the agricultural activities developed in the RUT Irrigation District form the main basis of the economy of the three important municipalities of the northern region of the Valle del Cauca Department, namely Roldanillo, La Union, and Toro and of the bordering regions of the departments of Risaralda and Quindio as well. Therefore, the local administrations of these municipalities are seriously concerned about the existing, critical situation of ASORUT, because it will contribute to an undesired social environment that is characterized by an increase in poverty, unemployment and insecurity. The latter developments will support the practice of several unwanted and illegal activities.

On the other hand, this concern of the local administrations is shared by the Administration of the Department, because the critical situation of ASORUT seriously affects the socio-economic environment of the whole Valle del Cauca Department. As main consequence awareness at local, regional and national level arose that has led to comprehensive discussions and political actions to find urgently solutions for the problems of the RUT Irrigation District. In this way, several meetings have been organized with the participation of the public and private institutions that are interested and willing to play a role in the search for alternative solutions. These events took place with the participation of INCODER and the Departmental Agriculture Secretary and the Municipality Units for Agricultural Technical Assistance (UMATA). Other participants in the discussions and meetings were coming from the environmental authority at regional level (CVC), the academic sector (Valle University), the private sector represented by the Chamber of Commerce and agro-industry enterprises of the region, and the farmers (especially the large farmers).

The possibilities to modernize the RUT Irrigation District through an International Cooperation were the past years a very important incentive that helped the discussions and analysis of the present situation. However, the low management capability of the ASORUT organization, the impact of serious threats by external, but unknown agents, and the recent judicial trouble of one of the most important large farmers form some of the main reasons why this international project will not achieve the expected results at short term.

The before mentioned actions and events are a clear example of the dynamics of the discussion environment that has been created around ASORUT during this step of the transition stage. However, it has to be mentioned that during these discussions and the decision-making process of this international project a traditional top-down approach was followed. Furthermore the subsidiary principle was not applied as the small and middle farmers were not taken into account and did not participate in the process. Only some large farmers participated in the meetings.

A process that clearly reflected a bottom-up approach was followed in the framework developed as part of this research. The different stakeholders and other actors involved got the opportunity to present their views and ideas in several participatory workshops and direct discussions. In this way, the key factors for the present critical situation of ASORUT, the wishes and expectations of the farmers and other actors, and the required interventions have been identified. As a result two scenarios for the management of ASORUT have been developed and considered. After a thorough analysis, one model is proposed that will strongly support an integrated and sustainable management for the RUT Irrigation District, namely the scenario that is based on a Participatory and Communal Approach.

The questions to be answered are: how is it possible to implement the selected scenario under the current situation of ASORUT; how long will it take to arrive at the desired management conditions and what are the requirements for the envisaged development? A vision of the near future under this management model shows that ASORUT will be an organization based on equity, efficiency, sustainability and competitive criteria with the mission to promote the welfare of the beneficiaries (farmers), to improve their socio-economic conditions through profitable agricultural practices and to be involved in all the stages of the production chain: production, post-harvest management, added value process (agro-industry), and marketing.

In this way a sound management of the irrigation and drainage system is one of the instruments to reach these social-economic goals, being the increase of the living standard of the community. The present irrigation system needs to be seen as a part of a more complex system; it will form the connection between the improvement of the natural, economic, human resources and the irrigated agriculture in order to promote profitable agricultural activities within a sound marketing environment, and at the same time guaranteeing the conservation and protection of the involved natural resources.

In this future vision ASORUT will be an organization that has a high credibility in the eyes of the farmers, because the future procedures for decision-making and management of the available resources will be clear and transparent. Moreover, ASORUT will be recognized by others as an organization with a high capability for inter-institutional relationships that will allow the establishment of a close cooperation with other institutions and the signing of agreements for the development of beneficial projects that are aimed at the improvement of the welfare of the farmers.

On the other hand, it is expected that the farmers will have much more confidence in ASORUT and will have a well developed sense of belonging to the organization, because they will find the new organization a reliable means to achieve better socio-economic conditions and to increase their standard of living. Therefore, they will be willing to support and implement the proposals that are reached in a mutual way for the development of financially viable projects in combination with irrigated agriculture and other parallel activities.

Passing from the current status (transition stage) to the desired condition will take considerable time. The ongoing deterioration process of the organization started a long time ago, which maybe is more evident when looking to the irrigation management. Reverting this situation of decline will take also a considerable time and a step-by-step process will have to be followed in order to reach the proposed management condition for a Participatory and Communal Approach.

The name "transition stage" has created a clear consciousness about the need and importance of an enhanced condition of ASORUT in view of an integrated and sustainable management. The inputs given during this transition stage will be the starting point for the formulation and implementation of actions needed to arrive at the new management model.

The formulation and implementation of some small productive agricultural pilot projects that will involve small groups of small and middle farmers will be used as a strategic approach. In this way ASORUT and the farmers will obtain more experience, confidence and knowledge step by step and it might be expected that a successful development of this strategic approach will result in the fact that more farmers and more projects will be involved over time. As a consequence a new organizational culture will be growing and the farmers will recognize its importance and benefits. In this line, it is expected that the organization will be consolidated and will reach maturity, autonomy and self-sufficiency at the short or middle term.

The knowledge of successful experiences based on this strategic approach form useful means. An example is the case of the Zulia Irrigation District. After a deep crisis caused by rural conflicts, the organization has now reached a degree of sustainable development where the farmers perform the central role, meaning that they lead and determine the development policy of the organization.

In order to embark on and to reach the new management conditions for ASORUT the key actors will have to play an important role. The Board of Directors will have to include representatives of the farmer's community along with the Government Agency represented by the Departmental Agriculture Secretary. Other supporting public and private entities will have to join their resources and efforts in a concerted way to contribute to the final realization of the new management model.

References

Abernethy Ch., 2001. *Managing irrigation systems in preparation for management transfer.* ICID 19th European Conference, June 2001

Agustin Codazzi Geography Institute (IGAC), 1973. *Inventario de suelos por clases agrológicas.* IGAC. Bogotá, Colombia

Alonso, B. D., 2004. *Informe de avance de actividades. Sistema de tratamiento de aguas residuales STAR.* Alcaldía Municipio de la Unión – Valle. Septiembre, 2004

Apollin, F., Eberhart, C., 1998. *Metodologías de análisis y diagnóstico de sistemas de riego campesino.* Riego andino. Camaren, Quito – Ecuador

Asociación de Usuarios del Distrito de Riego RUT (ASORUT), 2000. *Información general del distrito de riego RUT.* Seminario – Taller internacional: Control de flujo en redes de distribución de aguas superficiales (riego y drenaje). Evento Internacional Agua 2000. 19 – 27 Octubre, 2000. Universidad del Valle, Cali, Colombia

Asociación de Usuarios del Distrito de Riego RUT (ASORUT), 2003. *Reportes de la sección de sistemas del Distrito de Riego RUT.* La Unión, Valle

Blanco, N. H., 1996. *Diagnóstico de la situación actual del DAT RUT.* Contrato INAT 151/1995. Febrero, 1996

Bos, M. G., 1989. *Discharge measuring structures.* ILRI publication 20, Wageningen, The Netherlands

Bruns, B., Granjar, K., Tajidan, 1994. *Improving the performance of small irrigation systems: am impact survey in West Java and West Nusa Tenggara.* Appendix to final report of Asian Development Bank

Burbano, G. L; Forero, O. E., 1999. *Caracterización del Riego en el Distrito R.U.T.* Universidad del Valle, Universidad Nacional de Colombia Sede Palmira. Cali, Colombia

Brundtland, G. (ed), 1987. *Our common future.* The world commission on environment and development, Oxford. Oxford University Press

Calle et al., 1998. *Régimen jurídico del recurso agua. Normatividad, administración y sistematización para un aprovechamiento sostenible del agua.* Memorias del Seminario – Taller: Aprovechamiento integral del recurso hídrico con énfasis en riego, drenaje y sostenibilidad. Junio 1 – 12, 1998. Universidad del Valle, Cali, Colombia

Corporación Autónoma Regional del Valle del Cauca (CVC), 1986. *Diagnóstico del problema de salinidad y drenaje en el Distrito RUT.* Contrato CVC – HIMAT. División de Asistencia Técnica Agropecuaria. Cali, Colombia

Czech Committee of the International Commission on Irrigation and Drainage (ICID), 2001. *Workshop on transformation and rehabilitation of irrigation and drainage systems in central and eastern Europe.* Proceedings 19th European Regional Conference on sustainable uses of land and water, Brno and Prague, Czech Republic

Food and Agriculture Organization (FAO), International Network on Participatory Irrigation Management (INPIM), 2002. *Overview paper: Irrigation management transfer, sharing lessons from global experience.* International E-mail conference on irrigation management transfer. June – October, 2001

Garces-Restrepo, C., and Mora, L., 1997. *The Application of External Indicators to Assess Irrigation Management Transfer Impacts in Three Irrigation Districts in Colombia.* International Workshop on Irrigation Performance. Mendoza, Argentina.

Garces-Restrepo, C., 2001. *Irrigation management devolution in Colombia. IMT case study, Colombia.* International E-mail conference on irrigation management transfer. June – October, 2001

Gorriz C., Subramanian A., and Simas, J., 1995 *Irrigation Management Transfer in Mexico. Process and Progress.* Technical Paper Number 292. The World Bank. Washington, D. C., USA

Groenfeldt, D.; Sun, P., 2001. *Demand management of irrigation systems through user's participation.* Economic Development Institute. Handbook on participatory irrigation management. Washington, D.C: EDI/World Bank

Groenfeldt, D., 2001. *Moving upstream: changing roles for users and the state in irrigation management.* Economic Development Institute. World Bank

Hernandez, G. L; 1995. *Transfer of National Systems' Management from the Government to the Irrigator Association.* M. Sc. Thesis. IHE.

Hess, T., Leeds-Harrinson, P., Counsell, C., 2000. *WaSim technical manual.* Institute of water and environment. Crainfield University

Hofwegen, P. V., Jaspers, F. G. W., 1999. *Analytical Framework for Integrated Water Resources Management. Guidelines for Assessment of Institutional Frameworks.* IHE Monograph 2. A. A. Balkema, The Netherlands

Institute of Hidrology, Meteorology and Land Improvement (HIMAT), 1985. *Zonificación de Colombia y priorización de zonas para adelantar estudios de adecuación de tierras.* Himat. Bogotá, Colombia

International Conference on Water and Environment (ICWE), 1992. *Development issues for the 21th century.* The Dublin statement report of the conference. January 26 – 31, 1992. Dublin, Ireland

International Commission on Irrigation and Drainage (ICID), 1999. *Granada ICID Declaration.* 17th world congress on irrigation and drainage. Water for agriculture in the next millennium. Granada, Spain. September 11 – 19, 1999

International Commission on Irrigation and Drainage (ICID), 2001. *Important data of ICID members countries.* Databases. www.icid.org/index_e.html

International Commission on Irrigation and Drainage (ICID), 2002. *ICID Yalta Declaration.* 1st International workshop on irrigation management transfer in countries with a transition economy. Yalta, Crimea, Ucrania. Mayo, 2002. www.icid.org

International Institute for Infrastructural, Hydraulic and Environmental Engineering (IHE), 1998. *Duflow manual.* Delft, The Netherlands

Japanese National Committee of the International Commission on Irrigation and Drainage (ICID), 2000. *Proceedings Asian Regional Workshop on sustainable development of irrigation and drainage for rice paddy fields.* Tokio, Japan

Jiménez, H., Carvajal, Y., 1998. *Perspectivas de las fuentes de agua en Colombia.* Conferencia Internacional Agua 98. Junio 1 – 3, 1998. Universidad del Valle, Cali, Colombia

Johnson III, S.H., 1997. *Irrigation Management Transfer in Mexico: A Strategy to Achieve Irrigation District Sustainability.* Research Report 16. International Irrigation Management Institute, IIMI

Kloezen, W.H., Garces-Restrepo, C., and Johnson III, S.H., 1997. *Impact Assessment of Irrigation Management Transfer in the Alto Rio Lerma Irrigation District, Mexico.* Research Report 15. International Irrigation Management Institute, IIMI.

Kraatz, D. B., Mahajan, I. K., 1975.*Small hydraulic structures.* Irrigation and Drainage Paper 26. FAO, Rome

Malano, H., Burton, M., 2001. *Guidelines for benchmarking performance in the irrigation and drainage sector.* IPTRIP Secretariat. FAO, Rome 2001

Malano, H.M; Hofwegen, P.V., 1999. *Management of Irrigation and Drainage Systems. A Service Approach.* IHE Monograph 3. Balkema/Rotterdam/Brookfield. The Netherlands

Marin, R. R., 1992. *Estadísticas sobre el recurso agua en Colombia*. Ministerio de Agricultura, HIMAT. Segunda edición. Santa Fe de Bogota, Colombia

Meinzen-Dick, R., Knox, A., 1998. *Collective action, property rights, and devolution of natural resources management: a conceptual framework*. International Food Policy Research Institute

Merrey, D. I., 1996. *Institutional design principles for accountability in large irrigation systems*. Research report 8. International Irrigation Management Institute, IIMI

Ministerio de Agricultura y Desarrollo Rural, 2004. *Memorias 2003-2004. Manejo social del campo*. Santafé de Bogotá

Ministerio de Desarrollo Economico de Colombia, Financiera de Desarrollo Territorial (FINDETER), Instituto CINARA, 1999. *Servicios Sostenibles de Agua y Saneamiento. Marco Conceptual*. Segunda edición, Noviembre 1999. Cali, Colombia

Mora, L. A., Garcés-Restrepo, C., 1998. *Evaluación del impacto de la transferencia en tres distritos de riego en Colombia*. Memorias Seminario-Taller: Aprovechamiento integral del recurso hídrico con énfasis en riego, drenaje y sostenibilidad. Junio 1 – 12, 1998. Universidad del Valle, Cali, Colombia

Mott – McDonald International Ltda. UK, 1993. *Turnover evaluation report. Irrigation operation and maintenance and turnover component*. Irrigation subsector, project II. Jakarta

Muñoz, M. V., 1978. *Problemas geotécnicos en el distrito de riego del proyecto RUT*. Informe de consultoría. Cali, Colombia

National Administrative Department for Statistics (DANE), 2001. *Encuesta nacional agropecuaria*. Santafé de Bogotá, Colombia

National Administrative Department for Statistics (DANE), 2003. *División de síntesis y cuentas nacionales*. Santafé de Bogotá, Colombia

National Institute for Rural Development (INCODER), 2003. *Estructura organizativa. Decreto 1300*. Mayo, 2003. www.incoder.gov.co/organización/organigrama.asp

National Institute for Land Improvement (INAT), 1995. *La modernización del campo. Cartilla mejores proyectos*. INAT, Santafé de Bogotá, Colombia

National Institute for Land Improvement (INAT), 1996. *Situación de los distritos de mediana y gran irrigación*. Ministerio de Agricultura y Desarrollo Rural. Subdirección de Adecuación de Tierras. Grupo de Gestión de Distritos. Santafé de Bogotá, Colombia

National Planning Department (DNP), 1984. *National water study: availability of surface water*. Bogotá, Colombia

National Planning Department (DNP), 1991. *La Revolución Pacífica. Plan de desarrollo económico social 1990 – 1994*. Presidencia de la República. Santafé de Bogotá, Colombia

National Planning Department (DNP), 1997. La efectividad de las políticas públicas en Colombia: Un análisis neoinstitucional. Tercer Mundo Editores. Santafé de Bogota, Colombia

Oorthuizen, J., Kloezen, W.H., 1995. *The other side of the coin: A case study on the impact of financial autonomyy on irrigation management performance in the Philippines*. Irrigation and Drainage Systems, an International Journal. 9: 15 – 37, 1995. Kluwer Publishers; the Netherlands

Otero, M. F., 2002. *Optimization of water management in the RUT Irrigation District, Colombia*. MSc. Thesis. IHE, Delft, the Netherlands

Palacios E., 2001. *Benefits and Second Generation Problems of Irrigation Management Transfer in Mexico*. International E-mail Conference on Irrigation Management Transfer. June – October 2001

Peña, V. M., Muñoz, N., 2004. *Caracterización físico-química del agua residual doméstica y efluente del sistema de tratamiento con laguna anaerobia y facultativa existente en el Municipio de la Unión (Valle del Cauca)*. Instituto CINARA, Universidad del Valle, Cali, Colombia

Periódicos Asociados Ltda., 1997. *Nuevo atlas de Colombia*. Prensa Moderna Impresores, Cali, Colombia

Periódicos Asociados Ltda., 2003. *Colombia tierra de mil colores*. Tecimpre S. A. Cali, Colombia

Perry, C. J., 1995. *Determinants of function and dysfunction in irrigation performance, and implications for performance improvement*. International journal of Water Resources Development 11(1): 11 - 24

Ramirez, R. M., 1993. *Análisis de la política de adecuación de tierras en Colombia*. Planeación y Desarrollo. Volumen XXIX, Numero 2. Abril – Junio, 1998

Quintero, P. L., 1997. *Manejo participativo del riego: Beneficios y problemas de segunda generación, el caso de Colombia*. Taller Internacional Manejo Participativo del Riego. Cali, Colombia.

Savenije, H. H. G., 1997. *Lectures Notes on Water Resources Management. Concepts and Tools*. IHE, Delft, the Netherlands

Schultz, B., 2001. *Land and water development*. Lecture notes. International Institute for Infrastructural, Hydraulic and Environmental, IHE. Delft, the Netherlands

Schultz B., 2001. *Irrigation, drainage and flood protection in a rapidly changing world*. Irrigation and Drainage 50(4)

Schultz B., 2002. Economic and financial aspects of irrigation management transfer. Proceedings 1st International workshop *on irrigation management transfer in countries with a transition economy*. Yalta, Crimea, Ukraine

Schultz B., 2003. Irrigation and drainage. Present and potential role in food protection and sustainable rural development. In Proceedings Water Week, The World Bank, 4-6 March 2003, Washington, D. C.

Schultz, B., Thatte, C., and Labhsetwar, V., 2005. *Main contributors to global food production*. Irrigation and Drainage Journal. Managing Water for Sustainable Agriculture. 54: 263-278 (2005)

Small, L. A., 1990. *Irrigation service fees in Asia*. Irrigation management network. Paper 90. Overseas Development Institute, London

Small, L. A., Carruthers, I., 1991. *Farmer-financed irrigation: The economics of reform*. Cambridge: Cambridge University Press

Svendsen, M., Nott, G., 2001. *Irrigation management transfer in Turkey: Process and outcomes*. Country cases. International E-mail conference on irrigation management transfer. June – October, 2001

Tardieu, H., 2003. *Irrigation and drainage services. Some principles and issues towards sustainability*. A position paper of ICID. ICID Task Force 3, draft 21. January 2003

Tardieu, H., Préfol, B., 2002. *Full cost or sustainability pricing in irrigation agriculture. Charging for water can be effective, but is it sufficient?* Irrigation and drainage, Vol. 51 No. 2, 2002

Torres, O. J., 1999. *Hacia un marco institucional para la rehabilitación y construcción de proyectos de adecuación de tierras*. Seminario Hacia un programa para la rehabilitación y construcción de proyectos de adecuación de tierras. Santafé de Bogotá, Colombia

Torres, O. J., 2003. *El agua en los Distritos de Coello y Saldaña*. XXVIIICongreso anual de Induarroz. Cali, Colombia

Ukraine National Committee of the International Commission on Irrigation and Drainage (ICID), 2001. *Proceedings 1ˢᵗ International workshop on irrigation management transfer in countries with a transition economy.* Yalta, Crimea, Ukraine

UNESCO, 1978. *World water balance and water resources of the earth.* Series: Studies and reports in hydrology. Paris, UNESCO, 1979

Vermillion, D., 1992. *Irrigation management turnover. Structural adjustment or strategic evolution?* IIMI Review 6(2): 3 - 12

Vermillion, D., 1997. *Impacts of Irrigation Management Transfer: A Review of the Evidence.* Research Report 11. International Irrigation Management Institute, IIMI

Vermillion, D., Garces-Restrepo, C., 1996. *Results of Management Turnover in two Irrigation Districts in Colombia.* Research Report 4. International Irrigation Management Institute, IIMI

Vermillion, D., Garces-Restrepo, C., 1998. *Impacts of Colombia's current irrigation management transfer.* Research Report 25. International Water Management Institute, IWMI.

Vermillion D., 2001. *Irrigation Sector Reform in Indonesia: From Small-Scale Irrigation Turnover to the Irrigation Sector Reform Program.* IMT case study: Indonesia. International E-mail Conference on Irrigation Management Transfer. June – October 2001

Zhovtonog O., Dirksen W., Roest, K., 2005. *Comparative assessment of irrigation reforms in Eastern European Countries under transition laid down in the national reports.* 21st European Regional Conference on Irrigation and Drainage, 15-19 May 2005, Frankfurt (Oder), Germany; Slubice, Poland. Theme: Integrated Land and Water Resources Management: Towards Sustainable Rural Development

Appendix I - Physical and Chemical Characteristics of the Soils

The Soils in the RUT Irrigation District

The landscape of the RUT area presents 3 basic forms: namely the foothill alluvial plain, fluvial-lacustrine plain, and flood plain. The depositions have been caused by the streams coming from the West Mountain Range, which flow to the Cauca River. The main depositions resulted in the colluvial sediment basin, alluvium combined with colluvium, alluvial marsh, alluvial beds, natural fluvial dikes, and alluvial fans. The alluvial plain is the most important deposition formed by a series of alluvial cones, which comprises gravel, sand, clay, and pebble brought by the above-mentioned streams. The colluvial sediment basin is found as hillside deposit with its characteristic sloping form. The alluvial cones or fans are located at the foot of the hill in a downward direction. The alluvial beds consist of gravel and sand deposits, and are formed by the water flows during the rainy season. The natural fluvial dikes are relatively permeable deposits formed by sand and clay, and coarse particles next to the river. The alluvial marshes are formed by deposits mainly with clay; including muddy areas artificially dried up.

About 63% of the area of the RUT Irrigation District has clay loam soils, characterized by little limitations for agriculture and a low salinity; 14% presents problems with salts, water logging and a heavy texture, and 21% has a high content of gravel, sand, and no deep soils. The salinity problem is intensified by inappropriate irrigation methods, failure of the drainage system, and the use of the drainage water for irrigation. From a geological point of view the salinity is caused by:

- Release of the water contented in rocks due to weathering processes;
- Hydrolysis of mineral material which produces calcium and sodium salts that are carried in soluble form towards the plain; and
- Formation of alluvial marsh that contributes to the concentration of soluble substances coming from the mountain areas. These deposits cover large areas of the alluvial plain.

Some general characteristics of the soils in the RUT Irrigation District are a moderate – slow permeability, a good – fairly good available moisture, an infiltration rate of 0.2 – 1 cm/h, a content of organic matter of about 1.7% - 2.2%, a pH of 6.8 – 7.7, a medium – low fertility and a soil depth of 0.90 m.

Soil Unit	AREA ha	Unit Description	Physical Characteristics				Chemical Characteristics				Fertility
			Texture	Permeability	Infilt Rate cm/h	Moisture Content	PH	Salinity/Alkalinity	C.I.C (meq/100gr)	Effective Depth	
1	460	Colluvial soils, light texture along the profile, deep and light subsoil, no presence of carbonates, lightly rolling topography; good drainage	Loan, clay on sand loan	Moderate to moderately fast	1	Moderate to good	6.8 - 7.9	No salts and sodium < 4mmh/cm	Medium to high 30-40	Deep >90 cms.	High
2	20	Colluvial soils, medium texture along all the soil profile, light subsoil and deep, no presence of carbonates, concave topography with lightly inclined slope; poor drainage; periodic waterlogging	Clay loan on sand loan	Moderate to moderately fast	1	Moderate to good	6.8 - 6.9	No salts and sodium < 4mmh/cm	Medium to high	Deep >90 cms.	High
3	530	Colluvial soils, medium texture, deep and moderately deep soils, no presence of carbonates; light rolling topography, excessive drainage	sandy loan, loan, coarse loan on sandy loan, loan sand, and sand	Moderate to moderately fast	9,7	Low	5.8 - 7.5	No salts and sodium Light reaction with HCl on the soil surface	High	Deep >90 cms	Medium
4	1.170	Alluvial recent soils, medium texture, light subsoil; deep soils, no presence of carbonates flat relief; good drainage	Loan or clay loan on sandy loan or loan o franco.	Moderate to moderately fast	2,9	Moderate to good	6.0 - 8.0	No salts and sodium Light salinity more than 1.4 m Light reaction with HCl	Low to high 20-30	Deep >90 cms.	High
5	45	Alluvial soils, medium texture; light subsoil deep soils, no presence of carboantes, flat relief, poor drainage	Loan to clay loan on sandy loan o loan	Moderate to moderately fast	2,9	Moderate to good	6.0 - 8.5	No salts and sodium Light salinity more than 1.4 m Light reaction with HCl	Low to high	Deep >90 cms.	High
6	970	Alluvial soils, medium texture; deep soils no presence of carbonates; flat relief, good moderately drainage	Clay loan or loan on clay loan or sandy clay loan	Moderate to moderately slow	0,8	Good	6.2 - 8.2	No salts and sodium Light salinity more than 1.4 m Light reaction with HCl Moderate salinity	High	deep >90 cms.	Medium
7	445	Alluvial soils, heavy texture along all the soil profile, deep soils, light to moderately presence of carbonates; flat relief; moderately to no good drainage; light salinity more than 80 cm	Clay	Slow	0,2	Very good	6.1 - 8.0	Light to moderate presence of carbonates. Light salinity	High 36-46.8	Deep >90 cms.	High
8	595	Alluvial soils, medium texture along all the soil profile; sandy subsoil; no presence of carbonates; flat relief; imperfect drainage	Loan or clay loan on clay or clay loan	Moderate to moderately slow	0,9	Good	5.0 - 7.5	No salts and sodium. Light reaction with HCl. Light salinity	High 29.6 - 42.2	Deep >90 cms.	High
9	295	Alluvia soils, heavy texture along all the soil profile; deep soils, no presence of carbonates; flat relief; poor drainage and waterloging areas	Clay or clay loan on caly loan or clay	Slow to very slow	0,2	Good	7.5 - 8.4	No salts and sodium. Light salinity, < 4mmh/cm	Very high 47-54	Deep >90 cms.	Low
10	120	Alluvial soils, heavy texture, deep soils, no presence of carbonates; 10% of area with superficial saline-sodic soils; flat relief; imperfect and poor drainage	Clay	Slow	0.2 - 0.4	Very good	5.6 - 8.4	No salts and sodium 10% of area with very superficial saline - sodic soils. Presence of sodic and saline - sodic profiles at 70 - 120 cm depth	Very high	Deep >90 cms (10% very superficial)	Low

No.	Area	Description	Texture	Permeability	Slope	Drainage	pH	Salinity/Alkali	Severity	Depth	Hazard
11	3,06	Alluvial soils, medium texture; light subsoil; deep soils, no presence of carbonates; rolling relief; good drainage	Clay loam, silty loam, or loam on clay loam, or silty loam, sandy loam or loam sand	Moderate to moderately slow	0.8-12	Good	6.5 - 8.7	No salts and alkali. Light salinity. < 6mmh/cm at more than 80 cm depth	High	Deep > 90 cms.	High
12	865	Alluvial soils, heavy texture; medium subsoil; deep soils, no presence of carbonates; rolling relief; imperfect drainage	Clay on clay	Slow	0.2-0.4	Very good	6.0 - 7.5	No salts and alkali. Light salinity < 4mmh/cm	Very high 40.8-47.9	Deep > 90 cms.	Moderate
13	805	Alluvial soils, heavy texture; medium subsoil; deep soils; no presence of carbonates; rolling relief; imperfect and poor drainage; waterlogging areas	Clay on clay or clay loam	Slow	0.2-0.3	Very good	8.0 - 8.2	No salts and alkali. Light salinity and sodium at more than 40 cm depth	Very high	Deep > 90 cms.	Moderate
14	675	Alluvial soils, medium texture; medium subsoil; deep soils, no presence of carbonates; rolling relief; good moderately drainage	Clay loam on clay loam to fine sandy loam	Slow to moderately fast	1.0-1.2	Good	6.2 - 8.2	No salts and alkali. Light salinity and sodium (< 16 PSI)	High	Deep > 90 cms.	High
15	490	Alluvial soils, medium texture; light subsoil; no presence of carbonates; deep soils; good moderately drainage; deep soils. 10% of area with saline - sodic and sodic soils; medium texture, light subsoil; superficial soils, no presence of carbonates. 5% of the area with sodic soils, medium or heavy texture, light subsoil; deep and moderately deep soils; no presence of carbonates; rolling or flat, imperfect drainage	Clay loam on clay or clay loam	Moderate to moderately slow	0.2-0.6	Good	6.7 - 8.8	10% of area with very superficial soils; sodic and saline-sodic soils between 0-25 cm depth. 5% of area with moderately deep soils; sodic and saline-sodic soils more than 60 cm depth. Other deep soils with salts and sodium more than 90 cm depth. En general values of 40% for PSI and 8 mmh/cm for EC are found	Very high	Deep, moderately. Deep, and superficial	Moderate
16	305	Alluvial soils, heavy texture; light subsoil, deep soils; no presence of carbonates; imperfect drainage. 10% of area with sodic and saline - sodic soils, very superficial soils, heavy texture; no presence of carbonates. 5% of area with saline and superficial soils; heavy; saline - sodic subsoil; imperfect drainage; rolling relief	Clay on clay	Slow	0.2-0.6	Very good	6.3 - 7.5	10% of the area with very superficial soils; sodic and saline-sodic soils at the first 25 cms depth. 5% of th earea with superficial soils at the first 50 cm depth. For all cases: EC < 8 mmh/cm; and PSI between 15 - 30%	High	Deep, superficial and very superficial	Moderate
17	180	Alluvial soils; heavy texture; heavy subsoil; deep soils; no presence of carbonates; rolling relief; good moderately drainage	Clay or clay loam on clay or clay loam	Slow	0.5-0.6	Very good	6.5 - 7.5	No salts and sodium	Very high 32-64	Deep > 90 cms.	Moderate
18	70	Alluvial soils, medium texture; no presence of carbonates; rolling relief; poor drainage, waterlogging	Clay loam and loam on clay loam	Slow to moderately fast	1.0-1.2	Good	5.6 - 7.0	No salts but areas with moderate salinity	High	Deep > 90 cms.	High

Appendix II - Pumped Water Volume by Pumping Station

Monthly Pumped Volume (m³) of Water by each Pumping Station. Period: 1989-1995

Pumping Station: Tierrablanca

Year		Jan	Feb	Mar	Apr	May	Jun	Jul	Aug	Sep	Oct	Nov	Dec	Total
1989	Volume (m3)	1411826	1171103	3219588	4163597	2772918	2903958	2590901	1977437	1876918	2528488	4170345	2634038	31421117
	Pump. Hours	375	311	855	1154	761	834	752	597	541	677	1161	749	8767
	Q (m3/s)	1.05	1.05	1.05	1.00	1.01	0.97	0.96	0.92	0.96	1.04	1.00	0.98	1.00
1990	Volume (m3)	477828	3169256	3289867	3388535	4505991	3681774	1795409	1628082	4101271	565254	4114196	4174941	34892404
	Pump. Hours	135	869	890	936	1211	1050	530	520	1325	175	1160	1192	9993
	Q (m3/s)	0.98	1.01	1.03	1.01	1.03	0.97	0.94	0.87	0.86	0.90	0.99	0.97	0.97
1991	Volume (m3)	1995501	3785631	4678705	3290382	1730743	3329219	1638043	1715842	2696076	5743224	3238322	4255898	38097586
	Pump. Hours	575	1135	1374	919	482	955	496	514	838	1796	980	1119	11183
	Q (m3/s)	0.96	0.93	0.95	0.99	1.00	0.97	0.92	0.93	0.89	0.89	0.92	1.06	0.95
1992	Volume (m3)	2637432	2784050	5067867	2947464	4591826	4437684	2892960	1506438	4191372	4336222	3503804	4392007	43289126
	Pump. Hours	760	823	1555	930	1421	1475	980	509	1390	1470	1100	1287	13700
	Q (m3/s)	0.96	0.94	0.91	0.88	0.90	0.84	0.82	0.82	0.84	0.82	0.88	0.95	0.88
1993	Volume (m3)	1253517	3337963	4197240	1562202	807951	4671662	3503088	1506690	3099294	3476167	1231664	2112501	30759939
	Pump. Hours	357	938	1185	398	215	1312	1055	482	962	1064	335	561	8864
	Q (m3/s)	0.98	0.99	0.98	1.09	1.04	0.99	0.92	0.87	0.89	0.91	1.02	1.05	0.96
1994	Volume (m3)	1916691	1864825	2711232	305013	869674	2351887	2795936	778151	4186296	2434413	1010743	2729736	23954597
	Pump. Hours	509	499	720	81	231	652	804	804	1307	729	293	737	7366
	Q (m3/s)	1.05	1.04	1.05	1.05	1.05	1.00	0.97	0.27	0.89	0.93	0.96	1.03	0.90
1995	Volume (m3)	1620565	3144420	2903552	157212	659408	1779534	1418519	1339297	2219026	1314686	2595392	2885810	22037421
	Pump. Hours	466	994	892	43	178	497	415	379	659	365	691	785	6364
	Q (m3/s)	0.97	0.88	0.90	1.02	1.03	0.99	0.95	0.98	0.94	1.00	1.04	1.02	0.96
Total Period:														Period
	Volume (m3)	11313360	19257248	26068051	15814405	15935718	23155718	16634856	10451937	22370253	20398454	19864466	23184931	224452190
	% Volume (')	53.5	62.5	56.5	48.8	50.6	61.7	75.4	58.3	68.3	65.4	60.9	58.7	59.8
	Average Vol.	1616194	2751035	3724007	2259201	2276930	3307960	2376408	1493134	3195750	2914065	2837781	3312133	32064599
	Pump. Hours	3177	5569	7471	4461	4499	6775	5032	3805	7022	6276	5720	6430	66237
	Q (m3/s)	0.99	0.96	0.97	0.98	0.98	0.95	0.92	0.76	0.88	0.90	0.96	1.00	0.94

Pumping Station: Candelaria

Year		Jan	Feb	Mar	Apr	May	Jun	Jul	Aug	Sep	Oct	Nov	Dec	Total
1989	Volume (m3)	217872	312570	634572	933336	915732	929772	106560	905688	672732	528336	985968	620460	7763598
	Pump. Hours	33	57	259	373	170	162	19	168	133	89	167	107	1737
	Q (m3/s)	1,83	1,52	0,68	0,70	1,50	1,59	1,56	1,50	1,41	1,65	1,64	1,61	1,24
1990	Volume (m3)	373896	732906	853992	491346	924552	779904	240084	169920	656028	176796	655704	802872	6858000
	Pump. Hours	64	152	318	88	154	140	45	32	137	34	115	143	1422
	Q (m3/s)	1,62	1,34	0,75	1,55	1,67	1,55	1,48	1,48	1,33	1,44	1,58	1,56	1,34
1991	Volume (m3)	380304	554472	869580	784008	419436	848664	239976	464544	518004	975708	668376	1240452	7963524
	Pump. Hours	67	113	160	133	73	153	45	87	100	195	136	233	1495
	Q (m3/s)	1,58	1,36	1,51	1,64	1,60	1,54	1,48	1,48	1,44	1,39	1,37	1,48	1,48
1992	Volume (m3)	519768	553968	1127232	761724	911340	1463580	512856	169056	229842	655344	373806	896166	8174682
	Pump. Hours	92	102	219	154	178	317	125	62	85	248	114	191	1887
	Q (m3/s)	1,57	1,51	1,43	1,37	1,42	1,28	1,14	0,76	0,75	0,73	0,91	1,30	1,20
1993	Volume (m3)	598212	499986	821162	476442	260298	1505268	559278	475488	521928	620496	174096	694548	7207202
	Pump. Hours	119	105	162	85	45	302	138	142	142	271	42	123	1676
	Q (m3/s)	1,40	1,32	1,41	1,56	1,61	1,38	1,13	0,93	1,02	0,64	1,15	1,57	1,19
1994	Volume (m3)	455886	486846	1184706	236358	321084	1293102	900432	900432	1269396	755964	505620	1269720	9579546
	Pump. Hours	79	86	221	44	55	257	195	195	348	183	112	239	2014
	Q (m3/s)	1,60	1,57	1,49	1,49	1,62	1,40	1,28	1,28	1,01	1,15	1,25	1,48	1,32
1995	Volume (m3)	989532	1171729	1233101	184968	170442	713520	470628	436428	1091556	716760	1412604	954342	9545610
	Pump. Hours	223	338	328	61	33	146	109	92	256	142	252	183	2163
	Q (m3/s)	1,23	0,96	1,04	0,84	1,43	1,36	1,20	1,32	1,18	1,40	1,56	1,45	1,23
Total Period:														**Period**
	Volume (m3)	3535470	4312477	6724345	3868182	3922884	7533810	3029814	3521556	4959486	4429404	4776174	6478560	57092162
	% Volume (*)	16,7	14,0	14,6	11,9	12,5	20,1	13,7	19,6	15,1	14,2	14,6	16,4	15,2
	Average Vol.	505067	616068	960621	552597	560412	1076259	432831	503079	708498	632772	682311	925509	8156023
	Pump. Hours	677	953	1667	938	708	1477	676	778	1201	1162	938	1219	12394
	Q (m3/s)	1,45	1,26	1,12	1,15	1,54	1,42	1,24	1,26	1,15	1,06	1,41	1,48	1,28

Pumping Station: Cayetana

Year		Jan	Feb	Mar	Apr	May	Jun	Jul	Aug	Sep	Oct	Nov	Dec	Total
1989	Volume (m3)	2689200	2345760	3061800	2167856	2380889	1264212	1048795	3441160	805492	34466	744869	1320689	21305188
	Pump. Hours	830	724	945	756	809	551	419	108	358	26	346	533	6405
	Q (m3/s)	0,90	0,90	0,90	0,80	0,82	0,64	0,70	8,85	0,62	0,37	0,60	0,69	0,92
1990	Volume (m3)	90720	836518	2066709	2079439	1950783	513356	17222	0	456102	118015	235138	794074	9158076
	Pump. Hours	28	331	808	803	646	272	8	0	279	39	118	393	3725
	Q (m3/s)	0,90	0,70	0,71	0,72	0,84	0,52	0,60	0,00	0,45	0,84	0,55	0,56	0,68
1991	Volume (m3)	257353	870771	1612015	2014827	1394632	591624	221357	258732	836889	1646438	1268399	2058354	13031391
	Pump. Hours	158	326	690	672	507	332	107	135	423	716	458	648	5172
	Q (m3/s)	0,45	0,74	0,65	0,83	0,76	0,50	0,57	0,53	0,55	0,64	0,77	0,88	0,70
1992	Volume (m3)	10750068	591905	1246985	1158239	363528	1332288	259200	0	610016	1090102	1213974	651119	9592424
	Pump. Hours	383	236	423	352	204	514	100	0	342	588	675	361	4178
	Q (m3/s)	0,78	0,70	0,82	0,91	0,50	0,72	0,72	0,00	0,50	0,51	0,50	0,50	0,64
1993	Volume (m3)	88348	659016	1239816	1651954	2163380	1664514	430272	21384	551221	871413	1335204	2057961	12734483
	Pump. Hours	45	308	574	751	1159	831	166	12	298	445	774	1082	6445
	Q (m3/s)	0,55	0,59	0,60	0,61	0,52	0,56	0,72	0,50	0,51	0,54	0,48	0,53	0,55
1994	Volume (m3)	1940652	1042564	3211434	2662927	2351538	1453690	272981	235080	1463724	1164348	1343854	1457838	18600630
	Pump. Hours	986	528	1377	1148	1173	786	154	154	801	664	772	822	9365
	Q (m3/s)	0,55	0,55	0,65	0,64	0,56	0,51	0,49	0,42	0,51	0,49	0,48	0,49	0,55
1995	Volume (m3)	172854	900983	870318	988802	1022842	23657	156002	11038	695276	1415890	1832631	1473084	9563377
	Pump. Hours	97	533	489	753	777	15	84	6	374	802	1115	746	5791
	Q (m3/s)	0,50	0,47	0,49	0,36	0,37	0,44	0,52	0,51	0,52	0,49	0,46	0,55	0,46

Total Period:

	Jan	Feb	Mar	Apr	May	Jun	Jul	Aug	Sep	Oct	Nov	Dec	Period
Volume (m3)	6314195	7247517	13309077	12724044	11627592	6843341	2405829	3967394	5418720	6340672	7974069	9813119	93985569
% Volume (*)	29,8	23,5	28,9	39,3	36,9	18,2	10,9	22,1	16,5	20,3	24,4	24,9	25,0
Average Vol.	902028	1035360	1901297	1817721	1661085	977620	343690	566771	774103	905810	1139153	1401874	13426510
Pump. Hours	2527	2986	5306	5235	5275	3301	1038	415	2875	3280	4258	4585	41081
Q (m3/s)	0,69	0,67	0,70	0,68	0,61	0,58	0,64	2,66	0,52	0,54	0,52	0,64	0,64

All Pumping Stations: Values for the Period 1989 - 1995

	Jan	Feb	Mar	Apr	May	Jun	Jul	Aug	Sep	Oct	Nov	Dec	Period
Total Volume (m3)	21163025	30817242	46101473	32406631	31488987	37532869	22070499	17940887	32748459	31168530	32614709	39476610	375529921
Average Tot. Volum (m3)	3023280	4402463	6585925	4629519	4498427	5361838	3152928	2562984	4678351	4452647	4659244	5639516	53647132
% Volume (**)	5,6	8,2	12,3	8,6	8,4	10,0	5,9	4,8	8,7	8,3	8,7	10,5	
Total Pumping Hours	6381	9508	14444	10634	10482	11553	6746	4998	11098	10718	10916	12234	119712
Average Pumping Hours	912	1358	2063	1519	1497	1650	964	714	1585	1531	1559	1748	
Total System Q (m3/s)	3,13	2,89	2,79	2,81	2,81	2,94	2,81	4,68	2,56	2,50	2,90	3,07	
Average Q (m3/s)	0,92	0,90	0,89	0,85	0,83	0,90	0,91	1,00	0,82	0,81	0,83	0,90	0,87

Remarks:

(*): Related to Total Water Volume pumped by all pumping stations each month

(**): Related to Total Water Volume pumped by all the pumping stations during the period (7 years)

Appendix III – Pumped Water Volume for Irrigation and Drainage

Volume of pumped water for irrigation and drainage, charged water volume and energy consumption; period: 1991-2002

Year: 1991

Month	Pumped Water Volume (m3) Irrigation	Drainage	Total	Charged Water (m3)	Energy Consuption (KWh)	Relation: Irrigation Water/KWh	Efficiency (%)
January	2634333		2634333	1435179	118000	22.3	54.5
February	5536713		5536713	3028915	234600	23.6	54.7
March	7193942		7193942	4792438	371400	19.4	66.6
April	6269928		6269928	2889600	193200	32.5	46.1
May	3952255		3952255	1631651	175800	22.5	41.3
June	4561162		4561162	3206493	210600	21.7	70.3
July	2295375		2295375	828023	129400	17.7	36.1
August	2413937		2413937	851299	121200	19.9	35.3
September	4758926		4758926	1991355	190200	25.0	41.8
October	9834040		9834040	5691411	380400	25.9	57.9
November	5104465		5104465	3692806	143600	35.5	72.3
December	6623700		6623700	4045176	273600	24.2	61.1
Subtotal:	61178776	0	61178776	34084346	2542000		
Monthly Average	5098231		5098231	2840362	211833	24.1	55.7

Year: 1992

Month	Pumped Water Volume (m3) Irrigation	Drainage	Total	Charged Water (m3)	Energy Consuption (KWh)	Relation: Irrigation Water/KWh	Efficiency (%)
January	3975228	257040	4232268	1886396	241800	17.5	47.5
February	3734083	195840	3929923	1312823	136200	28.9	35.2
March	7442084		7442084	4830426	345000	21.6	64.9
April	4867427		4867427	2886644	289800	16.8	59.3
May	5866694		5866694	3489734	302400	19.4	59.5
June	7233552		7233552	4109952	331200	21.8	56.8
July	3665016		3665016	1747236	269400	13.6	47.7
August	1675494		1675494	1121843	88200	19.0	67.0
September	4990187	41043	5031230	3159157	281400	17.9	63.3
October	6053156	28512	6081668	5060246	355200	17.1	83.6
November	4769336	322248	5091584	2851515	319200	16.0	59.8
December	5939291		5939291	3738804	277800	21.4	63.0
Subtotal:	60211548	844683	61056231	36194776	3237600		
Monthly Average	5017629	168936.6	5088019.25	3016231.333	269800	18.6	60.1

Month	Pumped Water Volume (m3)			Charged Water (m3)	Energy Comsuption (KWh)	Relation: Irrigation Water/KWh	Efficiency (%)
	Irrigation	Drainage	Total				
March	7442084		7442084	4830426	345000	21.6	64.9
April	4867427		4867427	2886644	289800	16.8	59.3
May	5866694		5866694	3489734	302400	19.4	59.5
June	7233552		7233552	4109952	331200	21.8	56.8
July	3665016		3665016	1747236	269400	13.6	47.7
August	1675494		1675494	1121843	88200	19.0	67.0
September	4990187	41043	5031230	3159157	281400	17.9	63.3
October	6053156	28512	6081668	5060246	355200	17.1	83.6
November	4769336	322248	5091584	2851515	319200	16.0	59.8
December	5939291		5939291	3738804	277800	21.4	63.0
Subtotal:	60211548	844683	61056231	36194776	3237600		
Monthly Average	5017629	168936.6	5088019.25	3016231.333	269800	18.6	60.1

Year: 1993

Month	Pumped Water Volume (m3)			Charged Water (m3)	Energy Comsuption (KWh)	Relation: Irrigation Water/KWh	Efficiency (%)
	Irrigation	Drainage	Total				
January	1940077		1940077	814012	132600	14.6	42.0
February	4496965		4496965	2113885	152400	29.5	47.0
March	6153938	104280	6258218	3571711	321000	19.5	58.0
April	2249807	1440791	3690598	1049810	130200	28.3	46.7
May	1075541	2156088	3231629	329703	108600	29.8	30.7
June	7374456	466988	7841444	3958735	402600	19.5	53.7
July	4492638		4492638	1613809	233400	19.2	35.9
August	2003562		2003562	827267	133200	15.0	41.3
September	4172443		4172443	2098176	212400	19.6	50.3
October	4968076		4968076	3208327	261600	19.0	64.6
November	1435847	1337126	2772973	613104	115800	23.9	42.7
December	3072779	1812928	4885707	1028704	164900	29.6	33.5
Subtotal:	43436129	7318201	50754330	21227243	2368700		
Monthly Average	3619677	1219700	4229528	1768937	197392	18.3	48.9

Year: 1994

Month	Pumped Water Volume (m3)			Charged Water (m3)	Energy Comsuption (KWh)	Relation: Irrigation Water/KWh	Efficiency (%)
	Irrigation	Drainage	Total				
January	2371380	1941849	4313229	1082704	151800	28.4	45.7
February	3228399	165836	3394235	1231681	125400	27.1	38.2

Appendix III – Pumped Water Volume for Irrigation and Drainage

Year: 1995

Month	Pumped Water Volume (m3)			Charged Water (m3)	Energy Comsuption (KWh)	Relation: Irrigation Water/KWh	Efficiency (%)
	Irrigation	Drainage	Total				
January	2773317		2773317	1447998	135600	20.5	52.2
February	5217131		5217131	3165024	264600	19.7	60.7
March	5006971		5006971	3138945	254400	19.7	62.7
April	283080	1052977	1336057	76062	48600	27.5	26.9
May	814657	1038035	1852692	2372234	78000	23.8	29.1
June	2517711		2517711	1424698	111000	22.7	56.6
July	2013869	31280	2045149	1158700	111000	18.4	57.5
August	1775725	11038	1786763	668521	72000	24.8	37.6
September	4005858		4005858	2070406	181800	22.0	51.7
October	2872786	574550	3447336	1061859	108000	31.9	37.0
November	5322584	518043	5840627	3969925	152400	38.3	74.6
December	4999346	313890	5313236	3564945	255600	20.8	71.3
Subtotal:	37603035	3539813	41142848	21984317	1773000		
Monthly Average	3133586	505688	3428571	1832026	147750	21.2	58.5

Year: 1996

Month	Pumped Water Volume (m3)			Charged Water (m3)	Energy Comsuption (KWh)	Relation: Irrigation Water/KWh	Efficiency (%)
	Irrigation	Drainage	Total				
January	3065762	255343	3321105	2013380	157200	21.1	65.7
February	3817709	1443515	5261224	1793762	159000	33.1	47.0
March	4238413	1257624	5496037	2080665	151800	36.2	49.1
April	1577629	3061097	4638726	4739921	121800	38.1	30.0
May	872049	3151383	4023432	209031	96600	41.7	24.0
June	153852	1473568	1627420	264802	65400	24.9	172.1
July	3029496	1182638	4212134	832570	131400	32.1	27.5
August	2916069		2916069	943656	147600	19.8	32.4
September	3547858		3547858	2061837	174600	20.3	58.1
October	2159848	313293	2473141	708200	108000	22.9	32.8
November	6303978	320556	6624534	3602338	231000	28.7	57.1
December	4174516	747031	4921547	1963244	167400	29.4	47.0
Subtotal:	35857179	13206048	49063227	16947406	1711800		
Monthly Average	2988098	1320605	4088602	1412284	142650	20.9	47.3

Year: 1997

Month	Pumped Water Volume (m3)			Charged Water (m3)	Energy Consumption (KWh)	Relation: Irrigation Water/KWh	Efficiency (%)
	Irrigation	Drainage	Total				
January	985623	1152680	2138303	396800	78000	27.4	40.3
February	1980743	1514680	3495423	622099	122400	28.6	31.4
March	2628038	1155906	3783944	982760	111000	34.1	37.4
April	4253971	703606	4957577	2628991	194400	25.5	61.8
May	2842458	296704	3139162	1282092	114000	27.5	45.1
June	1909642	663860	2573502	853471	76200	33.8	44.7
July	4643111	69300	4711411	2762625	206400	22.8	59.5
August	3566800	11340	3578140	1680797	199800	17.9	47.1
September	3020457	31176	3051633	1381157	169800	18.0	45.8
October	5516517		5516517	4401575	299400	18.4	79.8
November	1507311	1118368	2625679	529269	81240	32.3	35.1
December	7744339	54768	7799107	6534289	322200	24.2	84.4
Subtotal:	40599010	6772388	47371398	24057925	1974480		
Monthly Average	3383251	615672	3947617	2004827	164570	20.6	59.3

Year: 1998

Month	Pumped Water Volume (m3)			Charged Water (m3)	Energy Comsuption (KWh)	Relation: Irrigation Water/KWh	Efficiency (%)
	Irrigation	Drainage	Total				
January	3060419	21384	3081803	2897398	169680	18.2	94.7
February	4000327	11448	4011775	3012807	252440	15.9	75.3
March	4753943		4753943	3272401	243720	19.5	68.8
April	981389	782267	1763656	392923	63600	27.7	40.0
May	648737	1694635	2343372	568774	78960	29.7	87.7
June	1450455	599976	2050431	1178310	97080	21.1	81.2
July	1768072		1768072	2027991	90240	19.6	114.7
August	3381456		3381456	2768206	189480	17.8	81.9
September	2094582		2094582	1891470	110400	19.0	90.3
October	2798519	214086	3012605	2432197	135600	22.2	86.9
November	1272178	1499082	2771260	1116825	84600	32.8	87.8
December	1162517	1565372	2727889	1215873	84960	32.1	104.6
Subtotal:	27372594	6388250	33760844	22775175	1600760		
Monthly Average	2281050	798531	2813404	1897931	133397	17.1	83.2

Year: 1999

Month	Pumped Water Volume (m3)			Charged Water (m3)	Energy Comsuption (KWh)	Relation: Irrigation Water/KWh	Efficiency (%)
	Irrigation	Drainage	Total				
January	1740406	1513454	3253860	1156803	83760	38.8	66.5
February	1589250	1926201	3515451	1348863	125880	27.9	84.9
March	1172750	3419887	4592637	1224599	185880	24.7	104.4
April	219226	3115861	3335087	96939	178560	18.7	44.2
May	1642290	2461985	4104275	1407941	167100	24.6	85.7
June	1043909	1439330	2483239	694114	106260	23.4	66.5
July	1931721	59235	1990956	1510772	76620	26.0	78.2
August	3385668		3385668	2803571	150300	22.5	82.8
September	1186125	233246	1419371	1054144	94380	15.0	88.9
October	933367	1448642	2382009	6559982	60840	39.2	70.3
November	880719	1729909	2610628	632113	87660	29.8	71.8
December	414340	1478717	1893057	359474	57540	32.9	86.8
Subtotal:	16139771	18826467	34966238	12945315	1374780		
Monthly Average	1344981	1711497	2913853	1078776	114565	11.7	80.2

Year: 2000

Month	Pumped Water Volume (m3)			Charged Water (m3)	Energy Comsuption (KWh)	Relation: Irrigation Water/KWh	Efficiency (%)
	Irrigation	Drainage	Total				
January	1638826	1396728	3035554	1371716	108540	28.0	83.7
February	2247661	1475353	3723014	2248941	124080	30.0	100.1
March	435370	2564082	2999452	362743	94920	31.6	83.3
April	1517738	2973398	4491136	1759027	119040	37.7	115.9
May	18684	3833072	3851756	21643	131820	29.2	115.8
June	342266	2412375	2754641	1224629	102660	26.8	36.4
July	1050122	221144	1271266	1291651	101710	12.5	123.0
August	3274584		3274584	3217667	108396	30.2	98.3
September	1349102	292746	1641848	1114641	97380	16.9	82.6
October	2788764	184810	2973574	2503517	139800	21.3	89.8
November	1691803	970288	2662091	1123620	75600	35.2	66.4
December	4334641		4334641	3870941	204000	21.2	89.3
Subtotal:	20689561	16323996	37013557	19010736	1407946		
Monthly Average	1724130	1632400	3084463	1584228	117329	14.7	91.9

Year: 2001

Month	Pumped Water Volume (m3)			Charged Water (m3)	Energy Comsuption (KWh)	Relation: Irrigation Water/KWh	Efficiency (%)
	Irrigation	Drainage	Total				
Jannuary	2233677		2233677	1665142	101400	22.0	74.5
February	3853057		3853057	2695555	204000	18.9	70.0
March	3050618	82929	3133547	2165350	137800	22.7	71.0
April	6422992		6422992	5996194	238767	26.9	93.4
May	2335382		2335382	1487872	170861	13.7	63.7
June	3583525		3583525	3152804	135000	26.5	88.0
July	3177803		3177803	2175053	241200	13.2	68.4
August	4464990		4464990	3328043	208800	21.4	74.5
September	2931797		2931797	2001992	211800	13.8	68.3
October	6777171		6777171	6347145	336794	20.1	93.7
November	3410662	35541	3446203	2593585	290400	11.9	76.0
December	421525	505470	926995	150785	35400	26.2	35.8
Subtotal:	42663199	623940	43287139	33759520	2312222		
Monthly Average	3555267	207980	3607262	2813293	192685	18.5	79.1

Year: 2002

Month	Pumped Water Volume (m3)			Charged Water (m3)	Energy Comsuption (KWh)	Relation: Irrigation Water/KWh	Efficiency (%)
	Irrigation	Drainage	Total				
Jannuary	4041306		4041306	2370773	202800	19.9	58.7
February	4965936		4965936	3639613	342000	14.5	73.3
March	4369636	77893	4447529	3301016	212400	20.9	75.5
April	1061449	1342206	2403655	534117	97200	24.7	50.3
May	3513171	487324	4000495	2472875	193800	20.6	70.4
June	2541782	63184	2604966	1530419	156000	16.7	60.2
July	4504468		4504468	3139950	240000	18.8	69.7
August	5728555		5728555	4459860	333600	17.2	77.9
September	4695911		4695911	3033047	208800	22.5	64.6
October	2072429	11847	2084276	866963	99000	21.1	41.8
November	1919496	826234	2745730	1513349	184200	14.9	78.8
December			0	2891855			
Subtotal:	39414139	2808688	42222827	29753827	2269800		
Monthly Average	3583104	468115	3518569	2479486	206345	17.4	69.2

Average Montly Values - Period: 1991 - 2002

Month	Pumped Water Volume (m3)			Charged Water (m3)	Energy Comsuption (KWh)	Relation: Irrigation Water/KWh	Efficiency (%)
	Irrigation	Drainage	Total				
Jannuary	2538363	934068	3472431	1544858	140098	23,2	60,5
February	3722331	961839	4684170	2184497	186917	24,8	59,8
March	4256532	1392161	5648693	2681017	218643	25,5	66,2
April	2505440	1923984	4429424	1580948	149247	27,7	55,5
May	2152763	1823365	3976128	1152777	144628	26,0	57,0
June	3085426	988145	4073571	1887186	165500	23,8	69,7
July	3045087	312719	3357806	1769212	167764	19,6	64,4
August	3021153	11189	3032342	1981019	153548	20,3	61,8
September	3619583	167177	3786759	2204440	189580	19,2	64,5
October	4203416	433018	4636434	2942463	207736	23,3	66,9
November	2974726	859935	3834660	1905797	157092	26,9	62,7
December	3981084	878818	4859902	2720470	184491	26,5	67,6
Total	39105903	10686417	49792321	24554685	2065245	18,9	62,8

Remarks:
Efficiency: Charged Water Volume / Irrigation Water Volume

Appendix IV - Annual Pumping Hours and Energy Consumption

Annual pumping hours and energy consumption of the pumping stations in rut irrigation district; period: 1990-2002

Year: 1990 - 2002

| Year | Pumping Station | | | | | | Total in RUT | |
| | Tierrablanca | | Candelaria | | Cayetana | | | |
	Hours	KWh	Hours	KWh	Hours	KWh	Hours	KWh
2002	8465	1525600	3018	355200	3737	204600	15220	2085400
2001	9056	1724400	3183	424983	3028	225600	15267	2374983
2000	4160	553706	1686	226800	7978	528600	13824	1309106
1999	3267	777300	1022	148800	8876	571800	13165	1497900
1998	5841	1140000	1497	159000	5312	351000	12650	1650000
1997	7147	1322880	2606	389600	6190	312000	15943	2024480
1996	5804	1046400	1968	265800	8939	439800	16711	1752000
1995	6360	1204800	2165	277200	5797	307800	14322	1789800
1994	6857	1228800	2082	279600	9363	521400	18302	2029800
1993	9017	1639200	1685	253800	6579	471600	17281	2364600
1992	13781	2555040	1887	265800	4145	314400	19813	3135240
1991	11275	1915200	1490	161300	5157	334300	17922	2410800
1990	10129	1898400	1505	213600	4445	364800	16079	2476800
Total	101159	18531726	25794	3421483	79546	4947700	206499	26900909
Average	7781,5	1425517,385	1984,2	263191	6118,9	380592,3	15884,5	2069300,692
%	49,0	68,9	12,5	12,7	38,5	18,4		

Appendix V - Irrigation Water Distribution

During the second participative workshop, the water users developed and proposed a plan to divide the area in sectors for an improved irrigation water distribution. This would allow for the establishment of an irrigation schedule for the whole area; and cropping schedules and growth calendars, especially for the grain crops. The establishment of the boundaries of the irrigation sectors or blocks will depend on several factors, for instance on the command area of an irrigation canal of the main system, namely the main canal, interceptor canal, and canal 1.0; the practical capacity of the pumping stations; the present division in five sectors in view of the management of the irrigation district. These sectors are mainly defined by management aspects and geographical conditions. Moreover, the boundaries of the new sectors will be influenced by the future envisaged development of the land use (cropping pattern) in the various sectors.

Several possibilities to divide the area in sectors can be considered for the current condition and without major modifications of the existing hydraulic infrastructure. For instance the whole area could be divided into two irrigation blocks at least, namely block one will be conform the grouping of sector 2 and 4 under the command of the main irrigation canal. The other block will be formed by the grouping of the sector 1, 3 and 5 under the command of irrigation canal 1.0 and the interceptor canal. Previously, the duration of the irrigation period had to be defined by taking into account the irrigation intervals (irrigation frequency) of the crops and next the irrigation period of each block is defined according to its command area. According to the cropping pattern for each sector and the related crop water requirements, the required discharges for each irrigation canal are determined and an operation schedule for the pumping stations has been established. The required discharges for each of the irrigation canals will be compared with their design discharges and from this comparison follows whether the division in sectors is a feasible alternative.

The water distribution block by block (sectors) means significant organizational changes for the water management. ASORUT as the irrigation agency will be responsible for the water management at the main irrigation level that will have to result in the supply of irrigation water on a timely, sufficient, equitable and efficient way; while the management at farm level or the level of a group of farms will be the responsibility of the farmers (water users organization at sector level). A proper management at this level will benefit from the establishment of storage reservoirs that are especially needed for the farms with highly sensitive cash crop (citric and fruit trees).

Alternatives for the Irrigation Water Distribution

The technical alternatives to be considered will be developed for the existing hydraulic infrastructure without considering major changes, improvements or enlargements from the physical point of view; only functional changes will be considered. In this way, the alternatives for the irrigation water distribution will be considered for two scenarios. The first one (S1) considers an irrigation water distribution based on a continuous flow to the whole area causing a division in irrigation sectors that are defined by the influenced area of each pumping station. The second one (S2) considers a rotational flow for the whole area causing a division of the whole irrigation district in irrigation sectors. This scenario is the most accepted by the water users and its establishment and implementation is expected for them, this expectation was identified during the development of the second participative workshop.

Starting Conditions

As was mentioned before functional changes will be only considered for the existing hydraulic infrastructure. The proposed scenarios will be considered under the following starting conditions:

Pumping Stations

Tierrablanca and Candelaria pumping stations will operate as they currently operate; this is irrigation and irrigation/drainage respectively. A functional change will be introduced for Cayetana pumping station which will operate only for drainage according to the design criteria established during the design stage; in this way its current function for irrigation will be suspended in view of to minimize environmental impacts caused by the reuse of drainage water for irrigation from the main drainage canal.

Tierrablanca pumping stations will operate with 75% of its total capacity (3 out of 4 pumping units) given the identified problems in the suction chambers when the four pumping units are operated. Its total installed capacity is 6.8 m^3/s (4 units x 1.7 m^3/s - unit) but only 5.1 m^3/s (3 pumping units) are available for irrigation. The Candelaria pumping station will operate at full capacity, 2.8 m^3/s (2 units of 1.4 m^3/s-unit). In this way, the total amount of available water for irrigation will be 7.9 m^3/s. It is assumed also that the pumping stations will operate at nominal capacity under condition of maximum efficiency. The operation of the pumping stations will be out of the "peak hours" (6 hours per day: 10-13 and 18-21) during which the present energy costs are very high; in this condition, the available irrigation time will be of 18 hours as maximum.

Main Drainage Canal

The function for irrigation of the main drainage canal will be suspended in view of to minimize environmental impacts by the reuse of drainage water. In this way, the main irrigation network is constituted by the main irrigation canal, conveyance canal, 1.0 canal and interceptor canal.

Main Irrigation Canal

The main irrigation canal will serve the zones 2 and 4 covering a total area of the 4,033.6 ha. Initially, the Tierrablanca pumping station will supply irrigation water to the first 18 km length of the main irrigation canal serving the sub-zones 2a, 2b, and 2c and covering an area of 1,867 ha. The remainder length will be supplied by the Candelaria pumping station serving the zones 2d and the whole zone 4 covering an area of 2,166.6 ha.

Efficiency

According to the field experience the value of the total efficiency of the irrigation system will be assumed equal to 40%. The value of the application efficiency depends on the crop type and the irrigation method: 70% for sprinkler irrigation (cotton, grains, and grass), 50% gravity irrigation (sugar cane, vegetables), and 95% for drip irrigation (fruits).

Design Capacity

The design capacity of the main irrigation canals is a limiting factor for the technical alternatives for the irrigation water distribution. According to the basic information the design discharges are the following: conveyance canal 6.8 m³/s, 1.0 canal 3.1 m³/s, interceptor canal 6.8 m³/s, and main irrigation canal 4.8 m³/s.
Figure V.1 shows a sketch of the hydraulic infrastructure of the RUT Irrigation District.

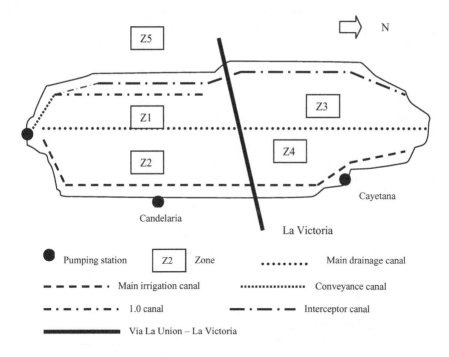

Figure V.1 Sketch of the basic hydraulic infrastructure of the RUT Irrigation District

Cropping Pattern

Taking into account the historic evolution of the cropping pattern in the RUT Irrigation District, the current situation, the future expectation and tendency, and the current land use per zones a scenario for the cropping pattern was proposed. The proposed cropping pattern is characterized by the increase of the crop area for sugar cane and the growing of fruit trees for whole the zone 5. Table V.1 shows the cropping pattern and land use proposed per zone

Table V.1 Proposed Cropping Pattern and Land Use in RUT Irrigation District

Zone	Zone Area		Crops - % Area per Zone					%
	ha	%	Fruits	Grains	Vegetables	Grass	Sugar Cane	Total
1	2,668	28.1	10.7	28.3	3.3	16.2	41.5	100
2	2,548	26.8	28.2	30.2	4.0	7.0	30.6	100
3	1,557	16.4	14.5		2.5	1.4	81.6	100
4	1,486	15.6	13.8		0.5	4.4	81.3	100
5	1,238	13.0	100.0					100
Total	9,498	100.0						

Irrigation Needs

The proposed cropping pattern, the crop water requirement, analysis of rainfall, determination of effective rainfall and efficiency of the irrigation system were considered in order to determine the crops irrigation needs. The crops water requirement was determined by using the evaporation tank method based on evaporation monthly records for a period of 33 years (1970 - 2002). The rainfall analysis was carried out by using rainfall monthly records for a period of 31 years (1967 - 1997). The effective rainfall was obtained by using USDA method. Finally the crops irrigation needs were determined at monthly level relating the crops water requirement and the effective rainfall. From analysis of results follows that the month of August presents the major irrigation water demand for the proposed cropping pattern. Table V.2 presents the irrigation needs and the corresponding required discharges for different available irrigation time for the peak month (August).

Table V.2 Required Discharges for the Peak Month (August)

Zone	Required Volume	Required Q		
	(m3/month)	(m3/s)		
		Available Irrigation Time (hours)		
		24	18	16
1	4,149,315	1.55	2.07	2.32
2	3,388,123	1.26	1.69	1.90
3	3,029,620	1.13	1.51	1.70
4	2,746,732	1.03	1.37	1.54
5	1,043,445	0.39	0.52	0.58
Total	14,357,235	5.36	7.15	8.04

Covering Area of Irrigation Canals

The main distribution system of irrigation water of the RUT Irrigation District is currently conformed by four canals namely: main irrigation canal, interceptor canal, 1.0 canal, and main drainage canal. The last canal was designed as drainage canal; however, in the course of time it has been used as irrigation canal covering 10.2% of the area under irrigation. According to the established initial conditions the main drainage canal will be considered as drainage canal only and its irrigation function will be suspended. As a consequence, the irrigation areas served currently by the main drainage canal will be redistributed among the other irrigation canals in a convenient way. Table V.3 shows the current influenced area of the irrigation canals in the different zones and the redistribution of the irrigation areas for the main drainage canal.

Table V.3 Current Covering and Redistribution of Areas Served by Irrigation Canals

Current Influenced Area of Irrigation Canals							
Canal	% Area per zone					Total area	
	Zone 1	Zone 2	Zone 3	Zone 4	Zone 5	(ha)	%
Main irrigation canal	0.0	92.7	0.0	78.3	0.0	3525.2	37.1
1.0 canal	70.1	0.0	0.0	0.0	0.0	1869.9	19.7
Interceptor canal	19.7	0.0	88.0	0.0	100.0	3134.3	33.0
Main drainage canal	10.2	7.3	12.0	21.7	0.0	967.4	10.2
Total						9496.8	100.0
Redistribution of Influenced Area of Irrigation Canals							
Canal	% Area per zone					Total area	
	Zone 1	Zone 2	Zone 3	Zone 4	Zone 5	(ha)	%
Main irrigation canal	0.0	100.0	0.0	100.0	0.0	4033.6	42.5
1.0 canal	78.6	0.0	0.0	0.0	0.0	2096.7	22.1
Interceptor canal	21.4	0.0	100.0	0.0	100.0	3366.5	35.4
Total						9496.8	100.0

Scenario S1: Continuous Flow Distribution

As was mentioned before Scenario 1 (S1) considers an irrigation water distribution based on continuous flow on whole area causing a division in irrigation sectors defined by the areas influenced by each pumping station. A key point in this scenario is to establish a balance between the required discharge by the irrigation system and the available discharge given by the pumping system. As was mentioned in the starting conditions the total capacity of the pumping system is 9.6 m³/s, but only 7.9 m³/s (83%) is available of which 5.1 m³/s are supplied by Tierrablanca pumping station and the remainder by Candelaria pumping station. The comparison between the required discharge by each irrigation canal and its design capacity is another key aspect to be taken into account. Combining the information from tables V.2 and V.3 the required discharges by each irrigation canal is obtained. Table V.4 gives the values for the required discharges. From Table V.4 follows that:

- This scenario is possible only for an available irrigation time (td) of 18 hours. A value of td equal to 24 hours is not considered because it would imply the operation of the pumping system during the "peak hours" which is not desired.

Values of td less than 18 hours will cause a required discharge of the irrigation system greater than the available discharge (7.9 m³/s); and
- The required discharges are less than the design discharges for all the irrigation canals.

Table V.4 Required Discharges for Irrigation Canals

Canal	Influenced Zone				Required Volume	Required Q (m3/s)				Design Q
	Zone	%Zone	Influenced Area		(m3/mes)	Available Irrigation Time (hours)				(m3/s)
			(ha)	%Total		24	18	16	%	
Main irrigation	2	100	2547.8			2.29	3.05	3.44	43	4.8
	4	100	1485.8							
Subtotal			4033.6	42.5	6134855					
1.0	1	78.6	2096.7			1.22	1.62	1.83	23	3.1
Subtotal			2096.7	22.1	3261362					
Interceptor	1	21.4	570.8			1.85	2.47	2.78	35	6.8
	3	100	1557.4							
	5	100	1238.3							
Subtotal			3366.5	35.4	4961018					
Conveyance					8222380	3.07	4.09	4.60	57	6.8
Total			9496.8	100.0	14357235	5.36	7.15	8.04		

According to Table V.4 the irrigation water distribution through the irrigation network is as follows (Figure V.2):

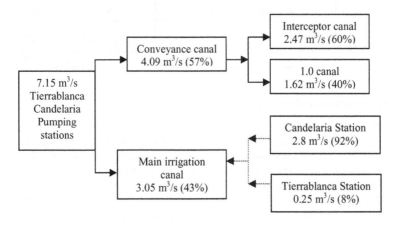

Figure V.2 Irrigation water distribution through the main irrigation network for scenario S1

The number of pumping units operating in the Candelaria pumping station is 2 (2.8 m³/s / 1.4 m³/s-unit) while 3 units will operate in the Tierrablanca pumping station ((7.15-2.8) m³/s / 1.7 m³/s-unit equal 2.6 units), meaning that is necessary to adjust the discharge of the pumping units for the Tierrablanca station.

On the other hand, the Tierrablanca station as the Candelaria one supply the main irrigation canal in a combined way. The irrigation area served by each station can be determined based on the total influenced area (4,033.6 ha) of the main irrigation canal and its total required discharge (3.05 m^3/s), and the individual discharge of each station (2.8 and 0.25 m^3/s for Candelaria and Tierrablanca station respectively). The unit supplying discharge is 0.756 liters/s-ha (3.05 m^3/s / 4,033.6 ha); then the irrigation area served by the main irrigation canal with irrigation water coming from Candelaria station is 3,704 ha (2,800 liters/s / 0.756 liters/s-ha), while 330 ha are irrigated by Tierrablanca station.

Scenario S2: Rotational Flow Distribution

Scenario S2 considers a rotational flow for the whole area of the RUT Irrigation District causing a division of the whole irrigation area in irrigation sectors. This scenario is the most accepted by the water users and its establishment and implementation is expected for them. For the development of this scenario the following considerations will be taken into account:

-    The whole area of the RUT Irrigation District will be divided in irrigation sectors taking as reference the grouping of the existing management sectors (5). Initially, two alternatives (S2.1, S2.2) will be considered, each of them corresponding to divide the total area in two irrigation sectors or blocks. S2.1 results of the grouping of the sectors 2-4 and 1-3-5; while S2.2 of the grouping of 1-2-4 and 5-3;
-    The current irrigation area under the command of the main drainage canal will be redistributed among the interceptor and 1.0 canal;
-    The duration of the irrigation period for each sector must be less than the irrigation interval. The irrigation interval will be assumed according to the agricultural practice for the grains and sugar cane crops, which have the largest irrigation intervals. Initially, an irrigation interval equal to 15 days will be assumed;
-    The duration of the irrigation period for each sector will be defined in a proportional way to the area size of each irrigation sector. In this way (theoretically), the value of the irrigation interval does not affect the magnitude of the required discharge for the irrigation canals.

Table V.5 shows the main characteristics of the alternatives S2.1and S2.2 for the division in irrigation sectors. On this table can be seen that irrigation sectors were conformed grouping existing zones served by a same irrigation canal and /or zones under the command area of an irrigation canal. It is the case of the irrigation sector 1-3-5, where the zones 3 and 5 are served by the interceptor canal while the zone 1 is served by the 1.0 irrigation canal.

Table V.6 presents the values of the required rotational discharge for two divisions in irrigation sectors and different value of available irrigation time. From the table can be seen that the alternatives S2.1and S2.2 present a total discharge requirement greater than the total capacity of the pumping stations (7.9 m^3/s) for an available irrigation time equal to 16 hours. The design capacity of irrigation canal 1.0 and the conveyance canal is exceeded 11 and 59% for the alternative S2.1 and 18 hours of the available irrigation time, while for the interceptor canal and the conveyance canal is exceeded by 12% in the alternative S2.2. The design capacity of the main irrigation canal is exceeded by 20% in S2.1 for the case of 24 hours of available irrigation time. From Table V.6 follows that the rotational distribution is only possible for the alternative S2.2 and an available irrigation time of 24 hours. However, the selection of this alternative will entail the operation of the pumping stations during the "peak hours", which is not desired.

Table V.5 Alternatives for the Division in Irrigation Sectors

Alternative	Irrigation Sectors				Required Q (24 hours)	
	Sectors	Zones	Area (ha)	%Zone	Canal	(m3/s)
S2.1	1-3-5	1	2.668,10	100	1.0	1,55
		3	1.557,40	100	Interceptor	1,13
		5	1.238,30	100	Interceptor	0,39
		Subtotal	5.463,80		Subtotal	**3,07**
	2-4	2	2.547,80	100	Main irrigation	1,26
		4	1.485,90	100	Main irrigation	1,03
		Subtotal	4.033,70		Subtotal	**2,29**
S2.2	5-3	5	1238,3	100	Interceptor	0,39
		3	1.557,40	100	Interceptor	1,13
		Subtotal	2795,7		Subtotal	**1,52**
	1-2-4	1	1947	73	1.0	0,83
		1	721,2	27	Interceptor	0,11
		2	2547,8	100	Main irrigation	1,26
		4	1485,9	100	Main irrigation	1,03
		Subtotal	6701,9		Subtotal	**3,23**

Table V.6 Required Rotational Discharge for Two Division Alternatives

Alter.	Irrigation Sectors			Required Q (m3/s)		Irrigation Period (days)	Required Rotational Q (m3/s) Available Irrigation Time (hours)		
	Zones	Area (ha)	%	Canal	24 hours		24	18	16
S2.1	1-3-5	5463,8	57,5	1.0	1,55	9	2,58	3,44	3,87
				Interceptor	1,52		2,53	3,38	3,80
				Conveyance	3,07		5,12	6,82	7,67
				Total	**3,07**		**5,12**	**6,82**	**7,67**
	2-4	4033,7	42,5	Main irrigation	2,29	6	5,73	7,63	8,59
	Total	9497,5		Total	**2,29**		**5,73**	**7,63**	**8,59**
S2.2	5-3	2795,7	29,4	Interceptor	1,52	4	5,70	7,60	8,55
				Conveyance	1,52		5,70	7,60	8,55
				Total	**1,52**		**5,70**	**7,60**	**8,55**
	1-2-4	6701,9	70,6	1.0	0,83	11	1,13	1,50	1,69
				Interceptor	0,11		0,15	0,21	0,23
				Conveyance	0,94		1,28	1,71	1,92
				Main irrigation	2,29		3,12	4,16	4,69
	Total	9497,6		Total	**3,23**		**4,40**	**5,87**	**6,60**

Conclusions

The division in irrigation sectors was an alternative suggested by the farmers and technical staff during the participative process developed through several workshops within the framework to formulate a management model for the RUT Irrigation District. The division in irrigation sectors was considered by the farmer as a way to optimize the irrigation water distribution whose current practice is considered inefficient and contributes to the very high pumping costs.

The farmers and technical staff realized that the division in irrigation sectors will have implications concerning the management of irrigation water at irrigation sector and farm level as at organizational one. From operational point of view, the division in irrigation sectors implies the consideration of storage reservoir especially for the areas where crops with short irrigation intervals are grown; this is the case of the fruits crop. From organizational point of view, the division in irrigation sectors will cause the need of an organization of farmers at sectoral level responsible for the distribution, control and regulation of irrigation water among farmers once the irrigation water has been delivered by ASORUT.

The division in irrigation sectors is restricted by the design capacity of the current hydraulic infrastructure and its characteristics for functioning. For instance, the conveyance capacity of the current irrigation network is a limiting factor to take in account for the division in irrigation sectors based on the distribution of a rotational flow. The available irrigation time is another factor to be considered. From the two alternatives considered for the irrigation water distribution by using a rotational flow only one of them was possible considering the conveyance capacity of the main irrigation network; however, it was possible considering an available irrigation time equal to 24 hours, which includes the operation of the system within the period of "peak hours".

The division in irrigation sectors results also in a division of the water distribution, which is a desired practice by the farmers and technical staff of the organization because it will cause an improvement of the irrigation water management, a more efficient use of the irrigation water and a reduction of the pumping costs. However, organizational and operational changes, and modification in view of the design capacity of the current hydraulic infrastructure are required.

Appendix VI - Water Management Optimization

Optimization of the water management in the RUT Irrigation District, Colombia

In the framework of this research a detailed study concerning the Optimization of Water Management in the RUT Irrigation District has been carried out (Otero, 2002). The study states that the present irrigation supplements the rainfall without defined criteria for the operation of the pumps and it is likely that more water than required is supplied to the area, affecting not only the pumping costs for irrigation but also those for drainage. To investigate these operation aspects the actual situation was evaluated based on an overall water balance at scheme level and an analysis of the fluctuation of the groundwater table in the area. Consecutively, a water balance at field level using the WASIM model was used to determine the irrigation requirements, yield reductions and drain flows based on a 20-year simulation. Hydrodynamic simulations for the main drain using the drain flow from the water balance and one-, two- and three-days rainfall with a return period of 5 years provided information on the required pumping capacity for drainage and the extent of the areas inundated. A summary of the mentioned study is presented as follows.

Groundwater Study

A water and salt balance study by Rodriguez and Delgado (1996) presents the observations in 97 wells from December 1995 to May 1996. These observations indicate that the average depth of the groundwater for this period was 1.60 m-surface, with a range between 1.47 and 1.88 m-surface. One of the conclusions of the study was that the District does not have an appropriate drainage system to control the groundwater table and salinity and it was recommended that the main drain and drains at field level should be deepened to lower the groundwater.

Otero (2002) transferred the observations of the wells to Arc-view, analyzed the spatial and temporal fluctuation of the groundwater table and identified the areas vulnerable to water logging (Figure VI.1). During April and October the groundwater table was less than 1.0 m-surface in 20.5% and 15.7% of the area, respectively. Moreover 57.1% of the area had a groundwater table between 1.0 and 1.5 m-surface during April. These periods correspond to the rainy seasons but also to the second month of the cropping seasons. Even though at these periods the crop roots have not yet reached their maximum depth, a shallow groundwater table can strongly affect the yields.

From Figure VI.1 follows that the months most affected by shallow groundwater are May and October, and the areas are the regions in the North near the drain, in the South near the pumping station Tierrablanca and some spots in the middle part. February and July are the months least affected and the groundwater is less than 1.5 m-surface in only 23.2% and 13.9% of the area, respectively. This classification was made based on a root depth of 1.5 m, which is the average for sugarcane (Allen, 1998). Shallow groundwater tables affect especially the area in the North of the District. This situation is aggravated due to the fact that this area is near the outlet, which is closed when the river level is high. The Central and the Southern region are also affected by water logging. The recharge comes from the Cauca River and the main irrigation canal, while in the western part the recharge mainly comes from underground flows from the mountains and seepage from the interceptor canal.

A comparison of the hydrographs of the wells and the rainfall distribution showed that the groundwater table and the rainfall have the same tendency. Hence, the fluctuation of the groundwater is not only affected by subsurface flows but also by rainfall.

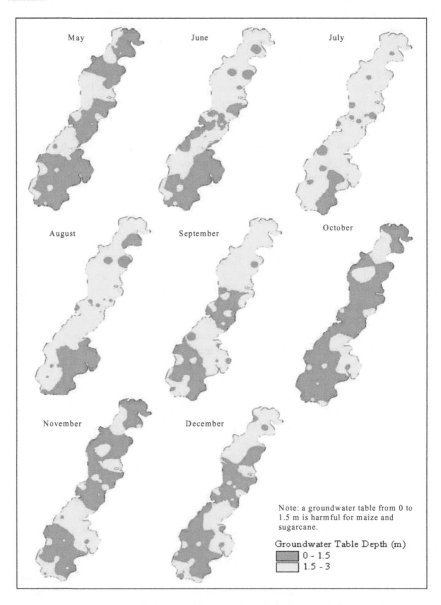

Figure VI.1 Areas with groundwater table depth shallower than 1.50 m (Otero, 2002)

Overall Water Balance of the RUT District

The total volume of water pumped in, is the volume pumped by the stations Tierrablanca and Candelaria. The stations Candelaria and Cayetana deliver the volume pumped out. The total average annual volume pumped is $45.4 \times 10^6 \, m^3$, of which 83% is for irrigation and 17% for drainage. For this type of district, where the irrigation is supplementary to the daily rainfall, a monthly analysis has been used to identify the periods with peak requirements for water supply and drainage. The value of ΔS, being the difference between the volumes flowing In and Out, is positive all over the year, which might be explained by the fact that the component for the recharge of the groundwater is not included in this water balance. Since there is not a defined criterion to supply water according to the daily precipitation and cropping calendar, it is possible that a larger amount of water than required has been supplied (pumped) to the system.

Water Balance at Field Level

The WASIM program has been used to simulate the daily water balance at field level for two main crops and three soil types. The main crops used in the simulation are maize and sugar cane, which cover large areas of the District. The three soil types are taken from a soil classification, which is based on their drainage parameter and physical characteristics. The by WASIM recommended soil parameters for irrigation and drainage design are presented in Table VI.1

Table VI.1 Soil Types (Types de sols)

Characteristic	Sandy Soil	Loamy Sand Soil	Sandy Clay Loam Soil
Saturation (%)	43.70	43.70	39.80
Field capacity (%)	11.50	16.80	24.10
Permanent Wilting Point (%)	3.30	5.50	14.80
Drainage Coefficient	0.69	0.51	0.17
Hydraulic Conductivity (m/d)	5.00	2.50	1.30
Area (ha)	3,939	4,439	1,190
Area (%)	41.10	46.40	12.50

The water balance has been studied for three scenarios, namely:
- Scenario 1: present situation in the District;
- Scenario 2: without any irrigation, to find the yields under rainfed conditions;
- Scenario 3: with irrigation for no yield reduction.

Some limitations in view of the results of the WASIM simulation are:
- Only two crops are used to represent the cropping pattern and to estimate the irrigation requirement;
- The daily water balance gives the daily-required water supply, which is difficult to supply from an operational point of view;
- The rainfall conditions during a cropping season can widely vary;
- Fixed planting dates are assumed, which in reality do not occur;
- The conveyance and distribution efficiency might be under-estimated since the supply is by gravity.

Scenario 1: Actual Situation

Scenario 1 takes the present irrigation supply from the Tierrablanca and la Candelaria pumping stations. The available data have been used to introduce in WASIM irrigation on a weekly basis during the two growing seasons, namely season A starting at early May and season B starting around September. For the calibration of the model the measured groundwater table was used to check the assumed hydraulic conductivity. The wells within the area were selected to compare the by WASIM calculated groundwater levels with the measured ones.

A frequency analysis of the drainage outflows, which were based on the percentage of each soil in the whole area, resulted in the total drain flow. Irrigation and effective rainfall that infiltrate into the soil result in an almost uniform sub-surface drainage flow, which ranges between 0 and 2 mm per day.

Scenario 2: No Irrigation

Scenario 2 is meant to determine the yield reduction of maize and sugarcane if there would be no irrigation. The yield reduction is calculated from,

$$\left(1 - \frac{Y_{actual}}{Y_{\max imum}}\right) = k_y \left(1 - \frac{ET_{actual}}{ET_{\max imum}}\right)$$

Where:

Y = the yield,

k_y = the yield response factor and

$ET_{actual}/ET_{maxumum}$ = the relative evapotranspiration

For this scenario $k_y = 1$ has been used; in general k_y for maize and sugarcane is higher than 1, mainly due to the sensitivity of these crops to water stress (Doorenbos and Kassam, 1979).

Table VI.2 Relative Yield for Maize and Sugarcane for Different Return Periods

Return Period (year)	Relative Yield for Maize (%)		Relative Yield for Sugarcane (%)
	Season A	Season B	
2	96.3	99.1	71.7
5	93.3	97.2	64.4
10	91.3	95.9	59.6
20	89.4	94.7	54.9
50	87.0	93.1	48.9

The yield reduction for maize for a return period of 2 year is less than 4% during season A and less than 1% for season B. According to these figures, if maize is not irrigated it can still give high yields. On the other hand, the yield reduction of rainfed sugarcane is about 30% for a return period of 2 year and higher than 35% for a return period of 5 year. The yield reduction on the sandy clay loam soil, which covers only

12.5% of the total area, is higher than for the other two soil types. Sugarcane requires a groundwater table below 1.6 m-surface and the crop water requirement is still high after the crop reaches its maximum root depth. From Table VI.2 follows that maize can be a rainfed crop, but sugarcane should be irrigated for an optimal yield.

Two other crops that are also important in the RUT District are sorghum and soybean. Under rainfed conditions sorghum will not show any yield reduction, but soybean requires more water and the production will be reduced by 8.2% and 13.3% in season A and B, respectively.

Scenario 3: Irrigation Requirements

The net irrigation requirement is derived from the field water balance and results from the difference in crop water requirements (output) and available water from groundwater, soil storage and rainfall (input). The total available moisture *TAM* is defined as:

Where D_r is the thickness of the root zone, θ_{FC} and θ_{WP} are the soil moisture content at field capacity and wilting point, respectively. The fraction p is the readily available moisture content ($RAM = p\ TAM$), which means that during the consumption of this water by the plant the actual evapotranspiration equals the potential evapotranspiration. The irrigation requirement follows from the water balance for the case that there is no yield reduction. This means that sufficient water is applied to the plants and that the available soil moisture is never below *(1-p)* of the total available moisture (De Laat, 1998). The *p* fraction is about 0.60 and 0.65 for maize and sugarcane, respectively.

In the case of sugarcane the irrigation requirements for sandy and loamy sand soil are almost the same (in mm), while for sandy clay loam the requirement is 200 mm more. A clay soil like sandy clay loam requires more water for the same cropping pattern, even when the water acquired by capillary rise can be considerable. Sugarcane requires irrigation from April until November, which coincides with the full maturity period. Maize requires irrigation at the end of both the seasons (A and B).

The groundwater table for the growing conditions of scenario 3 is adequate for the crops on the sandy soil; partly good in the loamy sand soil and harmful in the sandy clay loam soil. In the second soil type the groundwater is shallower than the root depth during April, May, and from September until the end of December, while for the third soil the groundwater is more than 40 cm above the root depth during the whole year. This comparison is made for a root depth of more than 1.5 m for perennial crops, such as sugarcane. Therefore it is recommended to plant crops with a smaller root depth or to manage the groundwater table in this soil type by drainage. The first alternative is recommended, since the area with this type of soil covers only 10% of the whole area.

Change in the Cropping Pattern

A change in the cropping pattern could be introduced as a measure to reduce the amount of pumped irrigation water. The following factors should be considered in these changes:
- *Admissible yield reduction*: some crops, like sorghum, maize or soybeans can still give high yields under rainfed conditions. The acceptable yield reduction should be

checked for the meteorological conditions and the viable amount of water to be pumped;
- *Market trends*: the question that has to be raised is: which crops are actually profitable and which crops could afford (support) the pumping costs of the District?
- *Groundwater table*: a shallow groundwater table affects some areas in the North near the outlet and in the centre near Cayetana. A change in crops that require less water and have smaller root depths could be recommended to reduce pumping costs as well as to improve growing conditions (Ritzema, 1994).

Effects of Irrigation on Drainage Flows

The water balances as simulated with WASIM clearly showed the impact of irrigation on the drain flow during season B and A for the year 1994 (being an average year for rainfall). As follows from Table VI.2 the relative yield for maize is almost 100% for an average year, which requires an irrigation supply of 160 mm for season B. When, for example, the irrigation supply is doubled to 320 mm about 2/3 of the extra water will flow almost directly to the drain and 1/3 will flow as deep percolation to the groundwater.

Hydrodynamic Computation

The hydrodynamic response of the main drain to different rainfall conditions and the pumping operations was simulated with the hydrodynamic model DUFLOW (IHE, 1998). The main drain collects water from runoff and sub-surface flows, and conveys the excess water to the northern part of the District where the outlet is connected to the Cauca River. The simulations with DUFLOW covered the following scenarios:
- Scenario 1: a one-day rainfall of 80 mm/d, which corresponds to a return period of 5 year, and the present agricultural practices without pumping operation;
- Scenario 2: a one-day rainfall of 80 mm/d, but now with pumping operation;
- Scenario 3: a two-day rainfall of 50 mm/d, which corresponds to a return period of 5 year and with pumping operation;
- Scenario 4: a three-day rainfall of 37 mm/d, which corresponds to a return period of 5 year and with pumping operation.

Scenario 1: Actual Irrigation without Pumping Operation

Scenario 1 was meant to find the water levels in the drain during peak rainfall without pumping and for different water levels in the Cauca River. For the first scenario, with a water level of 909.8 m+MSL (being the maximum river level measured during October 1994) the gate will be open and in some sections the water level in the drain is higher than the ground surface. The first 10 km in the southern region do not show too high water levels during the peak rainfall; while in some places in the centre and northern region the water levels are more than 1meter above the surface. During the rest of month the water level is everywhere below ground surface.

Two other scenarios, namely scenario 1b and 1c with a water level in the Cauca River at 911.5 and 912.5 m+MSL, respectively, were included in the simulation to estimate the extent of the inundations under even more extreme river water levels and in case the pumps are not operated. A comparison of the three scenarios shows that the water profiles become flatter for higher river water levels or for decreasing discharges at the outlet. In the last scenario (1c), the outlet gate remains closed and the drain water level is almost horizontal and higher than the surface; not only during the peak rainfall but also during the rest of the month.

The peak discharge through the flap gate is 11.5 m³/s at the highest measured river water level (scenario 1a), namely 909.8 m+MSL and keeps that capacity for three days. For smaller gate openings (scenario 1b), the peak discharge is 4.8 m³/s, in view of the head over the structure that is smaller than in the first scenario (1a). The gate is closed for water levels at 912.5 m+MSL or higher and many disturbances and backwater effects occur in the main drain during the maximum rainfall.

Scenario 2: One-Day Rainfall with Pumping Operation

In scenario 2 the two pumping stations are operated to analyze the behavior of the drain for a one-day rainfall of 80 mm/d and a return period of 5 year, which occurred in the middle of a simulation period of 10 days. In DUFLOW the pump operation is determined by a start and stop level, being the level at which the pump starts to operate for any level above that specific level and stops when the water level falls below that level. The start level for La Candelaria station has been determined at 910.5 m+MSL and the stop level is 910.0 m+MSL. For La Cayetana station the start and stop levels are 910.0 m+MSL and 909.5 m+MSL, respectively.

The present pumping capacity (2.8 m³/s) cannot lower the water levels below ground surface, only the duration of the peak is reduced from four days for scenario 1 to two days for scenario 2 and the pumps operate continuously for 42 hour in La Candelaria and for 38 hour in La Cayetana. Simulation with lower start levels, for example 909.5 m+MSL or 909.0 m+MSL did not achieve the desired water levels in a shorter time. For these low levels the pumps will operate during the whole period, also when the levels are below ground surface, which is not an economic application.

To compare the influence of the pumping capacity on the duration and height of the inundations the capacity of both the stations has been increased by 50% (scenario 2b). This increase resulted in smaller inundation areas (almost half) and the pump duration was about 2 - 3 hour shorter than with the present capacity. The discharge in the outlet is 11 m³/s, which is a little bit smaller than the discharge for the present situation.

Scenario 3: Two days Rainfall with Pumping Operation

For scenario 3 the 2 days rainfall is equal to 50 mm/d, which corresponds to a return period of 5 years. The water level in the drain near La Candelaria is 911.7 m+MSL for two and a half days; this level is about 0.70 m lower than in scenario 2. For La Cayetana the drain level is 0.60 m lower than the level without pumping. Although the level is lower, the pumping operation is longer, namely Candelaria pumps for 54 hours and La Cayetana for 52 hours, which is 12 hours more than for a one-day rainfall (scenario 2).

For the present pumping capacity the water level will not stay below the ground surface for a two-day rainfall of 50 mm/d. Increasing the pumping capacity of La Candelaria by 50% and keeping the capacity of La Cayetana the same (scenario 3b) resulted in a level below ground surface and the pump duration was 3 hour shorter than with the present capacity. The discharge in the outlet is 8.7 m³/s and the water profile is 0.50 m lower than the profile for a one-day rainfall.

Scenario 4: Three Days Rainfall with Pumping Operation

A three-day rainfall of 37 mm/d shows a water profile, which is lower than the ground surface; except at the two lowest point of the longitudinal profile (km 12 and 19). La Candelaria station functions for 78 hours and La Cayetana for 75 hours. The simulation shows that the present pump capacity is adequate to maintain the desired water levels. Increase of the pumping capacity of La Cayetana will only slightly lower the water levels at the lowest point as the draw dawn effect of the pumping station is limited due to the dimensions of the canal between the pumping station and the main drain.

According to the frequency analysis, a one-day rainfall with a return period of 1 year is 41 mm, which is comparable with the rainfall used in this scenario 4. This rainfall could occur anytime in a certain year meaning that for normal conditions the expected water levels in the drain are below ground surface along the whole drain length, as in scenario 4.

Area Inundated in the Different Scenarios

The extent of the inundated areas was estimated by overlaying the extreme water level, which occurs during the peak rainfall of scenario 1a, 1b and 1c, and the topographical map. For the first scenario, the inundation last for three days, for the second scenario the water level is higher and inundation in some spots occurs during the entire month, but the peak inundation will remain for almost 10 days. Finally, in the third scenario when the gate is closed, the water level is between 912.3 m and 912.5 m+MSL along the drain during the whole period and will only increase a few centimeters during the peak, reaching 912.8 m+MSL. The percentage of area inundated for the three scenarios is 8.4%, 15.8% and 51.2%, respectively. Table VI.3 gives the inundation depth for some scenarios

Table VI.3 Pumping Operations and Area Inundated in the Different Scenarios

Sc	Rainfall Intensity (mm/d)	Outflow Flap Gate (m³/s)	La Candelaria Station		La Cayetana Station		Area Inundated (%)
			Q (m³/s)	Operation Hours	Q (m³/s)	Operation Hours	
1a	80	12.0	0.0	0	0.0	0	8.4
1b	80	4.8	0.0	0	0.0	0	15.8
1c	80	0.0	0.0	0	0.0	0	51.2
2a	80	11.5	2.8	42	5.1	38	4.8
2b	80	11.0	4.2	40	7.6	35	2.7
3a	50	8.7	2.8	54	5.1	52	1.8
3b	50	7.3	4.2	51	5.1	52	0.6
4	37	4.4	2.8	78	5.1	75	0.2
Sc = Scenario							

The time of inundation for all scenarios is comparable with the operation time for La Candelaria, including some lag time. Comparison of the scenario 1a and 2a (same water level in the Cauca River but with the pumping stations operating in 2a) shows that the inundated area is reduced from 8.4 % to 4.8 % of the area. Even though the water level difference is small and the discharge difference at the outlet is only 0.5 m^3/s, the inundated area is reduced considerably. In the scenarios 3 and 4 the inundated area is relatively small, but still harmful in view of the fact that the groundwater is shallower than the root depth along some parts of the drain.

In the southern part of the District, near la Cayetana Station, a low area is always affected and has inundations of more than 1.5 m. This area of approximately 30 ha could store 450,000 m^3, which is about 65% of the volume that could be pumped out by both stations at full capacity during a complete day, being 682.560 m^3.

For a rainfall intensity of 37 mm/d, which is comparable with a one-day rainfall with a return period of 1 year, the inundation area covers about 0.2% of the whole area. This inundation can be expected any time during the year.

Conclusions

The most important conclusions and recommendations of the study were as follows:

- The present operation of the pumping stations can be improved. According to the water balance at field level, the areas planted with sugarcane and maize require on average 16% less water than the mean annual volume of water pumped into the system for this cropping pattern. This reduction in irrigation supply will also result in 23% less sub-surface water to be drained off;
- Shallow groundwater affects a large area of the district during the growing season. The areas most affected are located in the North near the outlet, in the South and in some spots in the centre of the district. The situation could be improved by a better drainage system at field level or by changing the cropping pattern with crops that have shorter root depths. Crops like maize, sorghum and even soybean can be a rainfed crop, but other crops like sugarcane should be irrigated;
- The groundwater levels calculated for the present irrigation conditions will be harmful, mainly in the sandy clay loam soils. This soil type should be clearly identified in view of possible improvement of the drainage at field level and the necessary periodical maintenance of the drains;
- Irrigation has a high impact on the drainage flows; the supply should be limited to or be less than the potential evapotranspiration. Any excess water will flow almost directly to the main drain or raise the groundwater table;
- Irrigation and effective rainfall result in an almost uniform sub-surface flow that ranges between 0 and 2 mm per day. The pumping operation for drainage is unpredictable since it depends mainly on the runoff and the water levels in the Cauca River. Pump operation at La Candelaria station should start from water levels at 910.5 m+MSL, while La Cayetana Station should start at 910.0 m+MSL. The present pumping capacity cannot prevent inundations and high groundwater for 1- and 2-day rainfall with a return period of 5 year (exceedance of 20%)

Appendix VII - Summary of Interventions

Interventions for the RUT Irrigation District

One of the results of the second participative workshop was the identification of the required physical and non-physical interventions in the RUT Irrigation District in order to reach a model for an integrated and sustainable management. Physical interventions are directly related to the improvement of the physical infrastructure (hydraulic works), operation and maintenance of the hydraulic infrastructure, and irrigation and drainage methods and practices. The non-physical interventions are related to the socio-economic, environmental, organizational and legal aspects. The proposed interventions are considered from technical, environmental, socio-economic, organizational and legal aspects

Interventions for the Technical Aspects

The most important interventions are related to the pumping stations, the marginal dike, the water distribution network, the hydraulic structures, and operation of the irrigation and drainage system. For each item several issues will be considered and the required interventions are proposed.

Interventions for the Environmental Aspects

From an environmental point of view, the interventions concern the preservation and improvement of the quality of the water, the appropriate disposal of solid waste, the protection and conservation of the watersheds and micro-watersheds, the control of the salinization process, and the formulation of plans for the environmental management.

Intervention for the Legal and Organizational Aspects

Based on existing agreements and resolutions for the Land Improvement Sector (Law 41/1993; Agreement 003/2004; and Resolution 1399/2005), ASORUT will have as one of its main tasks: the update of its internal regulations in line with the legal framework in force; to maintain a continuous and permanent verification of these regulations; and to evaluate the impact of the application of the rules. Moreover, ASORUT is responsible for the transfer of the knowledge and the sharing of the regulations with the farmers, the other users and actors involved with the purpose that they all know their rights and duties and that they can and will perform their respective role in view of a sound management of the irrigation district.

Interventions for the Socio-Economic Aspects

The interventions for socio-economic aspects concern the financial, economic and social components which were examined and discussed by the participants during the first participative diagnostic workshop.

Interventions in the RUT Irrigation District – Identification of Tendencies

Item	Topic	TECHNICAL ASPECTS	
		Modernization and finalization, operation and maintenance of hydraulic infrastructure	
		Intervention	Objectives/Impacts
Pumping Stations	Instrumentation	Installation of reliable discharge measurement devices; install in discharge pipes or canal outlets. Special emphasis: Tierrablanca Pumping Station	Monitoring of the performance of the pumping stations. Measurement of water pumped by the stations. Contribution to a more efficient water use.
	Alternative energy sources	Consider alternative energy resources: gas, wind, solar	Decrease operational costs for pumping
	Sediment control	Study, design, and construction of sand traps	Improve (physical) quality of water Decrease costs for canal maintenance Improve control of vegetation in canals Decrease costs for sediment management to be removed from canals (sediment disposal, sediment box) Improve conveyance capacity of canals Improve water distribution to secondary canals and water users Decrease conflicts between and with water users
	Operation during low water levels in the Cauca River	Lower the bottom of the suction chambers. Special emphasis: Tierrablanca pumping station	Improved and more reliable water supply in the dry season
	Rehabilitation and replacement	Changing the pump units	Improve efficiency of pump units. Reliable water supply to water users
Marginal Dike	Slide stability	Protection works (rip-rap, sheet piles) at critical places	Protection against flooding after dike breaching
	Heightening	Heighten the dike in critical points	Protection against flooding by overflow of dike due to too high water levels in Cauca River.
	Settlement	Re-location of settlements	Decrease the risk of dike failure due to inappropriate agricultural practices of the farmers.

TECHNICAL ASPECTS

Modernization and finalization, operation and maintenance of hydraulic infrastructure

Water distribution Marginal, conveyance, 1.0, and interceptor canal	Water losses	Canal lining. Repair joints. Fish control (corroncho). Control of cattle herds. Maintenance of hydraulic structures	Decrease water losses by seepage. More efficient water use. Increase irrigation area. Reduction of operation time of pump units. Improve stability of structures. Reduce or limit raise of water table and therefore reduce risk of soil salinization.
	Canal maintenance: Sedimentation control in interceptor canal	Especially sediment transport by small streams from the western mountain range Course correction of small streams. Reforestation of watersheds of small streams. Combine mutual solutions with local and regional entities involved: Umatas, CVC, Major Office	Improve operation of the distribution system Decrease operation and maintenance costs Improve water quality Environmental sustainability
	Weed control	Weed control in dilatation joints in canals. Aquatic vegetation control (especially calla lily in interceptor canal)	Improve operation of the distribution system Decrease operation and maintenance costs Improve efficient water use.
Hydraulic structures	Rehabilitation	Water level control structures in marginal irrigation canal (4); interceptor canal (3); conveyance canal (1); drainage canal (ceros) Water level control and discharge regulating structures in distribution canal (1.0 canal and others) Discharge measurement structures (wells in regulating gates)	Improve operation of the distribution system. Improve efficient water use; Reliable discharge measurement to the users and more realistic volumetric tariff.
Water Management	Irrigation	Irrigation scheduling and cropping plans. Mutual meetings by zones to establish cropping plans. Changes for the irrigation management: division in sectors of the water distribution process; implementation of water storage (at farm level, strategic points). Training, technological innovation, and technological transfer for the irrigation management (at farm level, water distribution).	More efficient water use and efficient operation of pump units. Decrease operation and maintenance costs. Decrease phyto-sanitary risks to establish cropping patterns.
	Drainage	Optimization of the operation of pumping stations Rehabilitation of well network for groundwater table control.	Contribution to a decrease of soil salinization risk. Improve operation of pump units for drainage.

ENVIRONMENTAL ASPECTS

Item	Topic	Intervention	Objectives/Impacts
Water quality	Disposal of waste water into interceptor canal	Optimization of the wastewater treatment plants of Roldanillo, La Unión and Toro municipalities. Alternatives for reuse of waste water in irrigation. Action to involve municipalities located within the influence area of RUT district in the solution of the optimal disposal of wastewater.	Improve irrigation water quality in the interceptor canal. Increase availability of irrigation water. Reduce health risks. Improve quality of agricultural products (fruits); competition and export opportunities for agricultural products. Reduce maintenance costs for canals by weed control (calla_ iris) and by presence of wastewater with high nutrients content.
	Disposal of domestic waste water	Alternative treatment of waste water (septic well, wetlands, lagoons, etc) for houses located along the canal network: interceptor, conveyance, 1.0, marginal, and drainage	Decrease risk of contamination of irrigation water. Reduce health risks. Improve quality of agricultural products (fruits). Reduce maintenance costs for canals by weed control and by presence of wastewater with high nutrients content.
	Cauca River	Contribution to the Integrated Management Plan for the Cauca River Basin	Improve irrigation water quality
Solid waste	Disposal of garbage	Promote: No disposal of garbage, construction rubble, etc into canal network by municipalities located within influence area of the RUT district. Good agricultural practices by farmers with land next to canal network.	Improve efficient operation of canal network. Conserve hydraulic capacity of canals. Reduce risk of canal overflowing. Reduce canal maintenance costs.
	Agro-chemical waste	Promote optimal disposal of agrochemical waste used by farmers at farm level: containers, bags, cans, etc	Landscape improvement especially at farm level. Reduce risk for soil contamination and water table by lixiviation.

		ENVIRONMENTAL ASPECTS	
Watersheds and micro-watersheds	Conservation of hillside zone: west side of the RUT district	Courses correction coming from west mountain range and discharging into interceptor canal. Reforestation of micro-watersheds of the streams before mentioned. Establishment of forest native species (abarco, balso, etc) for usable use	Cost reduction for interceptor canal maintenance. Improve irrigation water quality. Improve operation of interceptor canal. Landscape improvement. Possibility for income generation.
	Conservation of flat zone (inside RUT district)	Establishment of forest; native species for functional use in agricultural practices: fence construction, fertilizer preparation, animal nutrition, etc Specific project: reforestation by using bamboo and its bio-marketing: housing construction, agricultural facilities, crafts, etc	Erosion control by using living barriers. Establishment of micro-climate. Generation of additional income and jobs through bio-market projects.
Soils	Salinization	Land improvement at farm level for drainage (surface and subsurface) Optimization of irrigation practices: irrigation methods (salts leaching, efficient water use, water quality for irrigation)	Risk reduction for soil salinization. Contribution to agricultural profitability. Conservation of production potential of the soil resource.
Environmental management	Environmental education	Promote the clean production (organic agriculture). Rise environmental conscientious: development of specific projects for children taking advantage the presence of educational institutions in the region. Inter-institutional strengthening. Environmental education to users. Promote the environmental management and the development of the projects along with the leadership of the woman.	Improved life quality for the community. Contribution to the sustainability of the natural resources. Increase conscientious for sustainable human development: development focused on the welfare of the human beings. Creation of a culture with environmental vision.

LEGAL AND ORGANIZATIONAL ASPECTS

Item	Topic	Intervention	Objectives/Impacts
Legal framework	Regulating	Reviewing and spreading of the current regulation among users (land improvement: Law 41/1993, decree 003/2004, and Resolution 1399/2005). Reviewing, spreading and updating of statute for water users association Reform proposals for the current regulation. Specific case: ownership of the hydraulic infrastructure	Knowledge of organizations and supporting entities for the land improvement sub-sector. Weakness and strengths of the political and legal framework for the land improvement sub-sector. Organization mechanisms of the community base for the management and development of land improvement projects.
Organization	Supporting entities	Creation of a permanent consulting body for the study of technical, environmental, organizational, and socio-economic aspects which gives support for the decision-making	Strengthening and promotion of the sense of belonging to by users. Promotion for making decision in a communal and participative way. Promotion of efficiency, accountability, equity and transparency administrative.
	Inter-institutional strengthening	With: Public and private entities in order to share responsibilities for the solution of problems caused by external factors to the district. Specific case: disposal of waste water coming from municipalities within the area of influence of the district; conservation and recovery of watersheds; entities from local administration (Majors Offices) and environmental authorities with the purpose to declare part of the infrastructure like Public Works: dike, interceptor canal, roads, and others	Commitment and involvement of public and private; local, regional and national entities for the communal and participative solution for the problems of the district. Promotion and strengthening of the sense of belonging to by public and private, local and regional entities. Contribution of public and private, local and regional entities for the administration, and financing of maintenance and operation costs of the Public Works.
	Institutional strengthening	Motivation process of the users by which the association is considered as a means for economic and social progress. Promotion of others association forms inside the organization based on the mutual and team work: cooperatives. Training for technical, social, and environmental aspects in view of national and international marketing.	Contribution to increase the living standard of the users. Promote the economic-social development based on organizational reforms and mechanisms for a mutual and wide participation. Strengthening of the water users association in view of new challenges: market globalization, ALCA, TLC.

SOCIO-ECONOMIC ASPECTS

Item	Topic	Intervention	Objectives/Impacts
Financial	Tariffs	Definition of tariffs structure based on received benefits. Establishment of incentives for efficient water use. Establishment of penalties for non-efficient water use.	Difference between big and small users.
	Bills	Study and implementation mechanisms in order to collect debts from users. Promote efficiency for the recovery of debts.	Elimination of paternalism; contribution to self-sufficiency; financing of the association and financial sustainability.
	Negotiations for additional resources	Program for service offering: health, agricultural inputs, legal and accountant consultancy, formulation and management of agricultural projects, technical consultancy, agro-tourist (exhibition of production systems at farm level, tourist plans, etc)	Integral advantage of the production potential of the RUT district. Contribution to the socio-economic development of the organization. Contribution to the development of the sense of belonging to.
Economic	Marketing	Forward markets	Reliable marketing. Guarantee of prices and incomes.
	Strengthening for production systems	Strengthening of existing production chains: grains, fruits, vegetables, sugar cane, cotton, and grass. Promotion and development of new production chains: chilli, cacao, stumps, forest for functional use (abarco, balso). Promote crop diversification taking into account the agricultural suitable vocation of the RUT district being: permanent, semi-permanent, and finally short period. non-stimulate mono-crop. Promote production of cattle system and recover the traditional domestic practices. Introduction of adapting technologies for minifundio systems	Major efficiency for existing production systems. Enlargement of the agricultural frontier. Use advantage of the tropical environmental conditions. Reduce socio-economic risks caused by mono-crop (specific case: coffee crop). Contribute to the food security and improve the domestic economics.

SOCIO-ECONOMIC ASPECTS		
Social	Agricultural and cattle breeding	Recover traditional agriculture: small plot Give farmers alternatives for production projects through the practice of permanent and short period agriculture.	Contribute to social-economic stability of the users. Generate additional income for users.
	Land tenure	Strengthen the "minifundio" and don't stimulate the "latifundio"	Recover initial concepts of the RUT district. Recover traditional agricultural culture for the small farmers of the current regulation.
	Sense of belonging-to	Promote: Transparent and clear actions. Administrative efficiency. Generation change for the development of agricultural activities based on a profitable agriculture. Formulation and development of joint projects under efficiency, equity, and transparent criteria.	Strengthening and promotion of the sense of belonging to. Organizational strengthening.

Abbreviations and Acronyms

ACUAVALLE	Association of Water Supply Systems and Wastewater Disposal of the Valle del Cauca Department
ALCA	Free Trade Area for the Americas
ASODIMAR	Water Users Association of the Maria La Baja Irrigation District
ASORECIO	Users Associations of the Rio Recio Irrigation District
ASORUT	Users Associations of the RUT Irrigation District
ASUSA	Users Associations of the Samaca Irrigation District
ATP	Accelerated Transfer Program (Turkey)
ATPDEA	Andean Trade Promotion and Drug Eradication Act
BID	Inter - American Development Bank
BIRF	International Bank for Reconstruction and Development
C.C	Conveyance Canal
CAN	Nations Andean Trade
CAR	Autonomous Regional Corporation
CHO	Constant Head Orifice
CIS	Communal Irrigation System (Philippines)
CNA	National Water Commission (Mexico)
CONSUAT	Superior Council for Land Improvement
CORPOICA	Colombian Corporation for the Agricultural Research
CVC	Autonomous Regional Corporation of the Cauca Valley
DANE	National Administrative Department for Statistics
DAT	Land Improvement District
DNP	National Planning Department
DRI	Co-financial Fund for Rural Investment
FEDERRIEGO	Federation of Water Users Associations
FINAGRO	Agricultural Financial Fund
FONAT	National Fund for Land Improvement
HIMAT	Institute of Hydrology, Meteorology and Land Improvement
IA	Irrigation Administration (Philippines)
IBRD	International Bank for Reconstruction and Development
ICA	Colombian Agriculture Institute
ICEL	Colombian Institute for Electric Energy
ICID	International Commission on Irrigation and Drainage
ICR	Incentive for Rural Capitalization
ICWE	International Conference on Water and Environment
IG	Irrigation Group (Turkey)
IGAC	Agustin Codazzi Geography Institute
IIMI	International Institute for Irrigation Management (later IWMI)
IMT	Irrigation Management Transfer
INAT	National Institute for Land Improvement
INCODER	National Institute for Rural Development
INCORA	National Institute for Agrarian Reform
INDERENA	National Institute of Renewable Resources and Environment
INPA	National Institute for Fishing and Aquaculture

IWMI	International Institute for Water Management
M.D.C	Main Drainage Canal
M.I.C	Main Irrigation Canal
MSL	Mean Sea Level
NIA	National Irrigation Administration (Philippines)
NIS	National Irrigation System (Philippines)
RUT	Roldanillo – La Union – Toro
SECAB	Executive Secretary of the Andres Bello Agreement
SENA	National Apprenticeship Service
SIP	Small Irrigation Program
TLC	Free Trade Agreement with the United States of America
UMATA	Municipality Unit for Agricultural Technical Assistance
UNDP	United Nations Development Program
UNESCO	United Nations Educational, Scientific and Cultural Organization
USDA	United States Agriculture Department
USOMARIA	Users of the Maria La Baja Irrigation District
WUO	Water Users Organization

Samenvatting

Colombia is een tropisch land in Zuid-Amerika. Het landoppervlak is 114 miljoen ha. In Colombia zijn twee typen irrigatie: de kleinschalige en de grootschalige irrigatie. De kleinschalige irrigatie heeft zich vooral ontwikkeld in hellende gebieden, waar voedingsgewassen en andere marktproducten zoals graan, aardappelen en vooral groenten worden verbouwd. In deze gebieden heeft de landbouw zich niet goed kunnen ontwikkelen en daarom worden daar geen landbouwmethoden met geavanceerde technologieën aangetroffen. Deze landbouwgebieden met een traditioneel karakter worden gekenmerkt door kleine landbouwbedrijven, door boeren met weinig grond en een klein inkomen; er is een hoog niveau van analfabetisme en ondervoeding. De gebieden met grootschalige irrigatie worden gekenmerkt door grote landbouwbedrijven met veel grond, die vooral te vinden zijn in de vlakke gebieden en op zeer goede gronden voor de teelt van marktgewassen zoals suikerriet, katoen, sorghum, fruit, enz. In deze landbouwgebieden komen mechanisatie en geavanceerde irrigatie technologieën voor.

Het gebied van de grootschalige irrigatie en dat tot de publieke sector hoort is onderverdeeld in 24 irrigatiedistricten, die in totaal 264.802 ha beslaan. Ongeveer 133.000 ha hebben een infrastructuur voor irrigatie en drainage en 132.000 ha hebben alleen voorzieningen voor drainage. Alle irrigatiedistricten samen ontvangen ongeveer 499 m^3/s; het totale irrigatienetwerk is 1.905 km lang, 37% van de kanalen vormen de hoofdkanalen, 45% de secundaire kanalen en 18% de tertiaire kanalen; het drainagenetwerk is 2.132 km lang, ongeveer 44% van de afvoerkanalen zijn hoofdafvoerkanalen, 39% vormen de secundaire afvoerkanalen en 17% de tertiaire afvoerkanalen; het wegennet in de irrigatie districten is 3.382 km lang.

Ongeveer 9% van het land wordt gebruikt voor permanente gewassen, 51% voor niet-permanente gewassen en 40% voor gras. Ongeveer 62% van de landbouwgronden bestaan uit landeigendommen die kleiner dan 5 ha zijn en ongeveer 9% van het totale gebied beslaan; 17% van de landeigendommen hebben een oppervlak tussen de 5 en 10 ha en beslaan 13% van het totale landbouwgebied, ongeveer 16% van de eigendommen hebben een grootte van 10 tot 50 ha en 3% van de landbouwbedrijven zijn groter dan 50 ha en beslaan ongeveer 38% van het totale gebied.

Potentieel voor geïrrigeerde landbouw

Op wereldschaal bezien is Colombia een land met een overvloed aan natuurlijke rijkdommen, vooral met betrekking tot de aanwezige land- en watervoorraden, die samen een groot potentieel voor de verdere ontwikkeling van de geïrrigeerde landbouw vormen.

Gronden. Het Geografisch Instituut Agustin Codazzi (IGAC) heeft in 1973 een bodemonderzoek voor Colombia uitgevoerd. In Colombia kunnen twee grondsoorten voor de landbouw worden onderscheiden. De type A gronden komen overeen met de grondklassen I, II en III van de United States Department of Agriculture (USDA) en zijn geschikt voor mechanisatie, intensieve landbouw en veeteelt, en ook voor de ontwikkeling van geïrrigeerde landbouw. Deze gronden worden hoofdzakelijk aangetroffen in de vlakten langs de Caraïbische kust en in de gebieden langs de Orinoco en de Amazone. De meeste gronden van type B behoren tot klasse I - V; het grondoppervlak van de gronden type B varieert van vlak tot concaaf-vlak wat onder andere betekent dat grondbewerking nodig is om de gronden geschikt te maken, zoals bijvoorbeeld bescherming tegen hoogwater, drainage en zoutuitspoeling. Zodra deze gronden verbeterd zijn zal hun potentieel gelijk zijn aan dat van type A gronden. De typen A en B gronden beslaan samen een gebied van 10,6 miljoen ha die geschikt zijn voor gemechaniseerde en geïrrigeerde landbouw.

Watervoorraden. Volgens UNESCO heeft Colombia een specifieke oppervlakte wateropbrengst van 0,59 l/s per ha en bezet hiermee de vierde plaats op de wereldranglijst. Deze specifieke opbrengst is zes en drie keer groter dan de waarde voor het landoppervlak van de gehele wereld en van Latijns-Amerika, die respectievelijk 0,10 en 0,21 l/s per ha bedragen.

De gemiddelde jaarlijkse regenval in Colombia is 1.900 mm. Op 70% van het gebied is de gemiddelde jaarlijkse regenval 2.000 mm of meer en op bijna één derde van het landoppervlak is de regenval meer dan 4.000 mm (de vlaktes langs de kust van de Grote Oceaan en in het Oosten). Deze waarden liggen boven het gemiddelde van 1.600 mm per jaar voor Latijns-Amerika, en boven het gemiddelde voor de gehele wereld van 900 mm per jaar. In tegenstelling tot de rest van het land krijgt de Caraïbische kust weinig regenval en wordt als een droog gebied beschouwd.

Colombia heeft meer dan 1.000 het hele jaar stromende rivieren, een aantal dat veel groter is dan voor Afrika (60). Het heeft 10 binnenlandse rivieren, welke een gemiddelde jaarlijkse afvoer van meer dan 1.000 m³/s hebben. Colombia heeft een natuurlijk drainagenetwerk dat uit meer dan 740.000 micro-bassins bestaat en die ontstaan zijn mede door de hoge regenval, de topografie en de geologie. Deze natuurlijke drainage wordt verdeeld in 4 regio's of hoofdstroomgebieden.

Ontwikkeling van de irrigatie. Colombia draagt minder dan 3% van het geïrrigeerde landbouwgebied in Latijns-Amerika bij; het land bezet de zesde plaats onder de Zuid-Amerikaanse landen met zijn ratio van geïrrigeerd land over totaal gecultiveerd land (12,5%). Colombia heeft voor zijn irrigatieontwikkeling dezelfde tendens als de overige Latijns-Amerikaanse landen gevolgd. Het irrigatiegebied nam snel in oppervlak toe tot het eind van de jaren 1960, maar de groei daalde scherp tijdens de laatste twee decennia van de vorige eeuw. Gedurende de periode 1990 - 2000 werden de beschikbare middelen van de publieke sector vooral gebruikt voor de rehabilitatie, voltooiing en uitbreiding van de bestaande irrigatiedistricten.

Vooruitzichten voor de geïrrigeerde landbouw. Volgens de Nationale Water Studie van 1984 heeft Colombia 14,4 miljoen ha land dat geschikt is voor landbouw, hiervan is 6,6 miljoen ha (45,8%) geschikt voor gemechaniseerde landbouw en voor de ontwikkeling van geïrrigeerde landbouw door de aanleg van de benodigde irrigatie en drainage infrastructuur. Dit totale oppervlak zou kunnen toenemen tot 10,6 miljoen ha na de verbetering van sommige gronden, die op dit moment het gebruik van de gronden voor landbouw beperken. Van deze 6,6 miljoen ha heeft slechts 11,4% een infrastructuur voor irrigatie en drainage, waarvan 38% door de publieke sector en de rest door de particuliere sector is ontwikkeld.

De Colombiaanse Overheid ontwikkelde het "Programma voor de Landverbetering voor het Decennium 1991 - 2000" met als doel om 535.500 ha met een irrigatie en drainage infrastructuur te verbeteren. Tijdens de verschillende periodes heeft de Colombiaanse overheid dit landverbeteringsbeleid aangepast met als doel om zoveel mogelijk het oorspronkelijke programma voor de landverbetering te volgen dat voor deze periode door het Departement voor Nationale Planning was vastgelegd. Nochtans, de landverbeteringsprogramma's worden gekenmerkt door een trage ontwikkeling.

Beleid voor de landverbetering sector. De historische ontwikkeling van de landverbetering in Colombia werd in grote mate beïnvloed door het beleid zoals dat door de verschillende regeringen werd vastgesteld. De regeringen voerden hun beleid uit door de formulering van het vereiste juridisch kader en het beleid voor de landverbetering is na vele richtlijnen vaak veranderd, waardoor geen eenduidige ontwikkeling op de lange termijn mogelijk was.

De politieke richtlijnen van de Colombiaanse overheid voor de landverbeteringssector werden opgenomen in het Economisch-sociaal Ontwikkelingsplan voor de periode 1991 - 2000.

In 1991 was de Colombiaanse overheid van mening dat de publieke en particuliere investeringen een goed gereedschap zijn om tot modernisering van de landbouw te komen. De investeringen in de landverbeteringsprojecten zouden bij voorkeur gezamenlijk door de publieke en particuliere sector moeten worden gesteund en uitgevoerd. Om de investeringen op een eenvoudige manier terug te verdienen zou de overheid de deelneming van de gemeenschap aan het ontwerpproces van de projecten moeten bevorderen en na de aanleg zouden de projecten aan de Organisatie van Watergebruikers voor beleid, beheer en onderhoud moeten worden overgedragen. Om dit beleid door te kunnen voeren werd een nieuw institutioneel kader voorgesteld, waarin de rol van de overheid zal veranderen van alleen een uitvoerder naar meer een motor voor de investeringen en met een belangrijke betrokkenheid van de gemeenschap.

De wet 41 (1993) geeft de richtlijnen waarmee de landverbeteringssector werd georganiseerd. Op basis van deze institutionele regeling werd de Superieure Raad voor de Landverbetering (CONSUAT) de adviserende eenheid, die verantwoordelijk is voor de toepassing van het beleid voor de landverbeteringssector; het Nationale Instituut voor de Landverbetering (INAT) werd de uitvoerende eenheid om het landverbeteringbeleid met andere publieke en particuliere organisaties verder te ontwikkelen, en het Nationale Fonds voor de Landverbetering (FONAT) werd de administratieve eenheid voor de financiering van de irrigatie, drainage en projecten ter bescherming tegen overstromingen.

Na 7 jaar van landverbeteringsprogramma's binnen dit nieuwe institutionele kader nam de kritiek sterk toe in verband met de rol van de regeringsinstantie op grond van een niet efficiënt beheer van de economische middelen en een vrijwel tot stilstand gekomen landverbeteringsprogramma, dat vanaf 1990 in het kader van verschillende regeringsprogramma's was gelanceerd. De kritiek had onder andere betrekking op het beheer van de externe geldmiddelen voor het landverbeteringsprogramma, de kosten voor het functioneren van de instantie en de kwaliteit van de geleverde werkzaamheden door de regeringsinstantie.

Daarom werd aan het begin van 2003 een verandering in het institutionele kader geïntroduceerd en werden INAT en de andere publieke instanties binnen de landbouwsector opgeheven en geïntegreerd in een nieuwe organisatie, namelijk het Colombiaanse Instituut voor Plattelandsontwikkeling (INCODER). De overeenkomst van 3 Februari 2004 wordt beschouwd als een zeer belangrijk stap binnen het nieuwe institutionele kader waarbij de organisatorische richtlijnen voor de Associaties/Organisaties van Watergebruikers met inbegrip van de basisverordeningen voor het functioneren van kleine, middelgrote en grote irrigatiedistricten werden opgesteld.

Deze overeenkomst bevatte echter een aantal controversiële punten, die resulteerden in een nieuwe verordening voor het beleid, beheer en onderhoud van de irrigatiedistricten met inbegrip van de overdrachtvormen (delegatie, beleid, en concessie) en ervan uitgaande dat INCODER geen directe bemoeienis met de Organisaties van Watergebruikers moet hebben, maar alleen met het beheer van de infrastructuur (Resolutie 1399/2005).

Overdracht van het irrigatiebeheer

In het midden van de jaren 1970 begon de Colombiaanse overheid een programma om het irrigatiebeheer naar de Organisaties van Watergebruikers over te dragen. De overdracht van de irrigatiedistricten in Colombia is uitgevoerd gedurende drie perioden. In 1976 werden op verzoek van de watergebruikers de irrigatiedistricten Coello en Saldaña, overgedragen. De irrigatiedistricten Rio Recio, Roldanillo-La Union-Toro (RUT), Samaca, en San Alfonso werden overgedragen gedurende de jaren 1989 - 1993 en tenslotte werden van 1994 tot 1995 de meeste andere districten overgedragen in lijn met het beleid van de Colombiaanse overheid. Op dit moment zijn 16 van de 24 irrigatiedistricten overgedragen en de rest is nog onder het bestuur van een overheidsinstantie.

Toch werden tijdens en na het overdrachtproces ernstige problemen vastgesteld, onder andere met betrekking tot meningsverschillen over de waterrechten en andere wettelijke aspecten; ontoereikend beheer en onderhoud van de hydraulische infrastructuur; meningsverschillen over de rol van de overheidsinstantie en van de organisatie van watergebruikers; ten aanzien van de milieudegradatie; problemen met betrekking tot economische en financiële aspecten; ontoereikende participatie van de lokale gemeenschap; gebrek aan leiding en ontoereikend opleidingsniveau en irrigatiebeheer.

Voor een aantal irrigatiedistricten resulteerden de genoemde problemen en beperkingen in inefficiënte prestaties, verdere verslechtering van de fysieke infrastructuur en conflicten met betrekking tot het beleid, beheer en onderhoud van het irrigatie en drainagesysteem dat tot een niet-duurzame ontwikkeling van de irrigatiedistricten leidde.

In het kader van het huidige PhD onderzoek zijn de achtergronden en realisatie, de huidige status en effecten van het overdrachtprogramma in Colombia bestudeerd. Voor dit onderzoek zijn 5 irrigatiedistricten bezocht en tijdens het veldwerk werden uitgebreide gesprekken met de technische staf en de boeren gevoerd. Tevens zijn de lokale staf van de overheidsinstantie voor de Landverbetering en van publieke en particuliere instanties op lokaal en regionaal niveau gecontacteerd. Gedurende de onderzoeksperiode zijn de irrigatiedistricten van Rio Recio, Coello, Saldaña, Zulia en Maria La Baja bezocht.

Een uitgebreid literatuuronderzoek heeft geleid tot de vastlegging van de ervaringen met en kenmerken van de overdracht van het irrigatiebeheer in andere landen met vergelijkbare kenmerken als Colombia. Deze gegevens zijn voor het later te bespreken kader voor de overdracht van het beleid verder geanalyseerd.

De ervaringen met de overdracht van het irrigatiebeheer in Colombia tonen duidelijk aan dat de financiële last van de regering beduidend is afgenomen door de verschuiving van de totale kosten voor beleid, beheer en onderhoud van de overheid naar de Organisaties van Watergebruikers. Ook de gedeeltelijke en in vele gevallen de totale afschaffing van de subsidies en een reorganisatie van de institutionele structuur van de Landverbeteringssector heeft de totstandkoming van het decentralisatie- en delegatiebeleid in deze sector aanzienlijk vergemakkelijkt. Het effect van de overdracht van functies naar regionale en lokale instellingen kan op dit moment nog niet worden geëvalueerd, omdat de instellingen pas begonnen zijn met hun taken na deze overdracht. Toch bestaat er het nodige scepticisme onder de Organisaties van Watergebruikers betreffende de efficiëntie van deze overdracht.

De ontwikkeling van het overdrachtprogramma toont een duidelijk verschil tussen de eerste fase van de overdracht en de stadia daarna. Tijdens de eerste fase zijn het vooral de watergebruikers zelf geweest die de overdracht van het beheer bevorderde en toonde de overheid en de overheidsinstantie een duidelijke weerstand; in de daarop volgende fasen kwam de motivatie hoofdzakelijk van de overheid en niet van de gebruikers. De gebruikers zagen de nieuwe houding van de overheid als een regeringsstrategie om de fiscale crisis van de jaren 1990 onder ogen te kunnen zien en bovendien op deze wijze compromissen met de internationale bankinstellingen en daardoor nieuwe kredieten te verkrijgen.

Ongeveer 30 jaar na het begin van het programma van overdracht van het beheer tonen de Organisaties van Watergebruikers in de bezochte gebieden zeer verschillende stadia van ontwikkeling. Sommige hebben een uitstekend ontwikkelde en stabiele organisatie, voorbeelden zijn de irrigatiedistricten van Rio Recio, Coello, Saldaña, die financieel onafhankelijk zijn, omdat zij voorzien in een adequate en betrouwbare levering van irrigatiewater. Andere districten zoals Maria La Baja verkeerden in een ernstige organisatorische crisis, die uiteindelijk in een interventie van de regering resulteerde. Het irrigatiedistrict Zulia is het voorbeeld van een zeer goed ontwikkeld organisatorisch model met een dynamisch karakter dat is gebaseerd op een integrale benadering met een duidelijke inbreng van de gemeenschap rond enkele zeer winstgevende landbouwactiviteiten, die gericht zijn op de rijstproductie en steunen op een duidelijke visie voor de toekomstige ontwikkeling van de organisatie. Wat betreft het proces na de overdracht en vanuit een organisatorisch standpunt kan men drie belangrijke fasen onderscheiden: namelijk aanpassing, volwassen wording of volledige ontwikkeling, en consolidatie.

De verbetering van een efficiënt watergebruik is een andere doelstelling van de beheeroverdracht aan de Organisaties van Watergebruikers geweest. Echter waarnemingen in de onderzochte irrigatiedistricten tonen aan dat het watergebruik een kritische zaak is, die de duurzaamheid van de irrigatiesystemen in milieuopzichten kan gaan belemmeren. De financiële onafhankelijkheid (self-sufficiency) als resultaat van de levering van irrigatiewater tegen betaling, vooral voor gewassen die grote hoeveelheden water nodig hebben zoals rijst schijnt een factor te zijn die niet tot een meer efficiënt gebruik van irrigatiewater bijdraagt.

In verband met het milieu kan nog vermeld worden dat de toenemende ontbossing in de stroomgebieden van de belangrijkste rivieren van de onderzochte irrigatiedistricten ernstig gevolgen heeft voor onder andere de waterkwaliteit, de kosten voor het beheer en onderhoud en de beschikbaarheid van de watervoorraad .

Ten slotte kan gesteld worden dat de overdracht van het irrigatiebeheer in Colombia heeft geleid tot een andere kijk op de rol van een irrigatiesysteem. Deze kijk is in de afgelopen jaren in een ruimere context geplaatst. Een irrigatiesysteem moet niet als een eenvoudig hydraulisch netwerk worden beschouwd, dat alleen tot doel heeft om irrigatiewater aan de boeren te verstrekken, maar het moet vooral gezien worden als een belangrijke component in het gehele productiesysteem, die als uiteindelijke doelstelling heeft de verbetering van de levensomstandigheden van de boeren door geïrrigeerde landbouw volgens de normen van rentabiliteit, gelijkheid, efficiency en een nieuwe benadering van integraal beheer met deelname van de gemeenschap.

Duurzaam beheer van overgedragen irrigatiesystemen: conceptueel kader

Een conceptueel kader voor het duurzaam beheer van irrigatiedistricten die aan een Organisatie van Watergebruikers worden overgedragen, wordt gepresenteerd en is gebaseerd op algemene concepten en benaderingen van duurzaam waterbeheer en vooral van een duurzaam gebruik van land en watervoorraden en het beheer van land- en watersystemen. Dit conceptueel kader steunt op een aantal algemene principes, benaderingen en definities; vele van deze principes en benaderingen zijn uitgebreid toegelicht in tal van bijeenkomsten en conferenties betreffende de verbetering en behoud van de natuurlijke rijkdommen, speciaal de beschikbare land en watervoorraden.

De duurzaamheid van een irrigatie en drainagesysteem verwijst naar de fysieke en organisatorische infrastructuur voor beleid, beheer en onderhoud tijdens de levensduur van het systeem volgens duidelijk vastgestelde specificaties en na het geven van de verwachte voordelen aan de gebruikers zonder dat het milieu beschadigd wordt. Het concept duurzaamheid impliceert drie elementen, die ook een belangrijke rol in de definitie van het conceptuele kader voor een duurzaam beheer van een irrigatiesysteem spelen. De drie componenten zijn Gemeenschap (C = Communautair), Milieu (E), en Wetenschap & Technologie (S&T) en zij zijn onderling verbonden binnen bestaande institutionele en wettelijke contexten. Deze verbondenheid bepaalt de duurzaamheid van een irrigatiesysteem.

Communautair (C) vertegenwoordigt de verschillende belangengroepen in de irrigatie en drainagesector; in deze groep zijn de boeren de belangrijkste gebruikers. De gemeenschap (C) met zijn specifieke socio-economische en culturele kenmerken en organisatorische vormen is verbonden met de componenten Milieu en Wetenschap & Technologie om samen tot een duurzaam beheer van de systemen bij te dragen. Het milieu (E) verstrekt de natuurlijke rijkdommen, die nodig zijn voor geïrrigeerde landbouw als productiesysteem. Klimatologische omstandigheden, gronden, grondsoorten, ecosystemen en de watervoorraden zijn de belangrijkste gegevens die door de milieucomponent verstrekt worden. Factoren die het verband tussen Milieu (E) en Gemeenschap (C) in twee richtingen beïnvloeden vormen de risicofactoren. De Wetenschap & Technologie (S&T) vertegenwoordigen de wetenschappelijke en technologische kennis, methodes, hulpmiddelen en ervaringen, die beschikbaar zijn en door en voor de gemeenschap gebruikt kunnen worden met het doel om in het natuurlijke milieu in te grijpen om op deze manier aan zijn behoeften en wensen te voldoen. De Wetenschap & Technologie vormen het mechanisme waardoor de risicofactoren in de relatie tussen Milieu en Gemeenschap kunnen worden gecontroleerd of overwonnen. De technologische alternatieven voor de verstrekking van irrigatie en drainage moeten overeenkomen met de culturele, sociaal-economische, technische en financiële voorwaarden van de Gemeenschap. Het niveau en de kosten van de te leveren diensten, bereidheid om voor deze dienstverlening te betalen, de organisatorische vormen, technische mogelijkheden, deelname aan beheer en onderhoud, besluitvorming, conflict resolutie, en mobilisering van de benodigde middelen spelen een belangrijke rol in de relatie Wetenschap & Technologie en Gemeenschap.

Het RUT Irrigatiedistrict

Voor dit onderzoek is het RUT irrigatiedistrict als speciaal onderzoeksgebied gekozen. De historische achtergrond; beschrijving en kenmerken van de fysische, beheertechnische en sociaal-economische aspecten van het district; de kenmerken van en de effecten na de overdracht van het beheer zullen hier besproken worden. Het RUT irrigatiedistrict ligt in het Zuidwesten van Colombia, in het noordelijke deel van het Departement Valle del Cauca. Het district ligt binnen de drie gemeenten, die hun naam aan het district hebben gegeven, namelijk Roldanillo, La Union en Toro. Het district ligt op een hoogte van 915 tot 980 m+MSL (gemiddeld zeeniveau). Vóór de bouw van het RUT irrigatiedistrict stroomde het gebied regelmatig onder water; de overstromingen kwamen van zowel de Cauca Rivier als van de beekjes en stromen die van de bergen aan de westkant van het gebied komen. Hierdoor kwam een gebied van 1.500 tot 3.500 ha regelmatig, permanent of periodiek onder water te liggen. Tengevolge van deze hoge waterstanden heeft een gebied van ongeveer 2.500 ha een hoge bodemvochtigheid, die slechts de teelt van gras toestaat en een gebied van 4.000 ha, dat slechts voor extensieve veefokkerijen gebruikt kan worden.

In het algemeen kan gezegd worden dat voor het RUT district in de eerste plaats bescherming tegen hoogwater een rol speelt en pas daarna irrigatie en drainage.

Het lange en smalle komvormige gebied wordt omringd door een dijk langs de Oostgrens van het gebied, de Cauca Rivier, en een interceptiekanaal aan de Westkant. Een hoofdafvoerkanaal verdeelt het gebied vrijwel in tweeën en loopt in de laagste delen van het district. Parallel aan de rivierdijk loopt het belangrijkste irrigatiekanaal, dat door drie pompstations met een totale capaciteit van 13,8 m^3/s van water wordt voorzien. Het eerder genoemde interceptiekanaal doet ook dienst als irrigatiekanaal voor die watergebruikers, die kleine centrifugaalpompen gebruiken voor hun individuele waterbehoeften. Er is een aanvullend netwerk van zowel kleine irrigatie als drainagekanalen door het gehele gebied. Het belangrijkste afvoerkanaal loost vrij op de Cauca Rivier tijdens lage waterstanden en het drainagewater wordt opgepompt tijdens hoge waterstanden in de rivier.

Het RUT irrigatiedistrict is één van de belangrijkste irrigatiegebieden in Colombia vanwege zijn zeer goede en beschikbare natuurlijke rijkdommen (klimaat, grond en water) die gebruikt kunnen worden voor de ontwikkeling van geïrrigeerde landbouw. Verder zijn de strategische locatie van het district en zijn potentieel om nieuwe agro-industriële ontwikkelingen op te starten belangrijk voor de nationale en internationale markten.

Op dit moment vertoont het RUT district echter een toenemende gezagscrisis, die een ernstige bedreiging voor de duurzaamheid van het gehele systeem vormt; vooral wanneer de meest noodzakelijke maatregelen niet op korte termijn genomen worden. Externe en ook interne factoren met hun sociale, economische, milieutechnische en institutionele eigenschappen hebben in verschillende gradaties tot deze situatie bijgedragen.

Vanaf het prille begin heeft het beheer van het district een sterke weerstand van de lokale gemeenschap ondervonden, die zagen het project duidelijk als een project dat door de Nationale Overheid was opgelegd. Zoals zo vele projecten van de jaren 1960 en 1970 werd ook dit project ontwikkeld met een zeer beperkte tot minimale deelname van de lokale bevolking en onder bijna totalitaire verantwoordelijkheid van de overheid, die de uitvoering en ontwikkeling van dit landverbeteringsproject als een zuiver fysieke actie zagen, zonder enige of met beperkte aandacht voor de sociale, culturele en economische eisen binnen een aanvaardbare sociale context. Daarom toonden de boeren geen enkel saamhorigheidsgevoel en geloofden zij dat het alleen de verantwoordelijkheid van de Overheid was om het beleid, beheer en onderhoud van dit project uit te voeren. Tijdens de bestuursperiode van de overheid (1971 - 1989) had deze houding van de watergebruikers geen ernstige gevolgen, omdat de bestuursinstantie van de overheid voor alle aspecten van het bestuur, inclusief beheer en onderhoud en alle financiële middelen, die door de overheid verstrekt werden, verantwoordelijk was. Het was deze paternalistische houding van de landverbeteringssector; het negatieve effect van de ligging van de irrigatie en drainagekanalen op hun landbouwbedrijven en landbouwmethoden; de in hun ogen onvoldoende compensatie voor hun land dat door de overheid gebruikt werd voor de kunstwerken en het gebrek aan geloofwaardigheid van het investeringsproject, heeft geenszins de motivatie en de houding van de boeren verbeterd.

Na de overdracht in 1989 kreeg de Organisatie van Watergebruikers, ASORUT de volledig verantwoordelijkheid voor alle activiteiten op het gebied van beleid, beheer en onderhoud, inclusief alle financiële aspecten zonder enige subsidie van de overheid. De daadwerkelijke overdracht omvatte voornamelijk de activiteiten met betrekking tot het oprichten van de Organisatie van Watergebruikers, het innen van de waterbelastingen en de reparatie van de inlaatkunstwerken en het hoofdkanaal. Zelfs nu nog geloven de boeren dat de overdracht veel te snel is gegaan en zonder voldoende hulp van de overheid en dat de voornaamste reden van de overheid was om zo-snel-als mogelijk het beheer over te dragen om zodoende te voldoen aan de wensen van internationale financiële instellingen, die in de landverbeteringssector geïnteresseerd waren. Het belangrijkste aandachtspunt van het huidige beheer is de verstrekking van water, waarmee de organisatie de noodzakelijke inkomsten voor zijn financiering kan verkrijgen. Aangezien de huidige organisatie op dezelfde manier als de vroegere overheidsinstantie werkt, geloven de watergebruikers dat de overdracht slechts een verandering van acteurs binnen het beheer heeft gebracht.

Op deze wijze moest ASORUT na enige tijd de volgende en belangrijke negatieve aspecten onder ogen zien: lage opbrengst van de waterlasten; opschorting van de onderhoudsactiviteiten voor het irrigatienetwerk in tijd en plaats; een verminderde kwaliteit van de verleende diensten; een inefficiënt gebruik van de irrigatie infrastructuur en irrigatiewater; conflicten tussen de watergebruikers onderling en tussen hen en het administratieve en technische personeel van de organisatie; tenslotte een volledige mislukking van een overeenkomst betreffende de betaling van de waterlasten. Het voortdurend uitstellen van de nodige onderhoudsactiviteiten van de infrastructuur resulteerde in een verslechtering van de fysieke conditie van het systeem en een inefficiënte waterverstrekking.

De houding van het huidige beheer toont duidelijk aan dat zij alleen maar irrigatiewater levert zonder zich af te vragen of het water tot het succes van het productiesysteem van de boeren bijdraagt. De huidige situatie is het gevolg van de groeiende discrepantie in belangen tussen twee partijen: de organisatie is geconcentreerd op de verstrekking van irrigatiewater dat volgens het vastgestelde tarief moet worden betaald en de boeren verwachten juist acties van de organisatie om hen te helpen in de ontwikkeling van nieuwe projecten om zodoende hun levensstandaard te verbeteren.

Deze benadering van het beheer en een aantal externe factoren veroorzaakten meerdere problemen in het district, bijvoorbeeld het interceptiekanaal, de waterkwaliteit, de verzilting van de bodem, instandhouding van de infrastructuur en het afketsen van internationale samenwerkingsprojecten voor de modernisering van het district. De huidige crisis is een gevolg van de slechte conditie van ASORUT, vooral de technische, economische, sociale, milieutechnische en institutionele delen van de organisatie.

Het beheer van het district is tegenwoordig veel complexer dan vroeger en de watervoorziening wordt ongunstig beïnvloed door veranderingen in de te verbouwen gewassen, door het ontbreken van schema's voor de bevloeiing. De boeren beslissen zelf over hun gewassen, de datum van planten en het tijdstip waarop zij om irrigatiewater verzoeken. Een duidelijk voorbeeld is suikerriet dat op dit moment één derde van het district beslaat en veel meer water nodig heeft dan de traditionele gewassen. Ook de moderne landbouwpraktijken, die nodig zijn voor fruitbomen vereisen juist een watervoorziening met korte irrigatie-intervallen.

De belangrijkste milieuproblemen hebben betrekking op de waterkwaliteit. Het huishoudelijk en industrieel afvalwater en de ontbossing van de stroomgebieden beïnvloeden de waterkwaliteit van de irrigatiekanalen en van de Cauca Rivier. Op verscheidene plaatsen neemt de verzilting van de bodem toe door het hergebruik van drainagewater, inefficiënte irrigatiepraktijken, gebrek aan een goedwerkend drainagesysteem op veldniveau, en door verzilting, die door het grondwater wordt veroorzaakt.

Een factor die in ernstige mate de kritieke situatie van het district beïnvloedt is het feit dat er twee soorten boeren in het gebied zijn, namelijk de kleine en middelgrote boeren en de grote boeren. De kleine en middelgrote boeren zijn vanwege hun beperkte economische middelen samen met een beperkte technologie en een klein landbouwbedrijf niet in staat om gewassen met een hoge opbrengst te introduceren of te telen. De grote boeren kunnen wel gewassen met een hoge rentabiliteit, zoals fruitgewassen, introduceren. Dit leidt tot veel spanning en meningsverschillen tussen de twee groepen.

Duurzaam beheer voor het RUT Irrigatiedistrict: implementatie van een conceptueel kader

Een analytisch kader voor een integraal beheer van de beschikbare watervoorraden is gebruikt om een uitgewerkt voorstel voor een integraal en duurzaam beheersplan voor ASORUT te formuleren. De beoordeling van de huidige situatie, de formulering van het beheersplan en de vereiste acties zijn onder meer gebaseerd op gesprekken met boeren, administratief, technisch, en veld personeel van ASORUT, vertegenwoordigers van de gemeenten en vertegenwoordigers van publieke en particuliere instellingen, die nauwe contacten met de landverbeteringssector onderhouden.

De voorstellen voor scenario's voor een integraal en duurzaam beheer zijn gebaseerd op de potentiële, interne en externe factoren van het district en hun invloed op het beheersproces. Twee duidelijk verschillende scenario's zijn voor het RUT district ontwikkeld, namelijk een scenario gebaseerd op Directe Regeringsinterventie en een ander op een Benadering met Gemeenschapsdeelname: een Nieuwe Rol voor ASORUT.

Beide scenario's zullen beschreven worden vanuit de huidige, kritieke situatie, waarbij het eerste scenario is gebaseerd op de veronderstelling dat de situatie verder verslechtert als een gevolg van het huidige beperkte vermogen om het district goed te beheren en om de noodzakelijke veranderingen door te voeren, die tot een geleidelijke verbetering van de situatie zouden kunnen leiden. Verwacht mag worden dat sommige negatieve factoren zullen toenemen, bijvoorbeeld een toename van de onrendabele landbouwpraktijken waardoor de boeren niet bereid en niet in staat zijn om de prijs voor het water te betalen; een verhoging van de openstaande waterrekeningen omdat de watergebruikers niet voor de geleverde diensten kunnen betalen wat weer tot een afnemend financieel vermogen van de organisatie leidt. Andere factoren zijn de veranderingen in landgebruik, met name de toename van extensieve gewassen zoals suikerriet en van andere niet-landbouwkundige toepassingen; veranderingen in het landeigendom en het gebrek aan interesse om te investeren in gewassen voor de markt, agro-industrie en marketing; een toename van onwettige groepen in het district die werkloosheid en armoede zullen veroorzaken. In dit scenario zal ASORUT slechts in beperkte mate het district kunnen beheren en zal zijn geloofwaardigheid in de ogen van de boeren, de overheid en de ondersteunende publieke en particuliere ondernemingen tot een uiterst laag niveau dalen en dat zal tot een ernstige isolatie van de organisatie binnen de nationale, regionale en lokale context leiden. In dit scenario zal de kloof tussen de twee groepen boeren niet overbrugd worden, maar zal die eerder verdiept worden door het grote verschil in sociaal-economische omstandigheden. Deze situatie vormt het begin van de volgende fase waarin de overheid de totale controle van het district overneemt en de noodzakelijke acties voor een nieuw organisatorisch model zal opstarten. In dit geval kan de overheid tot een concessiemodel besluiten, welke nog niet zolang geleden, namelijk tijdens de voorstellen voor wijziging van het juridische kader voor de landverbeteringssector, is overwogen (Augustus, 2005). Volgens de wet omvat een Concessiecontract het beheer, gebruik, organisatie en gedeeltelijk of totaal beheer van een publieke dienst; of de bouw, het gebruik en gedeeltelijke of totale onderhoud van het werk of goed voor publiek gebruik. Het omvat ook alle noodzakelijke activiteiten die door de concessionaris moet worden ontplooid om de betreffende voorziening of een werk of publieke dienst te laten functioneren; de vergoeding voor de controle en de zorg van het werk of dienst door de organisatie van de concessionaris wordt geregeld door de vaststelling van rechten, tarieven, belasting, en waarde. Dit scenario zal leiden tot een beheersorganisatie, die zijn diensten als leverancier van irrigatiewater op de grote boeren zal concentreren, die al een sterke sociaal-economische eenheid zijn door hun winstgevende landbouwpraktijken. Dit beheersmodel zal tot een economische, financiële en technische duurzaamheid op korte en middenlange termijn leiden, maar zijn sociale duurzaamheid zal ernstig in gevaar worden gebracht door de verdieping van de kloof tussen de kleine en middengrote en de grote boeren. De verwachte voordelen zullen in sociale ongelijkheid resulteren, omdat de positie van de grote boeren zal worden versterkt en de klein en middelgrote boeren van elk voordeel zullen worden uitgesloten.

In het tweede scenario zullen de watergebruikers als gemeenschap in het district de belangrijkste rol spelen; de Organisatie van Watergebruikers zal als vertegenwoordiger van de gebruikers het ontwikkelingsproces gaan leiden. De vrees van de gemeenschap dat de overheid het beheer van het systeem aan een andere particuliere organisatie kan geven, vormt de belangrijkste aansporing om de huidige situatie zonder tussenkomst van de overheid te veranderen en verbeteren. Het ontwikkelingsproces naar duurzaam beheer zal een integrale benadering met deelname van de gehele gemeenschap volgen, waarbij alle interne en externe acteurs betrokken zullen zijn: namelijk de boeren als eerste begunstigden; de technische en administratieve staf; de vertegenwoordigers van de gemeenten binnen het district; de landverbeteringssector; overige ondersteunende publieke en particuliere organisaties. Als een wettelijk orgaan zal ASORUT namens alle boeren voor publieke en particuliere organisaties optreden om zodoende het project gezamenlijk verder te ontwikkelen. Binnen deze benadering met een uitgesproken gemeenschapsdeelname zal de organisatie een duidelijke verklaring ten aanzien van zijn missie moeten geven, die vooral gericht zal zijn op de verbetering van de leefomstandigheden van de boeren door een gezond beheer van het district volgens beginselen van gelijkheid, concurrentie, duurzaamheid en multifunctionaliteit volgens het wettelijke kader dat is opgesteld door de overheid Dit betekent dat activiteiten een breed terrein beslaan en gericht zijn op een verbetering van de socio-economische condities en de levensstandaard van de begunstigden op een integrale en duurzame wijze.

In dit scenario zal ASORUT zich naar verwachting zo goed ontwikkelen dat het vermogen heeft om ontwikkelingscoherente programma's te formuleren, te plannen en uit te voeren op de korte, middel en lange termijn. Deze programma's zullen sociaal-economische, technische en de milieucomponenten omvatten en zullen gebaseerd worden op een integrale, communautaire benadering met deelname van de gemeenschap; dit betekent de participatie van alle direct en indirect betrokken acteurs. De programma's zullen tot doel hebben om de geïrrigeerde landbouw te bevorderen en in praktijk te brengen enerzijds volgens criteria van rentabiliteit, efficiency, concurrentie en duurzaamheid en anderzijds rekening houdend met de bescherming en het behoud van de natuurlijke rijkdommen.

Sommige belangrijke verwachtingen naar aanleiding van de implementatie van deze benadering zouden kunnen omvatten: actieve participatie en een beter saamhorigheidsgevoel aan de kant van de landbouwer door een verbeterde levensstandaard als resultaat van nieuwe, succesvolle, commerciële projecten; de boeren zullen positiever denken over de organisatie; zij zullen zien dat de organisatie het welzijn van elk van hen nastreeft en daarom zullen de conflicten verminderen; de commerciële landbouw en de daarbij behorende activiteiten zullen armoede en werkloosheid verminderen, zullen de levensstandaard verbeteren en zullen leiden tot een gunstig milieu voor het behoud van de natuurlijke rijkdommen. Het vermogen van de organisatie om het district beter te besturen zal resulteren in een efficiënte watervoorziening aan alle boeren, maar ook in de formulering en de uitvoering van winstgevende landbouwprojecten; de gemeenschap in het district zal geleidelijk meer gezag verkrijgen en zal de wederzijdse besluitvorming verder ontwikkelen zodat de belangrijke verschillen tussen de twee groepen boeren kunnen worden geslecht in het voordeel van een harmonische ontwikkeling; ASORUT zal op lokaal, regionaal en nationaal niveau door alle publieke en particuliere organisaties worden erkend naar aanleiding van zijn goed ontwikkeld beheer en zal ook zijn mogelijkheden voor interinstitutionele contacten verder uitbouwen; de fysieke infrastructuur (pompstations en andere voorzieningen) zal op een zeer efficiënte manier worden gebruikt en ASORUT zal de marketing van landbouwproducten steunen en verbeteren.

Als eindconclusie kan worden gezegd dat dit laatste scenario het beste voor het district, de watergebruikers en boeren is in verband met zijn integraal en duurzaam beheer. De implementatie en succesvolle ontwikkeling van deze benadering zullen tot nieuwe voorwaarden leiden, die de gemeenschap meer macht zullen geven en hen in staat stellen om op een actieve manier deel te nemen in de besluitvorming, terwijl zij hun verantwoordelijkheid dragen omdat zij de belangrijkste acteurs in de noodzakelijke veranderingsprocessen zijn.

Fysieke en Niet-Fysieke Interventies in verband met het Nieuwe Beheerscenario

Vóór de implementatie van het nieuwe beheersscenario is het belangrijk dat de huidige infrastructuur van het RUT district op een hoger niveau wordt gebracht. De belangrijkste acties zullen gekoppeld worden aan de pompstations samen met de aanleg van een nieuwe zandvang; verbetering van het marginaal irrigatiekanaal, aanvoerkanaal 1.0 en het interceptiekanaal om de lekverliezen en de kosten voor onderhoud te reduceren; de modernisering van drie kunstwerken in het interceptiekanaal om de waterstand te regelen en om een betere werking van het kanaal te verzekeren. De nieuwe organisatie zal speciale aandacht aan een goed waterbeheer moeten besteden dat in een efficiënter watergebruik resulteert, in een juist gebruik van de infrastructuur en de pompstations om de hoge energiekosten te drukken. Op deze wijze zal de organisatie zijn aanzien en gezag terugwinnen.

De acties vanuit een milieustandpunt zouden vooral op de waterkwaliteit gericht moeten worden, de juiste verwijdering en afvoer van afval, de bescherming van de microstroomgebieden, de controle van de verzilting en de formulering van een uitgewerkt plan voor het milieubeheer. De beheerfunctie van de nieuwe organisatie zal zich op het gebruik, hergebruik en controle van water voor irrigatie moeten concentreren om de specifieke behoeften en eisen van de boeren te kunnen vervullen.

Één van de grootste problemen die door het nieuwe beheer moet worden opgelost is de financiering van het beleid, beheer en onderhoud. De financiering zou moeten komen uit een nieuwe tariefstructuur. Besprekingen met de boeren hebben aangetoond dat zij bereid zijn om voor de levering van water te betalen, omdat tot nu toe landbouw voor hen de enige winstgevende activiteit is. Het tarief zou moeten steunen op een aansporing tot een efficiënter watergebruik en hergebruik en op boetes voor inefficiënt watergebruik. Activiteiten om de economische situatie van de boeren te verbeteren zullen hun bereidheid vergroten om voor water te betalen, vooral de kleine en middelgrote boeren.

ASORUT zou andere mogelijkheden voor aanvullende economische bronnen moeten overwegen, bronnen die anders zijn dan de opbrengsten uit de levering van water. Op korte termijn kunnen nieuwe financiële middelen verkregen worden uit de deelname aan de identificatie en uitvoering van projecten op landbouwgebied; door het openen van nieuwe kredietlijnen; de ontwikkeling van technische diensten voor nieuwe landbouwactiviteiten; en uit marketing van landbouwproducten. Op midden en lange termijn zou ASORUT meer inkomen kunnen verkrijgen uit andere winstgevende projecten, zoals agro-tourisme, kleine landbouw ondernemingen, agro-industriële productverwerking; het opzetten van trainingsprogramma's voor andere irrigatiedistricten.

Productieve landbouw zou in de toekomst gebruik moeten maken van de bestaande natuurlijke situatie binnen het RUT gebied. Enkele gewassen, zoals granen, vruchten, groenten, katoen en gras bieden goede perspectieven in verband met internationale markten (vruchten en groenten) en kunnen extra toegevoegde waarde krijgen door agro-industriële verwerking binnen het gebied. Maïs biedt bijvoorbeeld goede mogelijkheden om aan de lokale en nationale vraag te voldoen en kan als grondstof voor de nationale industrie gebruikt worden. Suikerriet vormt de basis voor een belangrijke productieketen in het Departement van Valle del Cauca, maar het gewas is niet geschikt in verband met de bestaande condities in ASORUT; het heeft namelijk een zeer negatieve invloed op de fysieke infrastructuur.

ASORUT zou particulier vastgestelde termijncontracten moeten bevorderen, vooral voor kleine en middelgrote boeren zodat zij gunstige prijzen, het volume en kwaliteit van de landbouwproducten kunnen vastleggen. Voorbeelden van deze contracten zijn bijvoorbeeld: met grote warenhuizen, met grote boeren die een bepaald productievolume nodig hebben, met agro-verwerkingsindustrieën of met de Nationale Landbouwbeurs.

Het nieuwe ASORUT zou moeten proberen om het saamhorigheidsgevoel binnen de groep boeren te verbeteren en te versterken door de doelstellingen aan te passen aan de belangen en verwachtingen van de boeren die bijvoorbeeld een verdere ontwikkeling van winstgevende landbouwpraktijken en verbetering van hun levensstandaard op een efficiënte en billijke manier verwachten. Transparantie, verantwoordingsplicht, duidelijke acties, administratieve efficiency en gelijkheid zijn vereisten die door de boeren worden verlangd om het vertrouwen in de organisatie terug te brengen.

Verwachtingen voor een Nieuwe Rol voor ASORUT

Een proces met een duidelijk bottom-up benadering werd in de ontwikkeling van het kader als onderdeel van dit onderzoek gevolgd. De verschillende belanghebbenden en andere acteurs hebben de kans gekregen om hun meningen en ideeën in verschillende workshops waaraan de gemeenschap deelnam en in directe discussies te uiten. Op deze wijze konden de belangrijkste factoren voor de huidige kritieke situatie van ASORUT, de wensen en verwachtingen van de boeren en andere acteurs en de nodige interventies geïdentificeerd worden.

Twee scenario's voor het beheer van ASORUT zijn ontwikkeld en na een grondige analyse is één model, namelijk een Benadering met Gemeenschapsdeelname: een Nieuwe Rol voor ASORUT als de meest geschikte gekozen in verband met een integraal en duurzaam beheer van het RUT irrigatiedistrict

Het overgangsstadium zal beginnen vanuit het huidige lage organisatieniveau van ASORUT, dat een gevolg is van verschillende sociaal-economische en politieke gebeurtenissen tijdens de afgelopen periode van de organisatie en waarvoor ASORUT nog niet de meest adequate antwoorden gevonden heeft en ook niet het aanpassingsvermogen had om de nieuwe omstandigheden en verantwoordelijkheden na de overdracht van het beheer goed aan te pakken. Op dit ogenblik lijdt ASORUT onder een destructieproces op verschillende niveaus dat praktisch zal eindigen in een totale ineenstorting van de organisatie. Deze povere staat van de organisatie wordt bijvoorbeeld weerspiegeld in de slechte sociaal-economische situatie van de boeren, vooral de kleine en middelgrote; in het slechte vermogen van de organisatie om het district te beheren en in een lage geloofwaardigheid in de ogen van boeren en de maatschappij.

De vragen die moeten worden beantwoord zijn: hoe moet het geselecteerde scenario onder de huidige situatie van ASORUT uitgevoerd worden; hoe lang duurt het om de gewenste beheersituatie te verkrijgen en aan welke voorwaarden en eisen moet voldaan worden om de geplande ontwikkeling te krijgen? Een visie van de nabije toekomst onder het nieuwe beheersmodel toont aan dat ASORUT een organisatie is die op gelijkheid, efficiency, duurzaamheid en concurrentie criteria gebaseerd zal zijn met de opdracht om het welzijn van alle begunstigden (boeren) te bevorderen, om hun sociaal-economische situatie te verbeteren door winstgevende landbouwpraktijken en dat zij betrokken worden bij alle stadia van de productieketen: productie, beheer van de producten na de oogst, processing in verband met de toegevoegde waarde (agro-industrie), en op de markt brengen van de producten.

Daarom zullen de belanghebbenden van ASORUT en de verschillende acteurs zeer essentiële en verstrekkende besluiten op korte termijn moeten nemen om de organisatie en de fysieke infrastructuur in stand te houden en te beschermen. In het voorgestelde scenario wordt de huidige situatie beschouwd als een "overgangsstadium", die de huidige degenererende organisatie naar een nieuwe organisatievorm brengt, die gekenmerkt wordt door een dynamische houding en een vermogen om de wensen van de belanghebbenden te doorgronden en te voldoen in verband met hun streven naar verbetering van hun sociaal-economische levensomstandigheden. Bovendien zal door deze dynamische houding de nieuwe organisatie stabiele en duurzame contacten kunnen handhaven en verder ontwikkelen met de ondersteunende publieke en particuliere organisaties in de landverbeteringssector. Ook zal zij het grootste voordeel kunnen halen uit alle beschikbare mogelijkheden, die binnen het nieuwe beheerskader en het recente regeringsbeleid voor de landverbetering worden aangeboden. De overgang van de huidige status (overgangsstadium) naar de gewenste situatie zal veel tijd vergen, deze constatering is gebaseerd op het feit dat het nu aan de gang zijnde afbraakproces van de organisatie lang geleden begonnen is. De formulering en de uitvoering van een aantal kleine proefprojecten in de productieve landbouw voor een beperkte groep van kleine en middelgrote boeren zullen als een nieuwe strategische benadering worden gebruikt.

De hier beschreven ontwikkeling is gebaseerd op de zeer goede ervaringen in het Zulia Irrigatiedistrict in het noordoosten van Colombia, dat nu een staat van duurzame ontwikkeling heeft bereikt waar de boeren de hoofdrol spelen doordat zij het ontwikkelingsbeleid van de organisatie leiden en bepalen. Om de nieuwe voorwaarden voor ASORUT te bereiken zullen de belangrijkste acteurs een hoofdrol moeten spelen. De Raad van Beheer zou vertegenwoordigers van de gemeenschap van boeren samen met vertegenwoordigers van de Overheid, dat is vertegenwoordigd door het Secretariaat van de Landbouw, moeten opnemen. Bovendien zouden andere ondersteunende publieke en particuliere organisaties hun middelen en inspanningen op een gemeenschappelijke manier moeten bundelen om het nieuwe beheersmodel definitief tot stand te brengen.

RESUMEN

Colombia es un país tropical localizado en Sudamérica. Tiene un área total de 114 millones de hectáreas. En Colombia se distingues dos sectores de riego: el riego a pequeña escala o tradicional y el de la gran irrigación. El de pequeña escala se desarrolla en tierras localizadas en zonas de ladera donde se siembran cultivos de pan coger y algunos productos comerciales como el maíz, tomate y especialmente hortalizas. En estas áreas las técnicas de cultivo no están bien desarrolladas, y por lo tanto no es común encontrar agricultura con tecnología de avanzada. Pequeños agricultores con bajos ingresos, altos niveles de analfabetismo y desnutrición conforman gran parte de este sector tradicional. Por su parte, el sector de la gran irrigación corresponde a grandes propiedades localizadas en zonas planas con suelos de buena calidad y destinadas a la siembra de cultivos comerciales tales como la caña de azúcar, algodón, sorgo, árboles frutales, etc. En este tipo de sector prevalece la mecanización agrícola y se aplican tecnologías de riego avanzadas.

El área del sector de la gran irrigación desarrollada por el sector público se agrupa en 24 distritos de riego, los cuales cubren un área de 264,802 ha; de las cuales 132,918 tienen infraestructura de riego y drenaje y 131,884 tiene solo infraestructura de drenaje. El área de cobertura de los distritos de riego tiene una capacidad total de 498.8 m^3/s; la red de riego tiene 1,905.2 km de longitud de los cuales el 37% corresponde a canales principales, 45% a canales secundarios y 18% a canales terciarios; la red de drenaje tiene 2,132.6 km de longitud de los cuales el 44% son drenes principales, 39% drenes secundarios y 17% drenes terciarios; y la red de vías tiene 3,382.4 km. Cerca del 9% del suelo se usa para cultivos permanentes, 51% para no-permanentes y 40% para pastos. Cerca del 62% de las propiedades son menores de 5 ha y abarcan el 9% del área total, 17% de las propiedades están en el rango de 5-10 ha y cubren el 13% del área, mientras que el 3% son mayores de 50 ha y cubren el 38% del área total.

Potencial de la agricultura bajo riego

A escala global, Colombia se considera como un país de abundantes recursos naturales, especialmente en relación con los recursos de agua y suelo, lo cual ofrece un gran potencial para el desarrollo de la agricultura bajo riego.

Suelos. El Instituto Geográfico Agustín Codazzi (IGAC) llevó a cabo el estudio de suelos para Colombia en 1973. Como resultado dos tipos de suelo se identificaron. Los suelos tipo A corresponden a las clases I, II y III y son apropiados para el desarrollo de la agricultura bajo riego y mecanizada. Principalmente, estos suelos se localizan en las planicies del Caribe y en las regiones de la Amazonía y la Orinoquía. La mayoría de los suelos tipo B corresponden a la clase IV, su superficie varía de plana a cóncava requiriendo entre otros adecuación de tierras como lavado de sales, drenaje y protección contra inundaciones. Una vez hayan sido adecuados, su uso potencial es similar al de los suelos tipo A. Los suelos tipo A y B en conjunto constituyen un área potencial de 10.6 millones de ha para el desarrollo de la agricultura mecanizada y bajo riego.

Recursos hídricos. De acuerdo al Balance Mundial de Aguas y Recursos Hídricos de la Tierra de la UNESCO, Colombia tiene un rendimiento específico superficial de 0,59 l/s-ha, ocupando el cuarto lugar a nivel mundial. El rendimiento específico es 6 y 3 veces mayor que los valores correspondientes al mundo continental y América Latina, 0.10 y 0.21 l/s-ha, respectivamente.

Colombia tiene una precipitación promedia anual de 1,900 mm. El 70% del área tiene una precipitación promedia anual es de 2000 mm, y la precipitación es mayor de 4000 mm sobre casi una tercera del área (Costa Pacífica y Llanos Orientales). Estos valores están por encima del promedio anual para América Latina, y por encima del promedio global de 900 mm por año. En contraste, la Costa Caribe tiene poca precipitación y se considera como una región seca.

Colombia tiene más de 1000 ríos perennes, una cifra mayor que en África (60). Colombia tiene 10 ríos interiores, los cuales tienen una descarga anual promedia mayor a 1,000 m³/s. Colombia tiene una red de drenaje natural de 740,000 micro-cuencas distribuidas en 4 regiones o cuencas de ríos principales.

Desarrollo del riego. Colombia contribuye con menos del 3% del área de tierras bajo riego en América Latina, ocupando el sexto lugar entre los países latinoamericanos. Colombia ha seguido una tendencia similar a la región Latinoamericana en el desarrollo de la irrigación. El área de riego se incrementó rápidamente hasta el final de los años 60s, pero el crecimiento decayó considerablemente durante las dos últimas décadas. Durante el período 1990-2000 el sector público concentró sus esfuerzos en la rehabilitación, complementación y ampliación de los distritos de riego existentes.

Expectativas para la agricultura bajo riego. De acuerdo al Estudio Nacional de Aguas de 1884 Colombia tiene 14.4 millones de ha apropiadas para la agricultura, de las cuales 6.6 millones de ha (45.8%) son propicias para la agricultura mecanizada y el desarrollo de la agricultura bajo riego mediante infraestructura de riego y drenaje. Esta cifra podría incrementarse a 10.6 millones de ha después de la adecuación de algunos suelos, los cuales al momento limitan las condiciones para el uso en agricultura. De estos 6.6 millones solamente el 11.4% tiene infraestructura de riego y drenaje, de los cuales 38% han sido desarrollados por el sector público y el resto por el sector privado.

El Gobierno Colombiano estableció la "Década del Programa de Adecuación de Tierras para 1991 - 2000" para adecuar 535,000 ha mediante infraestructura de riego y drenaje. Durante diferentes períodos el Gobierno ha adaptado las políticas de adecuación de tierras con el propósito de cumplir el programa de adecuación de tierras preliminar, el cual fue establecido para el período en mención por el Departamento Nacional de Planeación. Sin embargo, los programas de adecuación de tierras han tenido un desarrollo lento.

Política para el sector de adecuación de tierras. El desarrollo histórico de la adecuación de tierras en Colombia ha sido ampliamente influenciado por la política establecida por los diferentes gobiernos de turno, los cuales han implementado sus políticas a través de la formulación de los respectivos marcos legales. La política para el sector de adecuación de tierras ha cambiado mandato tras mandato, lo cual no ha permitido un desarrollo consistente en el largo plazo.

Los lineamientos de política del Gobierno Colombiano para el sector de adecuación de tierras fueron incluidos en el Plan de Desarrollo Económico-Social para el período 1991 – 2000. En 1991el Gobierno Colombiano consideró que las inversiones públicas y privadas son buenos mecanismos para contribuir a la modernización de la producción agrícola. Las inversiones en los proyectos de adecuación de tierras preferiblemente tendrían que promoverse y desarrollarse de manera conjunta por los sectores público y privado. Para facilitar la recuperación de la inversión el Gobierno promoverá la participación de la comunidad durante el proceso de diseño de los proyectos, los cuales después de la construcción serán transferidos a las asociaciones de usuarios para su operación, administración y mantenimiento. En el nuevo marco institucional el

Gobierno cambiará su papel pasando de ser un simple ejecutor a un promotor de la inversión con una mayor participación de la comunidad.

La Ley 41 (1993) estableció las reglas por las cuales el sector de adecuación se organizó. En este arreglo institucional el Consejo Superior de Adecuación de Tierras (CONSUAT) fue la entidad consultiva encargada de la aplicación de las políticas parta el sector de adecuación de tierras, el Instituto Nacional de Adecuación de Tierras (INAT) era la entidad ejecutiva para el desarrollo de la política de adecuación de tierras en conjunto con otras instituciones públicas y privadas, y el Fondo Nacional de Adecuación de Tierras (FONAT) era la entidad administrativa de financiamiento de los proyectos de riego, drenaje y control de inundaciones.

Después de 7 años del programa de adecuación de tierras dentro del nuevo marco institucional, surgieron fuertes críticas al papel de la agencia gubernamental a causa del manejo no eficiente de los recursos económicos y al desarrollo casi nulo de los programas de adecuación de tierras lanzados desde 1990 por los diversos programas gubernamentales. Estas críticas entre otras estuvieron relacionadas al manejo de los recursos externos del programa de adecuación de tierras, los costos de funcionamiento de la agencia, y la calidad de su desempeño.

Como consecuencia, un cambio en el marco institucional se introdujo a comienzos del año 2003 e INAT y otras instituciones públicas fueron suprimidas e integradas a una nueva organización llamada Instituto Colombiano para el Desarrollo Rural (INCODER). El Acuerdo 003 de Febrero 2004 se consideró como un elemento clave dentro del nuevo marco institucional por el cual lineamientos organizacionales fueron establecidos para las organizaciones de usuarios incluyendo la regulación básica para el funcionamiento de los distritos de riego de pequeña, mediana y gran escala. Sin embargo, este acuerdo contenía puntos controversiales lo cual conllevó a la expedición de una nueva regulación para la administración, operación y mantenimiento de los distritos de riego, la cual incluyó formas de transferencia (delegación, administración y concesión) y la consideración de que el INCODER no debería tener ingerencia directa en la organización de usuarios pero si en el manejo de la infraestructura (Resolución 1399/2005).

Transferencia del Manejo del Riego

A mediados de los años 70s el Gobierno Colombiano empezó un programa de transferencia del manejo del riego a las asociaciones de usuarios. La transferencia de los distritos de riego en Colombia se ha llevado a cabo en tres períodos. En 1976, los Distritos de Coello y Saldaña fueron transferidos a solicitud de los usuarios del agua. Los Distritos de Riego de Río Recio, Roldanillo-La Unión-Toro (RUT), Samacá y San Alfonso fueron transferidos durante el período 1989-1993 y finalmente, durante 1994-1995, la mayoría de los restantes distritos de riego se transfirieron siguiendo la política establecida por el Gobierno Colombiano. Al momento, 16 de 24 distritos de riego se han transferido y los restantes permanecen aun bajo la administración de la Agencia Gubernamental.

Sin embargo, serios problemas se detectaron durante y después del proceso de transferencia relacionados con disputas sobre derechos de aguas y aspectos legales, inadecuada operación y mantenimiento de la infraestructura hidráulica, inconformidad por el papel de la Agencia de Gobierno y de las asociaciones de usuarios, degradación ambiental, problemas relacionados con aspectos económicos y financieros, insuficiente participación comunitaria, carencia de liderazgo, e insuficiente capacitación y manejo del riego. En algunos distritos de riego, estos problemas y restricciones dieron lugar a un desempeño ineficiente, deterioro de la infraestructura física y conflictos relacionados con la operación, mantenimiento y administración del sistema dando lugar a un desarrollo no sostenible de los distritos de riego.

En el marco de esta investigación se revisaron los antecedentes del programa de transferencia en Colombia, su implementación, estado actual e impactos. De esta manera, 5 distritos de riego fueron visitados, trabajo de campo y entrevistas con personal técnico y agricultores fueron llevados a cabo. Personal de la agencia Gubernamental del sector de adecuación de tierras, entidades públicas y privadas a nivel local y regional fueron también contactadas. Los Distritos de Riego de Rio Recio, Coello, Saldaña, Zulia y María La Baja fueron visitados. Una revisión de algunas experiencias y características de la transferencia del manejo del riego en otros países con características similares a Colombia fue también llevada a cabo.

La experiencia de la transferencia del manejo en Colombia mostró que la carga financiera del Gobierno se redujo significativamente al trasladarse los costos totales de administración, operación y mantenimiento a las asociaciones de usuarios, la eliminación parcial y después total de subsidios, y la reorganización de la estructura institucional para el sector de adecuación de tierras; facilitando así la implementación de la política de descentralización y delegación para el sector. El impacto de la delegación de funciones a instituciones locales y regionales no puede evaluarse aun puesto que estas instituciones recientemente han empezado a desempeñar las funciones de delegación; sin embargo existe escepticismo entre las organizaciones de usuarios acerca de la efectividad de esta delegación.

El desarrollo de la transferencia mostró una clara diferencia entre la primera fase del Programa de Transferencia y las siguientes. Durante la primera fase, la transferencia del manejo fue promovida por los usuarios mismos con una marcada resistencia por parte de la Agencia Gubernamental; mientras que para las fases siguientes la motivación surgió del Estado y se interpretó como una estrategia del Gobierno para enfrentar la crisis fiscal de los años 90 y atender compromisos con la banca internacional.

Después de 30 años de iniciada la transferencia, las organizaciones de usuarios investigadas presentaron diversos grados de desarrollo. Algunas de ellas presentaron una estabilidad organizacional relevante (Distritos de Riego de Rio Recio, Coello y Saldaña) basada en una autosuficiencia financiera producto de la prestación del servicio de agua de riego; otras como María la Baja presentaron una fuerte crisis organizacional dando lugar a la intervención del Estado; mientras que el Distrito de Riego Zulia presentó una gran dinámica organizacional en donde predomina un modelo de organización basado en un enfoque integral y participativo alrededor de una actividad agrícola rentable centrada en el cultivo del arroz, y con una clara visión del desarrollo futuro de la organización. Puede decirse entonces que en el proceso post-transferencia, y desde el punto de vista organizacional, se distinguen tres fases cruciales: adaptación, madurez o desarrollo pleno, y consolidación.

El incremento de la eficiencia en el uso del agua de riego fue otro de los objetivos del Programa de Transferencia. Sin embargo, de los distritos de riego investigados se deduce que el uso del agua de riego es un asunto crítico y amenaza la sostenibilidad ambiental de los sistemas. La autosuficiencia financiera como resultado de la venta de agua para el suministro de agua de riego, especialmente en cultivos que demandan grandes cantidades de agua como en el caso del arroz, parece ser un factor que no contribuye al uso eficiente del agua.

Desde el punto de vista ambiental merece mencionar el creciente proceso de deforestación que afecta las principales cuencas de las fuentes superficiales de los distritos. Esta situación afecta entre otros, la calidad y costos de las actividades de operación y mantenimiento y la disponibilidad del recurso hídrico.

Por último, la experiencia de la transferencia en Colombia permitió la consideración a un concepto más amplio del papel de los sistemas de riego. Estos no deben ser considerados solamente como una simple infraestructura hidráulica que suministra agua de riego a los agricultores, sino también como un componente importante de un sistema de producción cuyo objetivo final es contribuir al mejoramiento de las condiciones de vida de los agricultores mediante la práctica de una agricultura de riego bajo criterios de rentabilidad, equidad, eficiencia, y un enfoque de manejo integral y participativo.

Manejo sostenible de sistemas de riego transferidos. Marco conceptual

Un marco conceptual para el manejo sostenible de los distritos de riego transferidos a las asociaciones de usuarios del agua se formuló y desarrolló basado en conceptos generales y enfoques para el manejo sostenible del agua y especialmente el uso y manejo sostenible de los sistemas y recursos de agua y suelo. Este marco es soportado por ciertos principios generales, enfoques y definiciones, algunos de las cuales fueron enunciadas en numerosos eventos y conferencias relacionadas con el aprovechamiento y preservación de los recursos naturales con especial énfasis en los recursos de agua y suelo.

La sostenibilidad de un sistema de riego y drenaje se refiere a la infraestructura organizacional y física para el manejo, administración, operación y mantenimiento durante la vida útil del sistema de acuerdo a las especificaciones establecidas y para dar a los usuarios los beneficios esperados sin causar daño al ambiente. El concepto de sostenibilidad involucra tres elementos que juegan un papel importante en la definición del marco conceptual para el manejo sostenible de un sistema de riego. Los tres componentes son Comunidad (C), Ambiente (A) y Ciencia y Tecnología (C&T); ellos están Inter.-relacionados dentro del contexto legal e institucional existente. La interrelación determina la sostenibilidad de un sistema de riego.

La *Comunidad (C)* representa los diferentes grupos de interés en el sector del riego y el drenaje, en el cual los agricultores son los principales beneficiarios. La comunidad con sus características socio-económicas-culturales específicas y formas organizacionales interactúa con el Ambiente, la Ciencia & Tecnología para contribuir al manejo sostenible de los sistemas. El *Ambiente (A)* provee los recursos naturales para la agricultura bajo riego como sistema de producción. Condiciones de clima, suelos, ecosistemas, y recursos hídricos son los principales insumos suministrados por el componente ambiental. Los factores que afectan la relación entre el Ambiente y la Comunidad en ambas direcciones se constituyen en factores de riesgo. La *Ciencia y la Tecnología (C&T)* representa el conocimiento científico y tecnológico, métodos, herramientas, y experiencias los cuales están disponibles y pueden ser usadas por la comunidad para intervenir el ambiente natural y satisfacer sus necesidades y deseos. La Ciencia y la Tecnología constituye el mecanismo por el cual los factores de riesgo en la relación Ambiente y Comunidad pueden controlarse o superarse. Las alternativas tecnológicas para el suministro de un sistema de riego y drenaje deben estar en línea con las condiciones culturales, socio-económicas, técnicas y financieras de la comunidad. El nivel de especificaciones de servicio, costo de servicio, disponibilidad a pagar por el suministro del servicio, formas organizacionales, capacidad técnica, participación en la operación y mantenimiento, toma de decisiones, resolución de conflictos, y movilización de recursos juegan un importante papel en la relación Ciencia-Tecnología y Comunidad.

Distrito de Riego Roldanillo-La Unión-Toro: RUT

El Distrito de Riego RUT se tomó como caso de estudio para esta investigación. Se presentan los antecedentes históricos; descripción y características de los aspectos de manejo, físicos y socio-económicos del distrito; y finalmente las características e impactos del proceso de transferencia. El Distrito de Riego RUT se localiza en el suroeste de Colombia, en la parte norte del Departamento del Valle del Cauca. Descansa dentro de las tres municipalidades que dan su nombre, Roldanillo, La Unión y Toro. Su altitud sobre el nivel del mar varía entre 915 y 980 m.s.n.m. Antes de su construcción el área estuvo sujeta a inundaciones del Río Cauca y de las corrientes provenientes de las estribaciones de la Cordillera Occidental. Como consecuencia un área de 1,500 a 3,500 ha respectivamente estuvo bajo inundación permanente y periódica. Como resultado cerca de 2,500 ha mantenían un alto contenido de humedad permitiendo el cultivo de pastos, mientras que un área de 4,000 ha podía usarse para ganadería extensiva.

En términos generales, el sistema de riego RUT es principalmente un proyecto de riego/drenaje y control de inundaciones. La alargada y estrecha área en forma de tazón está rodeada por un dique de protección el cual corre a lo largo de la orilla izquierda del Río Cauca, y por un canal interceptor de inundaciones en su costado occidental. Un canal de drenaje principal divide el área casi a la mitad, circulando por los puntos más bajos del área. Paralelo al dique de protección se encuentra el canal de riego principal, el cual es abastecido por tres estaciones de bombeo con una capacidad total de 13.8 m^3/s. El canal interceptor también funciona como canal de riego para usuarios quienes usan pequeñas unidades de bombeo para satisfacer sus necesidades de agua. Hay una red complementaria tanto de riego como de drenaje a lo largo del área. El canal de drenaje principal descarga libremente al Río Cauca cuando este presenta niveles bajos, caso contrario el agua de drenaje es bombeada al río.

El Distrito de Riego RUT es uno de los mas importantes sistemas de riego en Colombia dado las muy buenas características de sus recursos naturales (clima, suelos, agua) para el desarrollo de la agricultura; su estratégica posición geográfica, y el potencial de desarrollo agroindustrial con miras a mercados nacionales e internacionales. Hoy en día, el Distrito de Riego RUT presenta una crisis creciente la cual es una seria amenaza para la sostenibilidad de todo el sistema. Factores externos e internos con sus características sociales, económicas, ambientales, técnicas e institucionales han contribuido in diferente grado a esta situación.

Desde sus comienzos, el manejo del Distrito de Riego RUT presentó una fuerte oposición por parte de la comunidad, la cual consideró el proyecto como una imposición por parte del Gobierno Nacional. Como muchos de los proyectos de los años 60s y 70s, este proyecto también fue desarrollado con una mínima participación de la comunidad y casi bajo la responsabilidad total del Gobierno, el cual concibió la implementación y desarrollo del proyecto de adecuación de tierras como una mera intervención física con limitada atención a requerimientos sociales, culturales y económicos dentro del contexto social. Por lo tanto, los agricultores mostraron un sentido de pertenencia muy bajo y consideraron que era solo responsabilidad del Gobierno llevar a cabo todas las actividades de administración, operación y mantenimiento. Durante la administración del Distrito por parte el Gobierno (1971 - 1989), esta actitud no ocasionó serios impactos puesto que la Agencia de Gobierno se encargó del manejo completo incluyendo la operación y mantenimiento del sistema con recursos financieros provenientes del Estado. Sin embargo, la actitud paternalista del sector de adecuación de tierras, el impacto negativo del alineamiento de canales en fincas, la insuficiente compensación de las tierras adquiridas por el Estado para la construcción de obras, y la falta de credibilidad en los proyectos de inversión del Estado no incentivaron la motivación y actitud de los agricultores.

Después de la transferencia en 1989, la Asociación de Usuarios del agua ASORUT fue completamente responsable de las actividades de administración, operación y mantenimiento asumiendo todos los costos y sin recibir subsidios por parte del estado. La transferencia fue centrada principalmente en actividades relacionadas a la organización de los usuarios, consideraciones acerca de las tarifas de agua y la reparación de la bocatoma y canal principal. Aun hoy, los agricultores piensan que la transferencia fue muy rápida y no tuvo la suficiente asistencia por parte del Estado, siendo el principal objetivo de este la transferencia del manejo a la mayor brevedad posible para cumplir compromisos con entidades financieras interesadas en el sector de adecuación de tierras. El objeto principal del actual manejo es el suministro del agua de riego del cual la organización obtiene los ingresos necesarios para su financiamiento. De esta manera, la actual organización de manejo actúa de una manera comparable o similar a la anterior Agencias Estatal, y por ello se piensa que el proceso de transferencia conllevó únicamente aun cambio de actores para el manejo.

Así, después de un tiempo la organización enfrentó las siguientes características negativas: bajo recaudo de tarifas, suspensión temporal y espacial de actividades de mantenimiento de la red de riego, baja calidad en la provisión del servicio de agua de riego, ineficiente uso del agua y de la infraestructura, conflictos entre usuarios y entre estos y el personal técnico y administrativo de la organización, y finalmente el incumplimiento de los acuerdos de pago por las tarifas de agua. El posponer continuamente las actividades de mantenimiento de la infraestructura de riego causó su deterioro e ineficiencia en su suministro.

La actitud de manejo actual claramente muestra que el manejo solo se ocupa del suministro del agua de riego sin plantearse la inquietud de si el agua contribuye al éxito del sistema de producción de los agricultores. La situación actual es el resultado de la discrepancia de intereses entre dos partes: por una parte la organización centrada en su actividad de suministro de agua de riego y por la cual recibe una remuneración a través de una tarifa de agua; y por otra, los agricultores a la espera de acciones de la organización que les apoye para el desarrollo de proyectos conducentes a mejorar su nivel de vida.

Este enfoque de manejo y algunos factores externos causaron varios problemas, como por ejemplo aquellos relacionados con el canal interceptor, la calidad del agua, procesos de salinización, mantenimiento de la infraestructura, y proyectos de desarrollo con cooperación internacional para la modernización del distrito. Las pobres condiciones de ASORUT son un reflejo de la presente crisis, especialmente en los aspectos técnicos, económico, social, ambiental e institucional.

La operación del distrito es hoy en día más compleja y el suministro de agua se ha afectado por cambios en el patrón de cultivos, carencia de una programación en el riego y planes de cultivo. Los agricultores deciden acerca del plan de cultivos, fechas de siembra y el momento para la solicitud del servicio de agua de riego. El cultivo de caña de azúcar cubre una tercera parte del área y demanda mucha mas agua que los cultivos tradicionales.

Los mayores problemas ambientales se relacionan con la calidad del agua. Las aguas residuales domésticas e industriales y la deforestación de las cuencas aledañas afectan la calidad del agua de los canales de riego y del Río Cauca. En varios sitios el proceso de salinización se ha intensificado a causa del reuso del agua de drenaje, ineficiencia en las prácticas agrícolas, carencia de sistemas de drenaje a nivel predial, y la salinización inducida por el agua subterránea.

Un factor que incide seriamente la crítica situación es la existencia de dos tipos de agricultores; los pequeños-medianos y los grandes agricultores. Dado sus limitados recursos económicos junto a su baja tecnología y las pequeñas áreas de sus predios, los pequeños agricultores no pueden, como los grandes, introducir y cultivar cultivos de alta rentabilidad como es el caso del cultivo de árboles frutales. Esto ha creado tensiones y desacuerdos entre los dos grupos de agricultores mencionados.

Manejo sostenible del Distrito de Riego RUT. implementación del marco conceptual

Un marco analítico para el manejo integral de los recursos hídricos se usó para formular una propuesta de un plan de manejo integral y sostenible para ASORUT. La evaluación de la situación presente, la formulación del plan de manejo y de las intervenciones requeridas se obtuvieron de entrevistas con agricultores, personal administrativo, técnico y de campo de ASORUT, representantes de los municipios involucrados y de las instituciones públicas y privadas estrechamente relacionadas con el sector de adecuación de tierras.

Las propuestas de escenarios para un manejo integrado y sostenible se basan en el potencial de factores internos y eternos al distrito y su influencia en los procesos de manejo. Dos escenarios se han desarrollado para el distrito a saber: un Escenario de Intervención Directa del Estado y otro de Enfoque Participativo y Comunitario: un Nuevo Papel para ASORUT.

Ambos escenarios parten de la condición actual del Distrito, pero el primer escenario se basa en el supuesto del deterioro continuado debido a la baja capacidad de manejo actual para producir los cambios necesarios que puedan conducir a una recuperación gradual de la organización. Se espera que algunos factores negativos continúen acentuándose, como por ejemplo el aumento de practicas agrícolas no rentables que den lugar a una baja disponibilidad y capacidad de pago de los agricultores para pagar las tarifas de agua, y un incremento de la cartera morosa dado que los usuarios no son capaces de pagar por los servicios lo cual conlleva al detrimento de la capacidad financiera de la organización. Otros factores incluyen cambios en el uso del suelo, especialmente la expansión de cultivos extensivos como la caña de azúcar y otros usos no agrícolas; cambios en la tenencia de la tierra y la carencia de incentivos para la inversión, especialmente en cultivos comerciales, agroindustria y mercadeo; aumento de la presencia de grupos ilegales que dará lugar a desempleo y pobreza. En este escenario ASORUT ofrecerá una baja capacidad de manejo y su credibilidad presentará el más bajo nivel ante los agricultores, el gobierno y las entidades de apoyo públicas y privadas, ocasionando el aislamiento de la organización dentro del contexto nacional, regional y local. En este escenario las diferencias entre los dos grupos de agricultores no serán salvadas, por el contrario serán ahondadas dado las grandes diferencias en sus condiciones socio-económicas.

Esta situación constituye el inicio de la siguiente fase de este escenario en el cual el Estado tomará el control del distrito y emprenderá las acciones necesarias para establecer un nuevo modelo organizacional. En este caso, el Estado puede decidirse por un modelo de concesión, el cual recientemente fue considerado durante los cambios al marco legal para el sector de adecuación de tierras (Agosto, 2005). De acuerdo a la Ley un contrato de concesión contempla la provisión, operación, explotación, organización y manejo total o parcial de un servicio público; o la construcción, explotación y conservación total o parcial de una obra o bien para uso o servicio público. También incluye todas las actividades necesarias a desarrollarse por el concesionario para la provisión apropiada o funcionamiento de la obra o servicio público; y la remuneración por el control y cuidado de la obra o servicio por la entidad concesionaria es acordada por medio de derechos, tarifas, impuestos, y valorización entre otras.

Este escenario dará lugar a una organización de manejo que concentrará sus servicios como proveedor de agua de riego a los grandes agricultores, los cuales presentan una favorable condición socio-económica debido a sus prácticas agrícolas más rentables. Este modelo de manejo proporcionará sostenibilidad económica, financiera y técnica en el corto y mediano plazo, pero su sostenibilidad social estará seriamente en peligro debido al ahondamiento de las diferencias entre los pequeños-medianos agricultores y los grandes; los beneficios esperados darán lugar a desigualdades sociales puesto que los grandes agricultores serán favorecidos y los pequeños-medianos serán excluidos de cualquier beneficio.

En el segundo escenario la comunidad constituida por los usuarios del agua del distrito jugarán el papel principal; la asociación de usuarios como representante de los usuarios estará a la cabeza del proceso de desarrollo. El temor de la comunidad de que el estado delegue el manejo del sistema a otra organización privada se constituye en el incentivo para cambiar la situación de manejo actual sin la mayor interferencia del estado. El proceso de desarrollo hacia un manejo sostenible obedecerá a un enfoque integral y participativo involucrando a todos los actores internos y externos: los agricultores como los primeros beneficiarios; el personal técnico y administrativo; representantes de cada una de las municipalidades del área de influencia del distrito, del sector de adecuación de tierras, y de las entidades de soporte públicas y privadas. Como organización legal, ASORUT actuará en representación de todos los agricultores ante las instituciones públicas y privadas para desarrollar programas de manera conjunta. Dentro este escenario participativo la organización tendrá una definición clara acerca de su misión, la cual apunta al mejoramiento de la calidad de vida de los agricultores mediante un buen manejo del distrito bajo principios de equidad, competitividad, sostenibilidad y multifuncionalidad en concordancia con el marco legal establecido por el Estado. Esto significa que sus actividades tendrán que ser amplias y enfocarse al mejoramiento de las condiciones socio-económicas y nivel de vida de sus beneficiarios de una manera integral y sostenible.

En este escenario ASORUT tendrá una muy bien desarrollada habilidad para formular, planear, e implementar programas coherentes de desarrollo en el corto, mediano y largo plazo. Estos programas incluirán componentes socio-económicos, técnicos y ambientales y se basarán en un enfoque integral, participativo y comunitario; significando con ello la participación de los actores directa o indirectamente involucrados. Los programas de desarrollo estarán basados en la promoción y práctica de una agricultura de riego bajo criterios de rentabilidad, eficiencia, competitividad y sostenibilidad; y al mismo tiempo en la protección y conservación de los recursos naturales.

Algunas de las mayores expectativas de la implementación de este enfoque comunitario y participativo incluyen: activa participación y mejoramiento del sentido de pertenencia de los agricultores debido a una mejor condición de su nivel de vida como resultado de nuevos proyectos comerciales exitosos; los agricultores tendrán una mejor opinión en relación a la organización; los agricultores percibirán que la organización persigue ahora el bienestar de todos ellos y como consecuencia los conflictos se reducirán; la agricultura comercial y los negocios correspondientes permitirán la reducción de la pobreza y el desempleo, incrementará el nivel de vida y propiciará un ambiente favorable para la conservación de los recursos naturales; la capacidad de manejo de la organización será mas fuerte dando lugar no solamente a un suministro eficiente de agua para riego a todos los agricultores sino también a la formulación e implementación de proyectos agrícolas rentables; la comunidad en el distrito obtendrá de una manera gradual mas autoridad y desarrollará un proceso mutuo de toma de decisiones en el cual las mayores diferencias entre los dos grandes grupos de agricultores puedan reducirse dando lugar a un desarrollo armónico; ASORUT será reconocido a nivel local, regional y nacional por todas las instituciones públicas y privadas como resultado del exitoso manejo de la organización y por lo tanto gozará de una gran habilidad para establecer relaciones interinstitucionales; la infraestructura física (estaciones de bombeo y otras instalaciones) se usarán de manera mas eficiente y ASORUT apoyará y resaltará el mercadeo de productos agrícolas.

Como conclusión final puede decirse que este escenario es el más apropiado para el distrito de riego, los usuarios y los agricultores en la perspectiva de un manejo integral y sostenible. Su implementación y desarrollo exitoso creará nuevas condiciones que permitirán el empoderamiento de la comunidad permitiéndole participar de manera activa en el proceso de toma de decisiones, puesto que ellos como actores centrales serán responsables de los procesos de cambio necesarios.

Intervenciones físicas y no-físicas para el nuevo escenario de manejo

Antes de la implementación del nuevo escenario de manejo es importante que la actual infraestructura física del Distrito de Riego RUT se llevada al mas alto nivel de servicio. Las intervenciones físicas necesarias mas importantes se relacionan a las estaciones de bombeo junto con la implementación de estructuras desarenadoras; adecuación de los canales marginal de riego, conductor e interceptor con el fin de reducir las pérdidas de agua y los costos de mantenimiento; y la modernización de las estructuras de control de nivel del canal interceptor para asegurar su correcta operación. El nuevo modelo de manejo debe prestar especial atención al manejo apropiado del agua con el fin de alcanzar un uso más eficiente, un uso apropiado de la infraestructura hidráulica y la correcta operación de las estaciones de bombeo para reducir los altos costos de energía.

Durante uno de los talleres participativos, los usuarios del agua propusieron dividir el distrito en sectores de riego con el fin de mejorar la distribución del agua. Las fronteras de estos bloques dependerán de un número de factores, como el área de comando de los canales del sistema principal y la capacidad de las estaciones de bombeo. Así, ASORUT será responsable por el suministro eficiente, oportuno y equitativo de agua a ser dirigido a cada uno de los bloques a nivel del sistema principal.

El manejo a nivel de bloque, siendo este un grupo de predios, será responsabilidad de los agricultores (una organización de usuarios a nivel de sector)

Las intervenciones desde el punto de vista ambiental deberían centralizarse en la calidad del agua, la apropiada disposición de los residuos sólidos, la protección y conservación de los micro-cuencas, el control del proceso de salinización, y la formulación de un plan para el manejo ambiental. La función operacional de la nueva organización se focalizará en el uso, reuso y control del agua de riego para satisfacer las necesidades específicas y demandas de los agricultores.

Uno de los mayores problemas a resolverse por la nueva organización concierne a la financiación de las actividades de administración, operación y mantenimiento. Esta financiación debería de provenir del establecimiento de una nueva estructura de tarifas. Las discusiones con los agricultores han mostrado que ellos están dispuestos a pagar por el servicio de agua para riego, puesto que para ellos la agricultura es hasta ahora, la única actividad económica rentable. La tarifa debiera contemplar incentivos por el uso y reuso eficiente del agua; y multas por el uso ineficiente de la misma. Las actividades tendientes a mejorar la condición económica de los agricultores ayudarán a aumentar su disponibilidad a pagar por el agua, especialmente en el caso de los pequeños y medianos agricultores.

ASORUT debiera considerar otras posibilidades para la obtención de recursos económicos adicionales a las tarifas de agua. En el corto plazo nuevos recursos económicos podrían obtenerse provenientes de la participación en la identificación e implementación de proyectos agrícolas productivos, obtención de nuevas líneas de crédito; provisión de servicios técnicos para el desarrollo de actividades agrícolas; y el mercadeo de productos agrícolas. En el mediano y corto plazo, ASORUT podría obtener otros ingresos económicos producto de su participación y desarrollo en proyectos paralelos rentables; como son los casos del agro-turismo, pequeñas empresas rurales, procesos agroindustriales y programas de capacitación a otros distritos de riego

En el futuro la agricultura debiera basarse en las condiciones naturales existentes dentro del área. Algunos de los cultivos, como los granos, frutas, hortalizas ofrecen buenas perspectivas con miras a mercados internacionales (frutas y hortalizas) y pueden adquirir valor agregado a través de procesos agroindustriales dentro del área. El maíz por ejemplo ofrece buenas posibilidades para satisfacer la demanda local y nacional y proveer de materia prima a la industria nacional. La caña de azúcar constituye la base de una importante cadena productiva del Departamento del Valle del Cauca, pero este cultivo no es apropiado para las condiciones que prevalecen en el área del Distrito RUT habiendo tenido ya un impacto negativo sobre la infraestructura física.

ASORUT debiera de promover los contratos de negociación a futuro, especialmente para los pequeños y medianos productores asegurando de esta manera precios favorables, volumen y calidad de los productos agrícolas. Estos contratos pueden llevarse a cabo con los grandes almacenes de cadena, los grandes agricultores quienes necesitan de ciertos volúmenes de producción para su comercialización, las compañías agroindustriales o la Bolsa Nacional Agropecuaria.

La nueva organización para ASORUT debe de recuperar y fortalecer el sentido de pertenencia de los agricultores adaptando los objetivos de la organización para estar en línea con los intereses y expectativas de los agricultores, quienes por ejemplo esperan el desarrollo de prácticas agrícolas rentables y el mejoramiento de su nivel de vida de una manera eficiente y equitativa. Transparencia, rendición de cuentas, acciones claras, eficiencia administrativa y equidad son los requerimientos demandados por los agricultores para crear confianza en la organización.

Expectativa por el nuevo papel de ASORUT

El marco desarrollado como parte de esta investigación refleja un proceso con un enfoque participativo de arriba–abajo (bottom-up). Los diferentes actores involucrados tuvieron la oportunidad de presentar sus ideas y puntos de vista en varios talleres participativos y discusiones directas. De esta manera, los factores claves de la actual situación crítica de ASORUT, los deseos y expectativas de los agricultores y otros actores, y las intervenciones necesarias se identificaron.

El capítulo 6 presenta los dos escenarios de manejo desarrollados para ASORUT y después mediante el análisis, un modelo, el "Escenario de Enfoque Comunitario y Participativo: un Nuevo Papel para ASORUT" se considera como el más apropiado con miras a un manejo integrado y sostenible para el Distrito de Riego RUT.

La etapa de transición comenzará con el actual bajo nivel de manejo de ASORUT el cual es el resultado de los diferentes eventos socio-económicos y políticos acontecidos a lo largo de la vida de la organización, y para los cuales ASORUT no tuvo capacidad de respuesta y adaptación para afrontar las nuevas circunstancias y responsabilidades adquiridas después de la transferencia del manejo del riego. En este momento, ASORUT sufre un proceso de deterioro a diferentes niveles que terminará en el estado de colapso total de la organización. Este estado precario se reflejará por ejemplo en la pobre condición socio-económica de los agricultores, especialmente los pequeños y medianos, en la pobre capacidad de manejo de la organización y su baja credibilidad ante los agricultores y la sociedad en general.

Las preguntas a responder son: cómo será posible implementar el escenario propuesto bajo la situación actual de ASORUT; cuánto tiempo tomará llegar a la condición de manejo deseada y cuales son los requerimientos para el desarrollo previsto? Una visión del futuro próximo bajo el nuevo modelo de manejo muestra a ASORUT como una organización basada en criterios de equidad, eficiencia, sostenibilidad y competitividad con la misión de promover el bienestar de todos los beneficiarios (agricultores) para mejorar su condición socio-económica mediante prácticas agrícolas rentables e involucrando todas las etapas de la cadena productiva: producción, manejo post-cosecha, proceso de valor agregado (agroindustria) y mercadeo.

Por lo tanto, los grupos de interés en ASORUT y los diferentes actores involucrados tienen que tomar a corto plazo, decisiones amplias y cruciales para salvaguardar la organización y la infraestructura física. En el escenario propuesto la situación actual se considera como una "etapa de transición", la cual permite el tránsito de una organización en decadencia a una nueva que se caracterizará por una actitud dinámica y con capacidad para interpretar y satisfacer los deseos de las partes de interés con miras al mejoramiento de sus condiciones socio-económicas de vida. Además la actitud dinámica de la nueva organización podrá mantener y desarrollar contactos duraderos y estables con entidades públicas y privadas de apoyo del sector de adecuación de tierras, y aprovechará las oportunidades disponibles para desarrollos adicionales mientras sean ofrecidos dentro del nuevo marco de manejo y la reciente política gubernamental de adecuación de tierras. El paso del estado actual (etapa de transición) a la condición deseada tomará un tiempo considerable, teniendo en cuenta que el proceso en curso de deterioro de la organización comenzó hace mucho tiempo. La formulación e implementación de pequeños proyectos pilotos agrícolas productivos que involucren grupos de agricultores pequeños y medianos se usará como estrategia.

Esta estrategia se basa en experiencias exitosas del Distrito de Riego Zulia, localizado en el nordeste de Colombia, el cual ha alcanzado un grado de sostenibilidad donde los agricultores desempeñan el papel central, donde son ellos los que lideran y determinan la política de desarrollo de la organización. Para emprender y alcanzar las nuevas condiciones de manejo de ASORUT los actores claves tendrán que desempeñar un papel de importancia. La Junta Directiva debiera incluir representantes de la comunidad de agricultores junto con la Agencia Gubernamental representada por la Secretaría de Agricultura Departamental. Otras entidades públicas y privadas de apoyo aunarán recursos y esfuerzos de una manera concertada para contribuir al establecimiento del nuevo modelo de manejo.

Curriculum Vitae

The author of this dissertation was born in the Municipality of Morales in the Cauca Department, Colombia on November 3, 1953. He followed the secondary school Santa Librada in Cali and started in 1972 his studies at the Faculty of Agricultural Engineering at the Valle University, Cali. He obtained his bachelor degree in Agricultural Engineering in 1977.

From 1978 till 1979 he worked as lecturer in the Universidad Technologica de los Llanos Orientales in the Agronomic Faculty, in Villavicencio in the Meta Department. Since 1979 he works for the Environment and Natural Resources School, Faculty of Engineering of the Valle University in Cali. At the university his main tasks include teaching, research and consultancy activities. He lectures in several topics, amongst others Fluid Mechanics, Applied Hydraulics, Irrigation and Drainage, and Pumping Stations. He has carried out various research projects in the field of irrigation and drainage and also consultancy activities in water resources management. Moreover, he has been in charge of a number of academic-administrative tasks such as head of the Fluid Mechanics Section, head of the Thermal Sciences and Fluid Mechanics Department, head of Fluid Mechanics Laboratory, and Academic Director of the Agricultural Engineering Program.

He is author of various publications in the field of hydraulics and irrigation of which are the most important: Irrigation in Small Communities in Colombia. Case study: Potrerito Project; Design and Management Optimisation of the Hydraulic Infrastructure for Irrigation and Drainage Systems by using Numerical Simulation; Flow Control for Irrigation Systems; Methodology for Water Resource Distribution in the Valle del Cauca Department; and Towards a Sustainable Management of Irrigation Systems by Water Users Associations in Colombia - Conceptual Framework. In addition, he has organized and coordinated national and international events and congresses in the field of water and water resources such as International Workshop: Integrated Water Resources Improvement with Emphasis in Irrigation, Drainage and Sustainability, International Event WATER 98 Colombia, 1998; International Course: Sediment in Irrigation and Drainage Canals, Colombia, 1999; and International Workshop: Flow Control in Distribution Networks for Surface Waters, Irrigation and Drainage, International Event WATER 2000, Colombia, 2000

He followed in 1998 the International Course for Irrigation Engineering in the International Experimentation Centre (CEDEX) in Madrid, Spain, where he obtained the title of Irrigation Engineering Specialist.

In October 1994 he came to the International Institute for Infrastructural, Hydraulic and Environmental Engineering (IHE) in Delft, The Netherlands and followed the International Course of Hydraulic Engineering, branch Land and Water Development and obtained the diploma in Hydraulic Engineering in 1995. After this Course in Hydraulic Engineering, he started his MSc research on the topic "Optimization and Evaluation of Small Irrigation Schemes. Case study: Potrerito Project". In 1997 he obtained his MSc degree in Hydraulic Engineering and in 2001 he started his PhD research in a sandwich construction at UNESCO-IHE, Institute for Water Education and at the Valle University in Cali. His research topic was "Sustainable Management after Irrigation System Transfer. Experiences in Colombia: the RUT Irrigation District". During his research he carried out detailed fieldwork and organized several workshops with the stakeholders from the RUT Irrigation District. Every year he stayed several months at UNESCO-IHE to present the progress of his research, to evaluate his research findings and to prepare fieldwork and other activities for the next period.

Printed and bound by CPI Group (UK) Ltd, Croydon, CR0 4YY

01/11/2024

01782618-0005